Education Innovation Series

Series Editors:

Wing On LEE
David Wei Loong HUNG
Laik Woon TEH

Executive Editor:

Siao See TENG

For further volumes:
http://www.springer.com/series/10092

Aims and Scope – Springer Education Innovation Book Series

Education holds the key to unlock human resources that a society needs to survive and flourish. This is particularly salient in a borderless knowledge economy. For the past decades, the sterling performance of economies such as Hong Kong, Finland, Japan, Singapore and Taiwan in international studies (e.g., TIMSS, PIRLS and PISA) has channeled much attention away from the traditional centers of education research in America and Western Europe. Researchers, policy makers and practitioners all over the world wish to understand how education innovations propel the emerging systems from good to great to excellent, and how different their trajectories were compared to the systems in America and Western Europe.

The *Education Innovation Book Series,* published by Springer, will delve into education innovations enacted by the Singapore education system and situate them in both the local and the boarder international contexts. Primary focus will be given to pedagogy and classroom practices; education policy formulation and implementation; school and instructional leadership; and the context and interface between education research, policy and practice. We believe that the latter is critical in making education innovations come to bear. Each volume will document insights and lessons learned based on empirical research (both quantitative and qualitative) and theoretical analyses. Implications to research, policy and professional practice will be surfaced through comparing and synthesizing Singapore's experience with those of successful systems around the world.

The audience of the edited volumes and monographs published in this series includes researchers, policy makers, practitioners and students in the fields of education and teacher education, and public policies related to learning and human resources.

Aik-Ling Tan • Chew-Leng Poon •
Shirley S.L. Lim
Editors

Inquiry into the Singapore Science Classroom

Research and Practices

Springer

Editors
Aik-Ling Tan
Shirley S.L. Lim
National Institute of Education
Singapore

Chew-Leng Poon
Ministry of Education
Singapore

ISSN 2211-4874 ISSN 2211-4882 (electronic)
ISBN 978-981-10-1339-3 ISBN 978-981-4585-78-1 (eBook)
DOI 10.1007/978-981-4585-78-1
Springer Singapore Heidelberg New York Dordrecht London

Softcover reprint of the hardcover 1st edition 2014

Printed on acid-free paper

Springer is part of Springer Science+Business Media (www.springer.com)

Series Editors' Foreword

We are very pleased to present *Inquiry into the Singapore Science Classroom: Research and Practices* co-edited by Aik-Ling Tan, Chew Leng Poon and Shirley S. L. Lim, as the fifth book of the Springer Series on Education Innovation. In this volume, the editors attempt to study and interpret Singapore's journey in embracing science as inquiry from the perspectives of multiple stakeholders – education policymakers, researchers, teachers as well as players who are not directly involved in formal education (e.g., journalists). They also situate Singapore's formulation and enactment of science curriculum in the broader sociohistorical context. This volume, as Mr. Siew Hoong Wong pointed out, is one of its kind.

Since Singapore's self-governance in 1959, a school curriculum with a substantial focus on science and technology has been a key strategy for Singapore to ensure its economic survival. Singapore's exit from Malaysia in 1965 which had resulted in a shift from import substitution to export-oriented industrialisation further heightened Singapore Education System's constant drive to ensure that its younger generation is well prepared for an economy that is propelled by science and technology. This effort has not relented in the twenty-first century where Singapore Education System progresses into the Value-Driven Student Centric Phase, even though the balance may have shifted from learning science as learning what is known to learning science as a way of knowing.

One recurring theme which the readers will read about in this volume is the inherent tension Singapore policymakers, researchers, teachers and students experienced in designing, implementing, teaching and learning science as inquiry. Some readers may find this unexpected because of Singapore's success in international benchmarking studies (e.g., TIMSS and PISA) and its widely cited tight coupling, or close nexus, between policymakers, teachers and researchers. The chapters contributed by policymakers, researchers and practitioners from within, and astute commentators from beyond, provide a pair of lens into how Singaporean stakeholders make sense of this inherent tension, and ameliorate it, albeit not always successful, whether collectively or individually. Readers (both local and overseas) are likely to find this useful in understanding and resolving their own challenges, even if the context is different.

As many contributing authors pointed out in this volume, learning and teaching of science as inquiry is not a new idea. Internationally, science education has gone through a number of reforms (e.g., during the post-Sputnik era) which appeared similar, and this could be viewed as evident failure of the earlier reforms. But perhaps the success of these reforms should be assessed collectively, instead of individually. If this is so, analogous to Bruner's concept of spiral curriculum, which emphasises curricular coherence across different levels of development, coherence in successive curricular reforms at an education system's different stages of development may also be a critical success criterion. The chapters in the book are important data for the readers to interpret as evidence and to create arguments, so that they can deepen their understanding of how science as inquiry can be enacted in a deep and sustained manner in all the classrooms throughout a jurisdiction.

National Institute of Education
Nanyang Technological University
Singapore

Wing On Lee
David Wei Loong Hung
Laik Woon Teh

Foreword

Science education has played an important role in the development of Singapore. Since the country's independence in 1965, a strong science education has helped equip Singaporeans with the knowledge and competencies that transformed Singapore into a modern and economically vibrant city-state. Fifty years on, it is befitting, therefore, that *Inquiry into the Singapore Science Classroom: Research and Practices* has been published and dedicated to a discourse on science education in Singapore.

There are a few other reasons why this publication is opportune. Firstly, students from Singapore have consistently performed well in science in international benchmarking studies. This has generated much interest and curiosity from science educators and scholars about the policies and practices surrounding science teaching and learning in Singapore's classrooms. This collection of essays will help all interested educators understand better the pedagogical practices and the curricular and assessment policies we developed in Singapore. In addition, through the history of science education in Singapore, readers will appreciate the contexts within which the teaching and learning of science has evolved.

Besides being the first comprehensive collection of essays on science education in Singapore, this volume is celebrated as a joint production between science educators from the National Institute of Education and senior specialists from the Ministry of Education. This reflects the traditionally strong relationship between the Ministry of Education and Singapore's teacher education institution – a partnership that has contributed in no small way to the quality of science teaching and learning in the classroom.

This book marks an important milestone as we reflect on the achievement and progress made in the past five decades of science education in Singapore and also reflects the challenges and opportunities available as we look forward to the next decade. Researchers and teachers will hopefully use this publication to strengthen their understanding of science education in Singapore and initiate conversations on how we want to move ahead together.

Deputy Director-General of Education (Curriculum) Wong Siew Hoong
Ministry of Education
Singapore

Preface

Aik Ling: We have decided to present this Preface as a metalogue to capture the active exchanges of our views, sometimes diverse and sometimes converging. The three of us (Chew Leng, Shirley and myself) come from different backgrounds and represent different perspectives about science education. Chew Leng works for the Ministry of Education and hence has a comprehensive idea about policies and their implementation. Shirley, on the other hand, is a natural scientist who has a keen interest in education. She offers the perspective of science from the angle of a practising scientist. Well, I am a science educator who works in the area of science inquiry. We came together because we have a passion for science education and for the progress of science education in Singapore.

Chew Leng: Indeed, we share a common passion for science teaching and learning in Singapore. Aik Ling, you kick-started this book project. Perhaps you could share with us the compelling reasons for launching this book project and why the focus on inquiry science?

Aik Ling: We have intended for this book to showcase science education research, particularly those focusing on science as inquiry in Singapore. There are compelling reasons why this book is important. Firstly, with stellar performance by Singapore students in international comparative studies, there are surprisingly few science education research publications coming from Singapore (Lee, Wu, & Tsai, 2009). Secondly, it has been 10 years since the Centre for Research in Pedagogy and Practice was set up. In these 10 years, we have seen the development, evolution and maturation of science education research in Singapore. It is timely to take stock of what forms of inquiry have taken place in our classrooms. Lastly, we hope that the ideas presented in this book will help to promote reflection in shaping the direction that science education and science education research can move forward. Chew Leng, how do you think this book will help policymakers?

Chew Leng: I thought the book is timely. We are five decades into science education in Singapore. We want to look back and see from where we came and look forward to what we aspire towards, but with the full knowledge of where we are now. The research findings are therefore helpful as they provide policymakers with honest and evidence-based feedback, particularly about the current reality of inquiry practice in the classrooms, the inroads made and the challenges to which we need to pay attention. Having said that, I must confess that I felt a little despondent when I read the chapters that describe the practice of inquiry in schools. There appear to be gaps and issues with our practices. Are they all telling us that we are not making any headway in the practice of inquiry in schools?

Aik Ling: I share the similar initial reaction with Chew Leng – I mean we do so well in international comparative studies. Surely, we must be doing good things in the schools! The processes must be sound before we can have the outcomes we can be proud of. Many chapters in the book painted a less than rosy picture about science as inquiry in the classrooms. But having reflected upon the chapters, I realise that they epitomise what we value in Singapore – critical self-reflection in our pursuit for continual improvement, change and excellence. This is likened to Freire's argument for the central role of dialogue and reflection. If we put the ideals and contradictions of science as inquiry in the centre as the object of our inquiry (the nucleus), with teachers, science education researchers and policymakers as different stakeholders of science education, then in order to establish a common understanding and goal of science as inquiry, all parties need to undergo many cycles of dialogue and reflection.

Shirley: Mindset changes in people are the most difficult to achieve as the acceptance of change is never easy. The most important point to note here is the awareness for the need to implement inquiry into the teaching of science in Singapore schools. As long as all the stakeholders concerned are open to and receptive to this need, there is hope!

Aik Ling: The more the different groups of people meet to engage in genuine dialogue, the more issues and contradictions they reflect upon. The more the groups divide and then reintegrate as a whole, the more closely we will approach the nucleus (which in this case, could be contradictions of the practice of science as inquiry in schools). It is only through these multiple interactions and exposition of the same reality that the level of awareness of all stakeholders can be raised. As such, the seemingly 'negative' situations described by some of our colleagues in the chapters should serve as platforms of dialogue and reflection such that the various groups can collectively move (rather than as individual groups representing only specific interests) towards the nucleus and develop a more holistic understanding of science as inquiry in Singapore.

Chew Leng: I think I see where you are coming from Aik Ling. If we can broadly categorise the chapters written by our colleagues into two groups – issues and suggestions – I think we have more suggestions than issues. Chapter 4 (describing teachers' concerns and assessment literacy in science), Chap. 6 (describing how teachers work around structural constraints to enact inquiry), Chap. 9 (describing contradictions between intended outcomes and practice) and Chap. 15 (describing the state of science-related article for general scientific literacy) can be described as chapters highlighting gaps in our practices of science as inquiry.

Shirley: And Chap. 5 (describing a programme to help physics teachers work with inquiry), Chap. 8 (describing a framework to facilitate teachers planning for science as inquiry), Chap. 10 (describing technology-enabled knowledge building in science), Chap. 11 (describing context for meaningful learning of inorganic chemistry), Chap. 13 (highlighting the possibility of homework as a means of informal learning) and finally Chap. 14 (describing the role of science centres in bringing science to the general population) are chapters that provide suggestions about how we can potentially move forward from where we are at the moment. They provide ideas about possible interventions and their likely outcomes.

Aik Ling: Well analysed. In fact, if we examine the ideas presented in the various chapters more closely, there are more authors presenting viable ideas and the way forward than those presenting the gaps in our current practice. Personally, I think we have a good balance of ideas and issues presented in this volume. The key issues raised by our colleagues with regard to the practice of science as inquiry in the Singapore classroom related to (1) teachers' ideas and their practices, (2) constraints posed at a systemic level, (3) students' competencies and readiness to learn through inquiry and (4) need for greater awareness of the role of informal learning avenues in science education.

Chew Leng: I concur with Aik Ling. Our passion is the unifying force amidst our differences in experiences. This is the first time I am involved in editing a book. Thanks, Shirley and Aik Ling for guiding me through the process. But, as usual, I have questions and questions. First off, why publish a book? Why not just send the articles to journals? Won't it save us a lot more work?

Aik Ling: It certainly will save us a lot of time. Books, however, serve slightly different purposes from articles in journals. Articles in journals are mainly aimed at a scholarly audience within a specialised knowledge domain. For example, articles published in a science education journal are targeted at researchers working in the field of understanding how people learn science. Even within science education, there are also subdomains of specialisation where some scholars examine only conceptual change, while there are others who delve into informal learning. Books, on the other hand, are likely to be less specialised (although a certain level of specialisation still exists). While still intended for scholarly readers, books such as these are broader and more divergent in the ideas that are expressed. Some scholars describe books as 'ideas playground' since they allow new and emerging ideas to be expressed and trialled. The format of books also offers greater depth of the ideas to be explored – this is a privilege and luxury that journal articles cannot afford. Hence, while both books and journal articles are important platforms for expression of scholarly ideas, their format and purposes differ to some extent. Both are hence necessary.

Shirley: While there is the perception that books may not be as 'specialised' as journal articles, they are 'specialised' in their own unique way. Let me explain by first distinguishing between an authored book and an edited volume, as well as comparisons with a special edition or proceeding published by a journal. Book authors are generally experts in their respective fields with a broad perspectives of the various topics featured in each chapter. In contrast, an edited book is conceptualised around a central theme by the editor(s) involved; after which, scholars and experts in the fields related to the theme are invited to contribute various chapters. This is quite similar to published proceedings of conferences and symposia that constitute special issues of journals, except that there is no preceding conference.

Chew Leng: Then how do you ensure high quality and robustness?

Aik Ling: Quality is indeed important in a piece of publication. Scholarly quality can be characterised by different aspects – (1) soundness of ideas that are expressed by the authors, (2) originality of the research that is conducted, (3) proposition of new insights into issues that are crucial for change and improvement and (4) robustness of research design that is employed in the study. Certainly, the characteristics of scholarly quality list are not exhaustive, but are fundamental for a good piece of work. In this edited volume, you would have realised that we read through each of the contributions by our colleagues with these conditions in mind.

Within the research community, the idea of blind (or anonymous) peer review or open peer review is also often taken as another measure of worthiness of a journal or book. While there are major criticisms (including that of differential power interplay) of blind peer review and open peer review process (Osborne & Brady, 2002), it is still used as a measure for worthiness of publications. Many universities rate peer review publications higher in terms of rigour and prestigious compared with non-peer-reviewed ones. This, of course, stems from the belief that ideas that are endorsed and accepted within the community of practising scholars are more valid than those that are not. As such, while we are cognisant of the pitfalls of a peer review process, we also recognise and embrace the process of peer review for every article in this book as we see the process as one that will help our peers develop their ideas further. Each article was subjected to a single blind peer review, and the authors of each chapter made amendments based on the recommendations. Their amended chapters were subsequently reviewed by the editors, and peer discussion ensued to ensure that the final product is something of which the authors can be proud. As an editor, while I am eager to suggest changes for each chapter to improve the quality, I am always cognisant to keep the authors' voices and the original ideas that they want to express. That, to me is one of the most important aspects of an edited book volume.

Shirley: And we are indeed grateful for friends in Singapore, Ken, Barbara, Benny and John, who have anchored each section with a commentary. They are busy people with numerous speaking engagements as well as heavy demands on research work and publication commitments – yet they have taken time to pen their views. Their commentaries help to consolidate the ideas presented by individual authors and also position the ideas and issues in Singapore against the international literature. We invited these four well-published science education academics because of their expertise in science education, as well as their familiarity with the science education scene in Singapore. We hope that the readers will read their views and reflect upon them. This is also an attempt to allow the authors, the commentary writers and the readers of this book to 'interact' and 'exchange' ideas with each other through reading the chapters.

Aik Ling: We must also not forget that Chaps. 1 and 2 set out important contexts for us to understand the state of science education in Singapore and the journey we have taken to arrive at this peak today. As a Singaporean actively involved as a teacher educator and directly with students, I am certainly very proud of how far we have come.

Chew Leng: Indeed, the ideas presented in all of the chapters are generally tangible and doable ideas that policymakers, teachers, students and school administrators can adopt, adapt and improve upon. After working on the chapter, tracing the development of science education in the last five decades, I am excited about how far we can all help make the science classrooms an exciting and engaging centre of learning and inquiry in the next 25 years of science education in Singapore.

National Institute of Education, Singapore — Aik-Ling Tan
Research and Curriculum Planning Ministry of Education, Singapore — Chew-Leng Poon
National Institute of Education, Singapore — Shirley S. L. Lim

References

Freire, P. (2009). *Pedagogy of the oppressed* (30th anniversary ed.). New York: Continuum.

Lee, M.-H., Wu, Y.-T., & Tsai, C.-C. (2009). Research and trends in science education from 2003–2007: A content analysis of publication in selected journals. *International Journal of Science Education, 31*, 1999–2020.

Osborne, M. D., & Brady, D. J. (2002). The room is long and narrow. *Research in Science Education, 32*, 163–169.

Contents

Chapter 1
Five Decades of Science Education in Singapore

Chew-Leng Poon

Abstract Beyond educational considerations, science education in Singapore has been shaped by historical, geopolitical and societal forces through more than five decades since Singapore's self-governance in 1959. This chapter, which presents a historical perspective of the development of science education in Singapore, is aimed at providing the social and educational contexts critical to understanding and appreciating the policies and practices discussed and debated in the subsequent chapters of this book. The narrative, pieced together by examining published and unpublished works of scholars, researchers and the Ministry of Education, traces the path of science education from an early focus on providing skilled manpower for an industrialising Singapore in the 1960s to an inquiry-centred science education to prepare students for living and working in the twenty-first century. This narrative also sheds light on the people and policies that have helped propel the small nation from its early days as a trading port to one respected internationally for the sustained achievement of its students in international mathematics and science studies.

Keywords Singapore science education • History of science education • Inquiry science • Education policies

Introduction

This book, *Inquiry into the Singapore Science Classroom: Research and Practices*, is a collaborative effort of a group of science educators and science policy makers in Singapore who are drawn together to document the historical, geopolitical and societal forces that shaped science education through more than five decades since

C.-L. Poon (✉)
Research and Evaluation, Planning Division, Ministry of Education, 285 Ghim Moh Road, Singapore 279622, Singapore
e-mail: poon_chew_leng@moe.gov.sg

A.-L. Tan et al. (eds.), *Inquiry into the Singapore Science Classroom*, Education Innovation Series, DOI 10.1007/978-981-4585-78-1_1,

Singapore's self-governance in 1959. It also captures the state of science teaching and learning in the classrooms that has propelled the nation from a little known trading port in the 1960s to one respected internationally for its sustained achievement in mathematics and science education (Darling-Hammond, 2010; Mourshed, Chijioke, & Barber, 2010; OECD, 2010c).

Four themes are developed in this book, with chapters on the science curriculum; teaching, learning and assessment practices; education of science teachers; and the range and roles of informal science education in Singapore. This chapter presents a historical perspective of science education in Singapore schools. It provides an overview of science education in Singapore and an introduction to some of the social and educational contexts that will be helpful for readers to understand and appreciate the policies and practices discussed and debated in the subsequent chapters. Detailed discussions of trends and issues specific to each area of science education are found in the other chapters of this book.

The five decades of education since Singapore's breakaway from British rule have often been marked out as three periods that corresponded to three distinctive philosophies or driving force underlying each of this period (Goh & Gopinathan, 2008):

1959–1978: The survival-driven phase
1979–1996: The efficiency-driven phase
1997–2011: The ability-driven phase

These three periods of Singapore's educational transformation are used to organise the science education narrative in this chapter. The rationale is that the same geopolitical and socio-economic forces that were shaping the Singapore education system were also impacting the development of science education.

I begin this narrative with a brief overview of Singapore, as size and geopolitical position play an important role in the development of Singapore's national psyche. A broad sweep of the current education system follows to provide the general education context. The rest of the chapter then focuses on the development of science education through the three periods.

Mode of Inquiry

One of the challenges faced in writing this chapter was the thin volume of publications that described the state of science education in Singapore, particularly those that referred to the earlier years following independence in 1965. Perhaps this difficulty only accentuates the urgency for this book to fill this gap in knowledge about the historical development of science education in Singapore.

Much of the evidence was pieced together by reviewing publications and documents that made references to science education in Singapore – including public documents from the Singapore Ministry of Education (MOE), speeches by Ministers for Education (as policies are often announced or explained through

Minister's speeches) and published and unpublished works of scholars and researchers related to science education in Singapore. While attempts were made to search for available local literature on science education residing at the National Institute of Education's database, it was inevitable that some could have been left out and I welcome other scholars to fill in the gaps in information.

To supplement the literature review, I conducted a face to face interview with five veteran science educators who, collectively, had wide knowledge of science teaching, curriculum development, teacher professional development and science education leadership from the late 1960s until their recent retirement. They filled gaps in knowledge, particularly of the early years of science education in Singapore and sent me publications that have long gone out of print. More importantly, they provided an additional dimension to this narrative – the names and faces of the pioneers of science education in Singapore and the roles that they played in shaping science education.

And finally, in presenting the events in this narrative, I am ever conscious of Roth's (2005) perceptive insight that at the same time I was 'gazing' (p. 256) back at these events from a distance of time, I was framing them in parallel with my own experiences as a science student (from 1971), a science teacher (from 1986), a school leader and a science curriculum researcher and policy maker (from 2000).

Overview of Singapore and General Education in Singapore

Singapore

Singapore is an independent island city-state of five million residents packed into all of 710 km^2 of land. It is about the size of the city of Edmonton in Alberta, Canada, or of Budapest in Hungary and smaller than Lake Geneva in Switzerland. Finland also has about five million residents but is 500 times the size of Singapore. The small island of Singapore is not blessed with any natural resources except perhaps for its strategic geographical location at the southern tip of the Malay Peninsula, right at the crossroads of major sea lanes in Asia. Having no other resources, there is immense attention paid to developing human capital. Education is therefore a key priority of both the government and her citizens.

Singapore's location at the major crossroads of Asia attracts a diverse group of immigrants since her early days as a seaport. Singapore is home to a multi-ethnic and multilingual society where 74 % of the population are Chinese, 14 % are Malays, 9 % are Indians and 3 % are of other ethnicities. As a reflection of its multi-ethnic heritage, there are four official languages in Singapore – English (which is the language of administration and the language of instruction in schools), Malay (which is also the national language), Chinese (Mandarin) and Tamil. Beyond these four official languages, the languages spoken at home and at other social spaces include Chinese dialects (such as Hokkien, Cantonese, Teochew) and

Indian languages such as Bengali, Gujarati, Hindi, Punjabi and Urdu. This diversity of ethnicity, culture and language is mirrored in the population of children that attends the schools.

Singapore's 2011 Gross National Income (GNI) exceeds S$300 billion, yielding a per capita GNI of S$61,692 (about USD 49,000) (Department of Statistics Singapore, 2012). Around 3.5 % of Singapore's Gross Domestic Product (GDP) is invested annually in education – to fund education in primary and secondary schools and preuniversity and post-secondary educational institutions (such as the Institute of Technical Education, Polytechnics and Universities). Although this proportion is small by international standards, the more than ten billion Singapore dollars (about USD eight billion) represents about 20 % of government expenditure (Ministry of Education, Singapore [MOE], 2011b) and is the second highest amount, after expenditure on defence. This again reflects the national priority given to education.

General Education in Singapore

Pre-school education in Singapore is not mandatory and can start as early as age four for nursery classes. Children can also enrol for kindergarten levels 1 and 2 at age 5 and 6, respectively. Pre-school education is largely provided by the private sector and community-based groups (such as the People's Association and religious organisations). Based on data reported in the 2009 Programme for International Student Assessment (PISA), the preprimary school participation rate for Singapore is 98 %, slightly higher than the average of 92 % for countries in the Organisation for Economic Co-operation and Development (OECD) and comparable to the 98 % in the United States (OECD, 2010b).

Formal schooling in Singapore begins at age 7. About half a million students are enrolled in public schools largely funded by the government – 173 primary schools, 155 secondary schools, 15 mixed-level schools (either schools with primary and secondary levels or schools with secondary and junior college levels) and 13 preuniversity institutions. A small proportion of students, about 2 %, are enrolled in private schools and schools for students who require more dedicated support to meet their special needs (MOE, 2011b).

Primary school education, Primaries 1–6 (generally for ages 7–12), is compulsory. Schools follow a national curriculum which emphasises broad-based holistic education. As students come from diverse home language backgrounds (with different levels of exposure to English and their mother tongue languages), more emphasis is given to the learning of the English language and mother tongue languages (could be Mandarin, Malay or Tamil) in the early primary school years. For this reason, science education begins only from Primary 3. Students also learn mathematics, social studies, art and music, in addition to health education, physical education and civics and citizenship education. Class size is generally 30 in Primaries 1 and 2 and about 40 in Primaries 3–6. The actual student to teacher ratio in the primary schools is smaller, at 19:1 (MOE, 2011b), as additional teachers are provided for the teaching of mother tongue languages, art, physical education and music. At the end of primary school (at Primary 6), students sit for the Primary School Leaving Examinations (PSLE[1]) in four subjects – English language, mother tongue language, mathematics and science. This reflects the importance of science education even though the learning of science begins only from Primary 3.

Secondary school education is almost universal, though it is not compulsory. Only about 1 % of students do not complete secondary education (Ministry of Education, Singapore [MOE], 2011a). Secondary education is characterised by a diversity of pathways – both in terms of courses offered and the diversity of schools which offer niche programmes that complement the national curriculum. Students can choose between various subjects and electives at the upper secondary levels (see curriculum framework in Fig. 1.1). There are also specialised schools that have an enriched programme for the arts, sports and science, mathematics and technology. Students offer a 4- or 5-year programme that culminates in the General Certificate of Education Normal or Ordinary (GCE 'N' or 'O') examinations, and there is flexibility for transfer between courses (MOE, 2011a). Since 2004, some schools have offered a 6-year integrated programme where students progress to preuniversity education without sitting for the national GCE 'N' or 'O' level examinations. These schools offer an enriched programme to broaden students' learning (Ministry of Education, Singapore [MOE], 2011c). The student to teacher ratio in secondary schools is 16:1 (MOE, 2011b), with actual class sizes of between 30 and 40 students.

General science is a mandatory part of the national curriculum up to the Secondary 2 level. Although science is an elective subject from upper secondary onwards, more than 90 % of upper secondary students offer at least one science subject.

About 3 in 10 of each Primary 1 cohort choose to study in a preuniversity institution (2-year junior college or 3-year Millennia Institute[2]), which prepares students for university education. Students in preuniversity institutions have a wide range of subject choices that include the languages, mathematics, sciences, humanities and the arts. About 86 % of preuniversity students offer at least one science subject (Ng, 2010).

Post-secondary education is characterised by greater diversity of pathways and flexibility of transfers between pathways (see chart on the educational pathways in Appendix). About 46 % (MOE, 2011b) of students enrol in 3-year diploma programmes offered by the five polytechnics in Singapore and the two tertiary arts institutions. These institutions offer diplomas that are closely linked to industries, in fields such as engineering, biosciences, business, digital media, the arts and design. Another 21 % (MOE) of secondary school graduates enrol in the Institute of

[1] The PSLE is a national examination taken by all students at the end of Primary 6.

[2] Millennia Institute offers a 3-year curriculum that is largely similar to that of junior colleges.

CCA: Co-curricular activities
CIP: Community Involvement Programme
CME: Civics & Moral education
NE: National Education
PCCG: Pastoral Care & Career Guidance
PE: Physical Education
PW: Project Work

Fig. 1.1 Curriculum framework for secondary school (*CCA* co-curricular activities, *CIP* community involvement programme, *CME* civics and moral education, *NE* National education, *PCCG* pastoral care and career guidance, *PE* physical education, *PW* project work) (Source: MOE, 2012, p. 7. Reproduced with permission from the Singapore Ministry of Education, 2012)

Technical Education (ITE) to read vocational courses in engineering, technology, business and hospitality services. Overall, about two-thirds of students in the polytechnics and Institute of Technical Education read courses in science and technology (MOE).

About a quarter of students from each Primary 1 cohort gains admission to one of the four government-funded autonomous universities in Singapore. Another one-fifth enrols in overseas or privately funded local universities (MOE, 2011a). Overall, about half of university degree students enrol in science and technology courses (MOE, 2011b).

Science Education in the Survival-Driven Phase (1959–1978)

The Context for Science Education

From self-governance in 1959 through the early years of post-independence in the 1960s, one of the key priorities of the government was to build up its fledgling economy through industrialisation (Goh & Gopinathan, 2008; Mackie, 1971). Industrialisation was seen as a viable option by the government at that time (Chua, 2010) to meet the challenges of high unemployment (about 14 % in 1959) coupled with a rapidly growing population. Science and technical education therefore became an area of focus right from the start of nation building to ensure that there was a pipeline of semi-skilled workers for the manufacturing industries that were attracted to set up shop in the country (Bhathal, 1982; Chua, 2010). This focus on science and technology (and mathematics) education has remained a key cornerstone of Singapore's education and economic strategy until this day even as the type of industries in Singapore has changed (e.g., from electronics manufacturing to high-end research and development industries). As the new industries demanded different skill sets and knowledge, the corresponding science and technology courses (particularly at the polytechnics and universities) also evolved to include the life sciences, precision engineering, clean energy, clean water technology and so on.

Science Teaching and Teachers

According to Yeow (1982), science first became part of the formal primary school curriculum in 1959, with a recommended syllabus and curriculum time (of 1.25 h per week for Primaries 1–3 and 2.25 h per week for Primaries 4–6). The aims of teaching science were to '(a) create an interest in nature and its working; (b) encourage the natural curiosity of children and to inculcate a spirit of inquiry; and (c) train the children to observe, to experiment and to seek further knowledge' (Yeow, p. 160).

Support for the syllabus came in the form of textbooks written by private publishers and approved by the Ministry of Education. Apart from the textbooks, there was little other support from the central Ministry in terms of teaching aids, teaching guides and in-service courses for science teachers (Yeow, 1982). The 1963 report by the *Commission of Inquiry into Education in Singapore* reported that science teaching consisted of verbal dictation of notes by teachers and that the lack of facilities and equipment meant students had limited exposure to laboratory activities (reported in Mackie, 1971). Yeow also characterised science lessons during this period as 'reading lessons' (p. 160) – literally reading and 'underlining' texts from textbooks. This was compounded by the fact that there were not enough science teachers and that a large number of the primary science teachers had neither the science background nor the science training to teach the subject well.

Cahill (1984) recorded that, during this period, science was taught in English or in the vernacular languages of Mandarin, Malay and Tamil, reflecting the education landscape where vernacular-language schools thrived alongside English-medium schools (Ministry of Education, Singapore [MOE], 2007). As part of its strategy of nation building through education, the government began building more English-medium schools. Vernacular schools were also encouraged to switch to teaching mathematics and science in English. One of the issues then was that students in these schools had to struggle with picking up the English language while learning science. The Ministry of Education tried to address these issues by posting teachers from English-medium schools to the vernacular schools and by offering the science PSLE papers in both the English language and in the vernacular language. By 1978, at the end of the survival-driven phase, science was taught only in English, beginning at Primaries 1–3 (Cahill).

With the rapid expansion of schools in the 1970s to cater to a growing student population, there was mass recruitment of teachers (Gopinathan, Wong, & Tang, 2008). The teaching of science was still regarded as undesirable by teachers, with Yeow attributing this to the lack of specialised training in science teaching:

> Many of these teachers, being unsure of themselves, discourage questions and discussion and tend to be authoritarian. The science lessons consist mainly of exposition of factual information through the talk and chalk method. Pupils memorise facts solely for regurgitation at tests and examinations. Scant attention is paid to the objective of arousing pupils' curiosity in their environment. Little is done to provide pupils with a variety of experiences through active involvement in conducting experiments and investigations which lead to the acquisition of the basic process skills and relevant concepts. (Yeow, 1982, p. 163)

It was only after the establishment of the Institute of Education in 1973[3] that specialised training on the teaching of science was made widely available (Stewart, 2010). Based on the accounts of science educators interviewed, the professional organisations worked in partnership with the Ministry of Education and the Institute of Education in upgrading the skills of in-service science teachers. Most notably, the Science Teachers Association of Singapore (STAS in short) was a community of science teachers who organised workshops, seminars and exhibitions for their own fraternity to improve their craft of teaching science. Some of the outstanding pioneer leaders who championed and promoted in-service professional development for science teachers included Sng Yew Chong, John Yip, Lim Jit Poh, Michael Liew, Mrs Ang Wai Hoong, Koh Peng Tiam, Ho Tat Kin and Chia Chye Kiat[4] (as mentioned by those interviewed).

Non-formal Science Education

The STAS also organised activities to promote the learning of science among students. The first competitive Science Fair was organised by STAS in 1969 (Bhathal, 1982) and has been held annually ever since. The Science Council of Singapore[5] also attempted to popularise the learning of science through sponsorship of related activities, including the Science Quiz that was first televised nationwide in 1972. This later became the 'Science and Industry Quiz' (Bhathal).

The Science Centre Singapore was another professional body that played a significant role in supporting the teaching of science in schools. The Science Centre

[3] Prior to the setting up of the Institute of Education in 1973, generic teacher training was carried out by the Teacher Training College.

[4] And probably many more names I have left out. I do apologise for the gap in knowledge and welcome information on other pioneer science teacher leaders I have left out.

[5] The Science Council of Singapore was first established in 1967 to drive government policies on science and technology. In 1991, it morphed into the National Science and Technology Board, the predecessor of the current Agency for Science, Technology and Research (A*STAR was established in 2001).

Singapore opened in 1977, promoting the informal learning of science through interactive exhibits, galleries, talks, demonstrations, film shows, publications, science camps and science competitions (Bhathal, 1982). More details of the role of the Science Centre Singapore in informal science education in Singapore, then and now, are provided in Chapter 14 of this book.

Summary Comments

In looking back at Singapore's journey from basic manufacturing (such as electronics) to high-end research and development in the biomedical and the physical science fields, Dr Ng Eng Hen, Singapore's Minister for Education from 2008 to 2011, shared his insights on how the early year decision to emphasise science and technology in education was an economic strategy borne out of limited choice but became a fortuitous policy decision:

> The journey for Singapore to venture into R&D was not an intuitive one, 45 years ago at our founding. Our forays into this field were as much a response to competition as it was a default choice when looking at limited options for our economic strategy. Fortuitously, we had started well by ensuring that our students had good capabilities in science and technology. This was a necessity because bereft of natural resources apart from human capital and without a hinterland, our economic strategies necessarily focused on providing highly skilled labour selling products to global markets. As we could not compete on size, we needed to leverage on science and technology to amplify our strengths and extend our reach. (Ng, 2010, pp. 1–2)

The survival-driven phase was indeed a challenging period in Singapore's history as the country sought to address pressing and immediate economic and social needs. Nevertheless, education policy decisions made during this period had a far-reaching impact than the immediate, as pointed out by Dr Ng. Whether it was good fortune or good foresight, the decision to develop capabilities in science and technology set the course of curricular emphasis and resource allocation for science education in Singapore schools and tertiary institutions for the next three decades.

Science Education in the Efficiency-Driven Phase (1979–1996)

The Context for Science Education

Education historians in Singapore have never missed a key milestone in Singapore's education history – the 1978 Goh Keng Swee report which marked the beginning of the efficiency-driven phase of education in Singapore. The report had a deep and far-reaching impact on the education landscape, with policies that left a strong legacy until today. Policies arising from this period included streaming, the gifted education programme, and the setting up of the Curriculum Development Institute of Singapore (CDIS) to develop textbooks and other teaching and learning resources.

The beginning of this efficiency-driven phase of education was motivated largely by the concern over educational 'wastage' during the survival-driven phase (Goh, 1978). The attrition rate at the time of the Goh report was 29 % for primary education and 36 % for secondary education. This worked out to about 55 % of a cohort who did not complete secondary school education (probably more if we took into consideration students who did not enter secondary schools even though they qualified for entry), compared to nearly zero attrition for Japan during the same period (Chua, 2010). Only 14 % went on to preuniversity and 9 % to university. Another related problem raised in the Goh report was the struggle some students faced in acquiring proficiency in both the English language and their mother tongue language (and for many, the 'mother tongue language' they offered in school may not be the mother tongue language they spoke at home).

The policies that were implemented to address these two key problems had an impact on science education. Under the New Education System in the post-Goh report, science was taught only from Primary 3 onwards as more curriculum time was allocated for language learning in Primary 1 and 2 (see Cahill, 1984; Tan, 1989). This policy is still in place today, as the dominant home language of students continued to remain diverse, with students coming to school with different levels of exposure to English. This distinguishes Singapore from most education systems where science education generally begins from grade 1, if not earlier (Mullis et al., 2012). One could perhaps argue that this policy is an illustration of how science education policies (or any other discipline for that matter) are more often shaped by societal forces than the educational needs of students. Often, as illustrated in this policy, the reality is that the two demands – societal forces and educational needs of the child – must and can be met. While it cannot be denied that the societal milieu is an important determinant of science education policies, in this case, the needs of the student did not play a diminished role. Some studies (e.g., Chall, 1996) have shown that students read to learn by the time they reach the age of about 10 and poor reading skills could pose some impediment to accessing learning in other disciplines. Analysis of the grade 4 (Primary 4) data from the 2011 Trends in International Mathematics and Science Study (TIMSS) and Progress in International Reading Literacy Study (PIRLS) also showed a strong relationship between reading literacy and science scores for Singapore students (correlation of 0.77). Given that a large proportion of Singapore students come from homes in which English is not spoken predominantly, allowing students to focus on language acquisition in their early years of primary education was therefore a way to also address the educational needs of the students.

Under the New Education System, students were also streamed into different ability groupings, with differentiated curricula that were intended to cater to the different pace of learning. At the primary level, only students in the normal and extended streams sat for the science examinations in the PSLE. Students from the monolingual stream did not offer science. At the secondary level, students in the normal course had an extra year (a total of 5 years in secondary school) to complete the same science syllabuses as students in the Express course before sitting for their GCE O-level examinations. However, those who did not go on to Secondary Five after their Secondary 4 Normal (Academic) examinations would experience a slimmer science syllabus compared to their counterparts from the Express course.

Teaching and Learning Science

During the efficiency-driven phase, steps were taken to address the challenge of effective science (and other subjects) teaching following the expansion of schools in the survival-driven phase. The curricular focus at the lower secondary level was on the development of 'inquiring minds' (Tan, 1989, p. 64), while the aim of the primary science syllabus was to engender 'learning of process skills, for intellectual growth and for nurturing of creativity' (Tan, p. 64). Despite these high aspirations of an inquiry-centric approach to the learning of science, the reality that had to be grappled with then was the readiness of teachers to deliver the contents of the syllabus. The CDIS was established in 1980 to address some of these issues.

Prior to CDIS, textbooks were generally not produced locally. CDIS brought teachers together to author primary and lower secondary science textbooks, student workbooks, teaching guides and materials and audiovisual aids that were aligned to the local science syllabuses. According to Gopinathan et al. (2008), CDIS textbooks and materials were an attempt to address the effectiveness of science teaching by 'teacher proofing' the curriculum – meaning that teachers who struggled with teaching the science curriculum could follow the textbook and teaching guide closely to maintain a certain minimum level of fidelity of curriculum implementation.

The CDIS teaching materials that were produced during this period were said to emphasise 'enquiry-based science learning' (Cahill, 1984). Examination of the first set of locally produced primary science curriculum materials (implemented from 1982) gave some indications that the focus of the 'enquiry' materials was on simple science process skills such as observation skills, collection and interpretation of data and classification and measurement skills (Singham, 1987). Despite the emphasis on 'problem-solving through observation, inquiry and discovery' (Tan, 1989, p. 65), Singham's study of classrooms showed that primary science teachers were unfamiliar with some of the basic process skills such as the setting up and testing of hypotheses. There was hardly any evidence of students raising questions in class or planning investigations. Indeed, teachers used the CDIS textbooks, but focussed on the science content more than the inquiry approach to teaching and learning as this mattered more in preparing students for the national examinations. A decade later, in a small study involving 97 students from two primary schools by Singh (1999), students were also found to be unfamiliar with the use of basic scientific instruments such as thermometers and stopwatches and had difficulties interpreting experimental results, corroborating with general impressions that primary science lessons were mostly textbook bound.

Learning science through experimental work and laboratory activities were emphasised in the science national curriculum. Cahill (1984) documented that in 1984, 60 % of primary schools had a science garden to support the learning of plant life and habitats in the syllabus. Today, every primary school is provisioned with an eco-garden, reflecting the enduring importance of the biological sciences in the primary science curriculum.

At the secondary school level, science laboratories were built to support the learning and assessment of science through laboratory activities. To emphasise that laboratory activities were integral to the learning of science through experimentation and investigations, the MOE recommended that between one-third to half of the curriculum time be devoted to laboratory-based activities. Science practical (usually laboratory-based investigations or activities) tests were also included as part of the national examinations for pure science subjects to signal the importance of laboratory activities. For General Science and Combined Science (various combinations of biology, chemistry and physics in a single subject), questions on laboratory activities were incorporated into the written papers in place of science practical tests. There were no records, however, of the extent that laboratory activities were an integral part of science teaching and learning.

Science Teacher Development

One of the pressing issues that had to be dealt with during the early years of the efficiency-driven phase was teacher preparation and development that went beyond the 'hardware' of textbooks, workbooks, science gardens and laboratories. In the 1980s, many of the science teachers in the primary schools did not have a strong background in science – of the approximately 7,000 science teachers, only 15 % had a degree in science and these were deployed to the upper secondary and preuniversity levels (Cahill, 1984). Some of the primary science teachers did not read any science or science-related subjects beyond the O levels.

The issue was not easily addressed, especially at the primary school level. Primary school teachers were not subject specialists and teacher preparation had to give attention to many other areas, particularly in the teaching of English language and mathematics. In fact, science was not offered by all prospective teachers in primary school teacher preparation courses. In schools, however, these teachers could still be deployed to teach science.

During my interview of the five retired science educators, I was told that the Science Teachers' Association supported the onerous work of the Institute of Education through in-service training. These trainings took place on weekends and after-school hours, augmenting teachers' content knowledge and enhancing their science pedagogical skills. The trainers were science teachers themselves who were passionate about science education and therefore voluntarily spent their weekends and after-school time to train other teachers. Schools also organised school-based workshops to share effective teaching practices. Tan (1989) mentioned workshops such as ‘activity-based science lessons’, ‘skilful science teaching’ and ‘use of questioning techniques to promote higher-order thinking in science teaching’. There was a strong sense of ‘ownership’ and ‘community’ in the way that the science teachers banded together to help themselves – reminiscent of the culture and spirit of teacher leadership and ownership that today’s Academy of Singapore Teachers is actively cultivating.

Non-formal Science Education

During the efficiency-driven phase, significant programmes were launched to enthuse and engage students in science through non-formal settings. In 1982, the ‘Young Scientist Badge’ was jointly launched by STAS, the Singapore Science Centre, Singapore Association for the Advancement of Science and the Singapore National Academy of Science. Primary school students could earn a series of ‘young scientist’ badges by conducting various simple investigations in different fields such as ecology, zoology and astronomy. Students could do this independently or through their school science clubs or societies. Thirty years on, this programme remains popular among primary school students.

For the older youths, there were science camps that brought urban students closer to nature. A science research programme was also launched where selected junior college students were mentored by science researchers from the universities. The Singapore Youth Science Fortnight offered a range of activities such as science forums and demonstrations, science congress where students presented their research findings, film shows and opportunities to ‘meet the scientists’ (Bhathal, 1982). Students also participated in overseas enrichment programmes such as the International Summer Institute at Weizmann Institute of Science in Israel, the London International Youth Science Fortnight, Thailand Youth Science Week and the Centre for Excellence in Education Summer Program in the USA. Singapore students also started to pit their talents against their counterparts from all over the world in competitions such as the International Chemistry Olympiad and the International Physics Olympiad.

Summary Comments

The survival-driven phase was marked by responses to immense challenges that reflected the economic and political challenges of the period. During the efficiency-driven phase, there was more time to pause, take stock and review student learning outcomes and put in place pragmatic policies and programmes which laid the foundation for the years ahead – a foundation that might sometimes be seen as more system efficiency driven than student centric (e.g., 'teacher-proof' textbooks, streaming) but nevertheless brought about lower student attrition (by 1981, the attrition rate for primary education had dropped to 14 %, see Chua, 2010) and built strong fundamental knowledge in science.

Towards the end of this phase, Singapore participated in the 1995 TIMSS for the first time. Singapore's students ranked 7th in grade 4 (Primary 4) science and first in grade 8 (Secondary 2) science. This was a respectable performance for students who only started learning science from Primary 3 in a language that was not predominantly used at home. It was definitely a confidence booster for all the teachers who taught to the best of their ability, regardless of how much science they had in their own educational background.

Science Education in the Ability-Driven Phase (1997–2011)

The Context for Science Education

The GDP per capita for Singapore was S$1,310 in 1960, was more than doubled to $2,832 in 1970, grew eight times to $10,685 in 1980 and by 1997 reached $38,840 (Department of Statistics, Singapore, 2012). Within 40 years, or what most would term a generation, Singapore had moved from third world to first world economically. Other countries, however, were not standing still and were building similar growth models and chasing the same jobs (e.g., tourism, multinational electronics firms, ports) with much better competitive edge than Singapore – abundance of workers, low wages, access to natural resources (Singapore even has to buy its water) and cheaper land.

With keen competition, hastened along by globalisation and technological advancements, Singapore chartered new economic directions – shifting from a more labour-intensive, lower-skill economy into a higher-end knowledge-based economy that would pay people well for the higher level of skills and expertise they brought to the job. The call was for education to once again play a key role in this economic transformation, by equipping students with the dispositions and habits of mind that would help them thrive in such an economy. Against this backdrop, the *Thinking Schools, Learning Nation* (or TSLN) vision of an ability-driven education was conceived in 1997 after much consultation with teams of educators – a more consultative ground-up effort that contrasted with the top-down clinical approach that launched the efficiency-driven phase of education.

The hallmarks of TSLN included greater autonomy for schools (e.g., self-evaluation through the School Excellence Model, school leadership and collaboration through the school cluster system) and provided multiple education pathways in recognition of the diverse strengths and aspirations of students. Most significantly, a specialised school (the NUS High School for Mathematics and Science) was set up to nurture talents in mathematics and science for students in Secondary

1 through to preuniversity without the encumbrances of the national curriculum and examinations. Curricula were reviewed and content trimmed to infuse thinking skills, national education (akin to citizenship education) and the use of ICT (Sharpe & Gopinathan, 2002). Later, in 2003, efforts were made to nurture more creative, innovative and entrepreneurial aptitudes and skills through school programmes (Ng, 2004). The *Teach Less, Learn More* movement initiated in 2005 sharpened the focus on the quality of teacher-student interaction that would nurture inquiring minds and more independent learners.

Science Curriculum and Pedagogy

Initiatives under the *TSLN* movement had an impact on the direction of science curriculum, pedagogy and assessment during this period. The science syllabuses were trimmed to free up curricular time for thinking skills, national education and the use of ICT. At the same time, they were refreshed to reflect new developments in science and technology. At the threshold of the twenty-first century, great strides were being made in the fields of computing technology, telecommunications, biotechnology, nanotechnology and clean energy. Scientific developments were occurring at the boundaries of traditional science disciplines. For instance, developments in the life sciences centred around the molecular level, aided by advancements in biotechnology and biophysics, which in turn fuelled applications in the medical and pharmaceutical fields as well as food industries. Nationally, there was a push into life sciences research and applications, spearheaded by A*STAR.

Biology and other science syllabuses were reviewed and changes implemented in the year 2000 to ensure that concepts and skills fundamental to students' access to the new fields of life sciences were incorporated (e.g., the study of simple structures of plant and animal cells and of hereditary traits was introduced into the primary science syllabus). The move towards greater emphasis on molecular biology at the higher level, however, was not without its detractors who were concerned that there was too far a swing towards molecular biology at the expense of curriculum time for learning about classical macro-biological systems such as ecosystems and animal physiology. Such debates also reflected the ongoing tensions in deciding on the scope of the national curriculum. MOE has often maintained the need for the curriculum to remain trim such that any additional contents introduced to update the curriculum have to be accompanied by the corresponding removal of topics, which often becomes a difficult decision to take as it has implications for the different stakeholders.

Beyond the life sciences, a significant shift was made in the syllabuses to give greater emphasis to teaching and learning science by inquiry to nurture students' curiosity, inquiring minds, higher-order reasoning skills and problem-solving skills (see Lee, 2008). There was greater emphasis on using hands-on and more open-ended investigations to encourage students to move away from mechanically following instructions to more self-directed learning and creative thinking (see

Chin & Kayalvizhi, 2002; Lee, Tan, Goh, Chia, & Chin, 2000). Darling-Hammond (2010), in describing her visit to Nan Chiau Primary school, observed that students conducted science projects in an 'experience/investigate/create cycle that was repeated throughout the year' (p. 6). She was impressed with the Ngee Ann Secondary school students undertaking mini research science projects:

> Science classes also support inquiry and invention. For example, a set of students from one biology class undertook a project to create an insect repellent that is 100 % natural, safe, environmentally friendly, convenient, and effective. They discovered that common spices such as cinnamon, cloves, and star anise have insect repellent properties and extracted oils to create an effective product in paper, liquid and solid form. (Darling-Hammond, 2010, p. 187)
>
> Another class of much younger students shared an inquiry they had conducted to find out, using a tachometer, what shapes of blades produce the most revolutions per minute. Their 'action research', presented by PowerPoint, featured careful questions and controls, and students were able to answer additional questions about how to go further in their investigation to sort out whether weight or shape was the key variable. They also explored applications to wind power for a greener approach to energy. (Darling-Hammond, 2010, p. 187)

Such anecdotal accounts (see also Poon, Lee, Tan, & Lim, 2012) indicate that there were schools and classrooms that were adopting more inquiry-based teaching and learning of science. However, to date, there has been no published studies of the pervasiveness of such practices across schools and classrooms in Singapore.

Science Resources to Support Teaching and Learning

Schools received additional funding and equipment to support the teaching of the life sciences and the use of ICT. Primary schools were provisioned with an additional science room (from one to two science rooms) from 2001 to support more hands-on investigative approaches towards the teaching and learning of science. Under the first Information and Technology (IT) Masterplan (1997–2002), school laboratories were also equipped with dataloggers (Tan, Hedberg, Koh, & Seah, 2006). However, the provision of more equipment and resources did not necessarily transform pedagogy immediately. In a 2003 study on the use of dataloggers in schools, Tan et al. (2006) found that more than 8 in 10 of the 593 science teachers surveyed (teaching grades 7–12) used dataloggers for more teacher-centred or teacher-directed activities such as demonstration or experiment. Fewer were using them for student-directed investigations such as fieldwork or project work. This is consistent with studies which found that changes in pedagogy often require time for teacher learning (Richardson & Placier, 2001) and are usually not accomplished immediately with the introduction of new tools and resources.

In the case of the revamped life sciences curricula, a DNA centre was set up at the National Institute of Education in 2003 to provide teachers with hands-on experience of investigations that students could do as part of the life sciences curriculum. Two guidebooks were also published, providing suggested activities that teachers could use (Teo, 2003). Teachers also requested for support on the laboratory protocols and on how they could deal with biosafety issues. In response to this, a book with biosafety guidelines was published by the Curriculum Planning and Development Division to provide schools with information on the handling of biological samples and contamination, proper and safe usage of life sciences equipment and the release and disposal of organisms and chemicals. This project was accomplished with the help of the science community from various institutions, including the universities, hospitals and research institutions. This episode illustrates once again the important role that the larger science community played in supporting science education in schools.

Science Practicals and Assessment

The ability-driven phase saw changes to assessment to align it with the vision of *Thinking Schools, Learning Nation.* More thinking questions were introduced (Sellan, Chong, & Tay, 2006), with students required to demonstrate their ability to transfer learning into unfamiliar contexts and communicate their arguments, reasoning and inferences coherently. One of the most significant changes to the science national assessment was the introduction of the School-based Science Practical Assessment (SPA) to the GCE A-level examinations from 2004 (for junior college year 1 students) and to the GCE O-level pure science examinations from 2006 (first batch of Secondary 3 students in selected schools).

Leading up to the introduction of SPA, the traditional practical tests at the GCE O and A levels were subjected to much criticisms. The science practical tests were criticised as 'close-ended, content-dominated and leading only to verification of known physical laws or constants' (Toh, Diong, Chia, Boo, & Yeo, 1991, p. 13). As it stood, the practical tests were not consistent with or measuring critical thinking skills, with students often 'entering his results in a provided table, followed by analysis and deductions made from it' (Toh et al., 1991, p. 12).

With SPA, students were graded on their demonstration of practical skills (not just on paper) on an ongoing basis. Students were required to go beyond making and recording observations to planning investigations (see Ministry of Education, Singapore [MOE], 2008), selecting techniques, apparatus and materials that fit their experimental design rather than following a sequence of instructions. Students evaluated methods and suggested improvements. This was more in line with the higher-order thinking skills and student-centric learning that the Singapore education system was promoting during the ability-driven phase. However, in its implementation, SPA was inundated with challenges. Teachers found it difficult to reconcile their roles as both teacher and assessor since SPA was school based, with teachers meant to design their own assessment and grade their own students. Towndrow, Tan, Yung, and Cohen (2010) also suggested that teachers did not have adequate levels of assessment literacy to implement SPA to its original intent. This led to subsequent reviews and modifications to SPA, including the current ongoing review.

Whatever the new direction that science practical assessment will take, SPA symbolised an important step taken in reforming national examinations. In its original intent, SPA reflected efforts for examinations to move in step with changes to curricular objectives – that is, it recognised that there was little resonance between twentieth-century assessment approaches and twenty-first-century pedagogy and curriculum. The idea (and ideal) behind SPA was the highly professional science teacher who made decisions not only on their students' learning (Liu et al., 2009) but also on their assessment, not unlike his/her well-lauded Finnish counterpart (Hargreaves & Shirley, 2012). In that sense, SPA was an important advancement of the professionalisation of science teachers in Singapore.

Non-formal Science Education

With the move to a high-value knowledge-based economy, the Singapore government set in motion plans to build its own indigenous R&D capabilities, especially in areas such as the biomedical sciences, environmental science and interactive and digital media (Ng, 2010; National Research Foundation [NRF], Singapore, 2012). Programmes were therefore stepped up, particularly by A*STAR and the National Research Foundation, to enthuse students to pursue a career in science and technology R&D. Students with a strong interest in science spent 6–8 weeks (sometimes even more) in the laboratories of research scientists from the universities and research institutions. Students were encouraged to publish their research findings and to send their works to local and overseas (e.g., the International Science and Engineering Fair and the Young Physicists' Tournament) science fairs and competition to benchmark their projects against some of the best from around the world. Singapore students came back with medals and prizes from these competitions and from the science Olympiads. Participation in the international competitions also spawned the organisation of local competitions (e.g., local Olympiads and Science and Engineering Fair) which engendered wider student participation and involved a larger community of science professionals volunteering their time to support these activities.

Efforts were also not spared for the younger students. In 2003, the DNA Leaning Lab was opened at the Singapore Science Centre. The centre collaborated with the DNA Learning Centre at Cold Spring Harbour in New York to distil cutting-edge research work in the life sciences (e.g., DNA extraction, bacteria transformation) into simple experiments accessible to young students.

Summary Comments

At the end of the ability-driven phase, there were improvements in the quality of science education, both broad-based and for students who had a strong interest in science. Singapore participated for the first time in the 2009 PISA conducted by the OECD. Singapore students did very well in science (among top performing education systems), showing that her 15-year-old students who were nearing the end of their 10–11 years of primary and secondary education could apply knowledge and skills and analyse, reason and communicate effectively as they solve problems in a variety of real-life contexts (OECD, 2010a). One-fifth of Singapore's students were described as top performers[6] in science compared to 8.5 % among OECD students. At the top end (top 5 %) of the PISA scale, Singapore also overtook first-placed Shanghai with a score of 704 (compared to Shanghai's score of 700).

In the most recent cycle of TIMSS (2011), Singapore was again among the top performing countries in both Primary 4 and Secondary 2 science. Singapore had the highest proportion of students attaining the highest Advanced TIMSS Benchmark for science (40 % of Singapore S2 students compared to international average of 7 % and 33 % of Singapore P4 students compared to international average of 7 %) (Martin, Mullis, Foy, & Stanco, 2012). What was more significant was the substantial gains made in the higher-order *Reasoning* domain in 2011 over that of 2007 (see Table 1.1), signalling perhaps some impact of the shift towards more inquiry-based teaching and learning in the classrooms.

In addition, TIMSS also found that Singapore students liked learning science, more than their international peers (Martin et al., 2012). Indeed, during the ability-driven phase, there were opportunities for students with a passion in science to pursue it – both in and beyond school. I remembered one occasion where I sat on a panel to interview some junior college students. Beyond the usual engineering, biomedicine courses, a few shared their interests to pursue their interests in quantum physics and space science. There is no space programme in Singapore and our national curriculum hardly covers space science – but these students have been cultivating their interest through school clubs, school-based integrated programme modules and research attachment (both local and overseas). So although I started this section with what seemed as a very utilitarian driver for ability-based education, in effect students' interests were also served – not only for their livelihood but also in meeting some of their interests and strengths. A large part of this credit goes to the teachers who have taught their students well and worked closely with the

---

[6] PISA describes top performers as students who achieved the highest proficiency levels of 5 and 6. Top performers in science 'clearly and consistently demonstrate advanced scientific thinking and reasoning, and they demonstrate willingness to use their scientific understanding in support of solutions to unfamiliar scientific and technological situations….can use scientific knowledge and develop arguments in support of recommendations and decisions that centre on personal, social or global situations' (OECD, 2010a, p. 147).

Table 1.1 Comparison of Singapore Primary 4 and Secondary 2 students' performance in the three cognitive domains, 2007 and 2011

	Mean scores in the three cognitive domains of					
	'Knowing'		'Applying'		'Reasoning'	
Grade/subject	2011	2007	2011	2007	2011	2007
P4 science	570	599	590	587	597	576
S2 science	588	561	589	570	592	568

Source: Extracted from Martin et al. (2012)

community to open up opportunities for students to pursue their interest in various fields of science.

Final Remarks

> Public education has served us well, in large part because so many educators have worked so hard . . . All that is wonderful, and should be celebrated, but it does not earn educators any free pass into the future. Education, like every other public institution, must find ways to cope with the new challenges and new contexts that are an essential part of the human condition. (Levin, 2008, pp. 1–2)

This chapter is very much an attempt to paint with very broad strokes the development of science education in Singapore over the last five decades. It is in some ways a celebration of our past and recognition of the hard work of science educators in Singapore. But as Levin (2008) has prophetically warned us, these successes do not earn us a 'free pass into the future'. Singapore needs to be alert to and even anticipate the challenges, and opportunities, ahead – not just within science education but also in the changing context of education.

While we celebrate the high enrolment of students in science and engineering courses at the polytechnics and universities (almost half of total enrolment in 2011), there has been some concern of declining interest and enrolment in engineering and physics courses. For example, engineering's share of enrolment in first degree courses in the public universities declined from 43 % in 2001 to 38 % in 2006 and then to 30 % in 2011 (Department of Statistics Singapore, 2012). Even if we consider absolute numbers of engineering students, there is still a decline from 2006 (18,481) to 2011 (17,321). These numbers also did not take into account the number of students who graduated from our local schools against those who came from outside Singapore. I am not advocating that it is only healthy if we have high numbers of students pursuing science and engineering courses. The declining trend, however, can help science educators here reflect on a few things: Do students find science fascinating and intellectually stimulating? Do students find learning science engaging? How has the teaching of science help students go beyond the abstract and conceptual levels to experiencing science as hands-on and practical in solving real world problems?

Singapore's past successes, including students' high performance in international benchmarking studies, like TIMSS and PISA (and national examinations), must also not lull us into thinking that we have the right mix of success factors and should do more of the same. We need to venture out of our comfort zone, particularly in teaching approaches and in assessment, to capture the imagination and passion of students and to harness the curiosity of students so that they will sustain their interest in science.

Science education in the next lap would also need to help students acquire life skills and twenty-first-century competencies valued in the workplace, such as creativity and innovation, decision making, learning to learn, collaboration, strong communication skills and personal and social responsibility (Binkley et al., 2012). But beyond the competencies, living in the twenty-first century would increasingly demand knowledge and values to make informed decisions – today it is about choices on genetically modified food; tomorrow it could be about genetically modified human embryos. Competencies and values are not only for top students in science but also for students offering Foundation Science[7] in primary schools and secondary school students offering the Normal (Technical) syllabus. These students would need fundamental scientific knowledge and skills and proficiency in the twenty-first-century competencies so that they can make informed choices and participate actively in the economic and social arenas. Every syllabus review is an opportunity for science educators to think creatively and courageously about what science education can be for all students, in the context of the different science syllabuses offered.

In this chapter, I have cited examples of how the wider community of science educators and researchers at the NIE, universities, science research institutes, the Science Centre and science-related associations (e.g., STAS, Singapore Institute of Biology, Singapore National Institute of Chemistry, Institute of Physics Singapore and Institute of Engineers) has played important roles in the history of science education to enthuse students in science. This close partnership presents even greater opportunities in the years to come, and future generations of teachers and curriculum specialists would do well to continue working hand in hand with the science community towards a shared mission of inspiring students in science and enthusing some of the best minds to pursue science-related careers.

Finally I would like to thank all who have contributed to this chapter, particularly the interviewees (Betty Larowe, Brenda Yeow, Gan Chee Hau, Chng Meng Seah and Tan Siok Cheng) who provided me with rich oral accounts of the earlier years. The history of Singapore's science education continues to be written and I look forward to reading the next chapter of the journey.

[7] Students in Primary 5 and 6 can be offered either 'Standard' science or 'Foundation' science. Foundation science is a trimmer syllabus than Standard science and is offered to about 10 % of students. Normal (Technical) science is offered as an elective to students in the Normal (Technical) course.

Appendix: Multiple Educational Pathways in the Singapore Education System

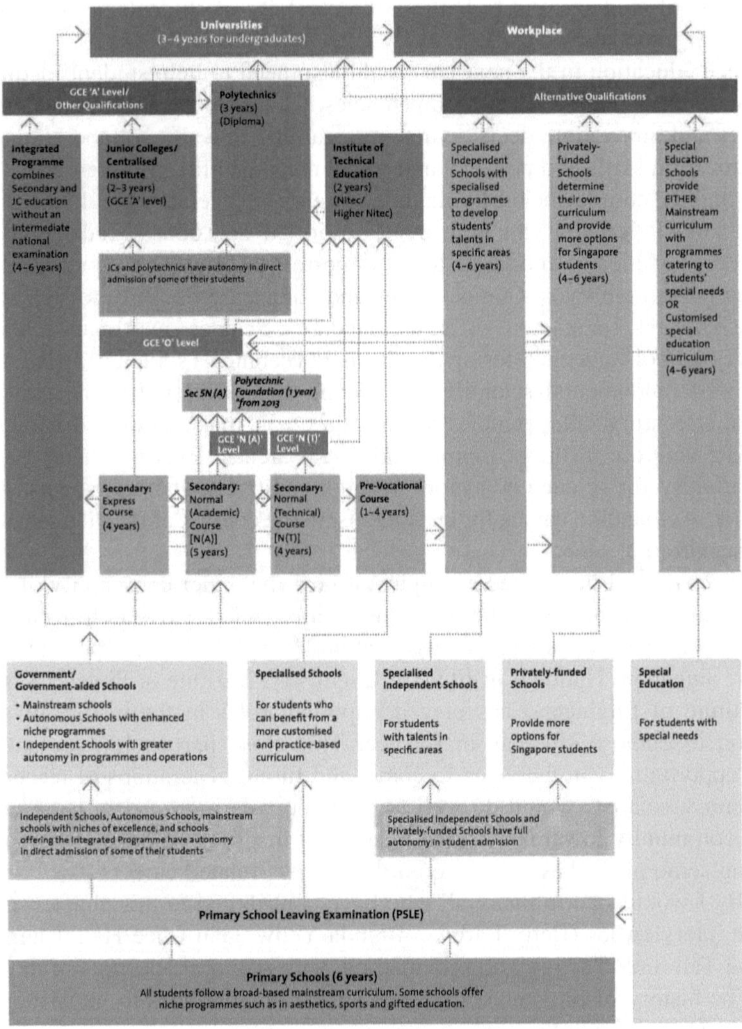

Source: MOE (2011a, p. 18). Reproduced with permission from the Ministry of Education

References

Bhathal, R. S. (1982). Singapore. In *UNESCO report*: *Out-of-school science education in Asia and the Pacific* (pp. 196–210). Bangkok, Thailand: UNESCO Regional office for Education in Asia and the Pacific.

Binkley, M., Erstad, O., Herman, J., Raizen, S., Ripley, M., Miller-Ricci, M., et al. (2012). Defining twenty-first century skills. In P. Griffin, B. McGaw, & E. Care (Eds.), *Assessment and teaching of 21st century skills* (pp. 17–66). Dordrecht, the Netherlands: Springer.

Cahill, B. (Ed.). (1984). *Science education in Asia and the Pacific*. Bangkok, Thailand: UNESCO Regional Office for Education in Asia and the Pacific.

Chall, J. S. (1996). *Stages of reading development* (2nd ed.). Fort Worth, TX: Harcourt Brace.

Chin, C., & Kayalvizhi, G. (2002). Posing problems for open investigations: What questions do pupils ask? *Research in Science & Technological Education, 20*, 269–287.

Chua, M. H. (2010). *Pioneers once more: The Singapore public service, 1959–2009*. Singapore: Straits Times Press.

Darling-Hammond, L. (2010). *The flat world and education: How America's commitment to equity will determine our future*. New York: Teachers College Press.

Department of Statistics, Singapore. (2012). *Yearbook of statistics*. Singapore: Department of Statistics, Ministry of Trade & Industry.

Goh, K. S. (1978). *Report on the Ministry of Education*. Singapore: Ministry of Education.

Goh, C. B., & Gopinathan, S. (2008). The development of education in Singapore since 1965. In S. K. Lee, C. B. Goh, B. Fredriksen, & J. P. Tan (Eds.), *Toward a better future: Education and training for economic development in Singapore since 1965* (pp. 12–38). Washington, DC: The World Bank.

Gopinathan, S., Wong, B., & Tang, N. (2008). The evolution of school leadership policy and practice in Singapore: Responses to changing socio-economic and political contexts (insurgents, implementers, innovators). *Journal of Educational Administration and History, 40*, 235–249.

Hargreaves, A., & Shirley, D. (2012). *The global fourth way: The quest for educational excellence*. Thousand Oaks, CA: Corwin.

Lee, K.-W. L., Tan, L. L., Goh, N. K., Chia, L. S., & Chin, C. (2000). Science teachers and problem solving in elementary schools in Singapore. *Research in Science and Technological Education, 18*, 113–126.

Lee, Y.-J. (2008). Thriving in-between the cracks: Deleuze and Guerilla science teaching in Singapore. *Cultural Studies of Science Education, 3*, 917–935.

Levin, B. (2008). *How to change 5000 schools: A practical and positive approach for leading change at every level*. Cambridge, MA: Harvard Education Press.

Liu, X., Zhang, B., Liang, L. L., Fulmer, G., Kim, B., & Yuan, H. (2009). Alignment between the physics content standard and the standardized test: A comparison among the United States-New York State, Singapore, and China-Jiangsu. *Science Education, 93*, 777–797.

Mackie, D. K. (1971). *A comparative study of some aspects of science education in the secondary school of Scotland and Singapore*. Unpublished master's thesis, University of Singapore, Singapore.

Martin, M. O., Mullis, I. V. S., Foy, P., & Stanco, G. M. (2012). *TIMSS 2011 international results in science*. Boston: Boston College/IEA.

Ministry of Education [MOE], Singapore. (2007). *Many pathways, one mission: Fifty years of Singapore education*. Singapore: Ministry of Education.

Ministry of Education, Singapore [MOE]. (2008). *Singapore 2008 GCE O-level exam syllabus: Physics (New syllabus with SPA)*. Retrieved October 28, 2008, from http://www.seab.gov.sg/SEAB/oLevel/syllabus/2008GCEOLevel Syllabuses/50582008.pdf

Ministry of Education, Singapore [MOE]. (2011a). *Education in Singapore: Findings from international benchmarking studies*. Singapore: Ministry of Education.

Ministry of Education, Singapore [MOE]. (2011b). *Education statistics digest, 2011*. Retrieved March 2, 2012, from http://www.moe.gov.sg/education/education-statistics-digest/files/esd-2011.pdf
Ministry of Education, Singapore [MOE]. (2011c, November 18). *Implementation of integrated programme on track*. Press release. Retrieved March 20, 2012, from http://www.moe.gov.sg/media/press/2011/11/implementation-of-integrated-programme-on-track.php
Ministry of Education, Singapore [MOE]. (2012). *Education in Singapore* (MOE's corporate brochure). Singapore Ministry of Education. Retrieved December 14, 2012, from http://www.moe.gov.sg/about/files/moe-corporate-brochure.pdf
Mourshed, M., Chijioke, C., & Barber, M. (2010). *How the world's most improved school systems keep getting better*. London: McKinsey & Company.
Mullis, I. V. S., Martion, M. O., Minnich, C. A., Stanco, G. M., Arora, A., Centurino, V. A. S., et al. (2012). *TIMSS 2011 Encyclopedia: Education policy and curriculum in mathematics and science* (Vol. 1). Boston: TIMSS & PIRLS International Study Center, Lynch School of Education, Boston College.
National Research Foundation (NRF), Singapore. (2012). Retrieved December 14, 2012, from http://www.nrf.gov.sg/nrf/strategic.aspx?id=134
Ng, E. H. (2010, July 22). *Opening speech by Dr Ng Eng Hen*, Minister for Education at the "2nd World Conference on Research Integrity", Pan-Pacific Hotel, Singapore, July 22. Retrieved September 17, 2012, from http://www.moe.gov.sg/media/speeches/2010/07/22/opening-speech-by-dr-ng-eng-he.php
Ng, P. T. (2004). Students' perception of change in the Singapore education system. *Educational Research for Policy and Practice, 3*, 77–92.
OECD. (2010a). *PISA 2009 results: What students know and can do – Student performance in reading, mathematics and science* (Vol. I). Paris: OECD Publishing.
OECD. (2010b). *PISA 2009 results: Overcoming social background – Equity in learning opportunities and outcomes* (Vol. II). Paris: OECD Publishing.
OECD. (2010c). *Strong performers and successful reformers in education: Lessons from PISA for the United States*. Retrieved July 22, 2011, from http://www.oecd-ilibrary.org/education/lessons-from-pisa-for-the-united-states_9789264096660-en
Poon, C.-L., Lee, Y.-J., Tan, A.-L., & Lim, S. S. L. (2012). Knowing inquiry as practice and theory: Developing a pedagogical framework with elementary school teachers. *Research in Science Education, 42*, 303–327.
Richardson, V., & Placier, P. (2001). Teacher change. In V. Richardson (Ed.), *Handbook of research on teaching* (4th ed., pp. 905–947). Washington, DC: American Educational Research Association.
Roth, W.-M. (2005). *Doing qualitative research: Praxis of method*. Rotterdam, the Netherlands: Sense Publishers.
Sellan, R., Chong, K., & Tay, C. (2006, May). *Assessment shifts in the Singapore education system*. Paper presented at the 32nd annual conference of International Association for Educational Assessment, Singapore.
Sharpe, L., & Gopinathan, S. (2002). After effectiveness: New directions in the Singapore school system? *Journal of Education Policy, 17*, 151–166.
Singh, J. (1999). *Performance of primary school pupils in science investigations*. Unpublished master of education thesis, Nanyang Technological University, Singapore.
Singham, J. K. (1987). *An investigation of the science process skills in the intended and implemented PSP of Singapore*. Unpublished Ph.D. thesis, University of Liverpool, UK.
Stewart, V. (2010). Singapore leads the way in changing teacher education. *Kappan, 92*, 92–93.
Tan, K. C. D., Hedberg, J. G., Koh, T. S., & Seah, W. C. (2006). Datalogging in Singapore schools: Supporting effective implementations. *Research in Science & Technological Education, 24*, 111–127.

Tan, T. W. (1989). School science: The Singapore scene. In M. H. Kuok & N. P. Das (Eds.), *Proceedings of the seminar on trends in science education* (pp. 60–74). Singapore: Singapore National Academy of Science.

Teo, C. H. (2003, March 29). *Address by RADM Teo Chee Hean*, Minister for Education and second Minister for Defence, at the opening of the DNA Learning Lab at the Singapore Science Centre. Retrieved March 29, 2003, from http://www.moe.gov.sg/media/speeches/2003/sp20030329_print.htm

Toh, K. A., Diong, C. H., Chia, L. S., Boo, H. K., & Yeo, T. K. (1991). *A state-of-the-art review on science education in Singapore*. Singapore: Institute of Education, Singapore.

Towndrow, P. A., Tan, A.-L., Yung, B. H. W., & Cohen, L. (2010). Science teachers' professional development and changes in science practical assessment practices: What are the issues? *Research in Science Education, 40*, 117–132.

Yeow, B. C. C. (1982). Development of primary science education in Singapore. *Hong Kong Science Teachers Journal, 10*, 160–165.

Chapter 2
Design and Implementation of the National Primary Science Curriculum: A Partnership Approach in Singapore

Tan-Ying Chin and Chew-Leng Poon

Abstract Singapore has a national science curriculum that spans from the primary levels, where it is mandatory, to the pre-university levels where science subjects are electives. The science curriculum undergoes a 6-year review cycle, with an intermediate third-year review to ensure currency of the curriculum. In this chapter, we present a partnership framework for curriculum design and implementation that fosters close collaboration among curriculum developers, school leaders, science educators and practitioners. Exploring this partnership specifically through the development of the primary science curriculum, we describe the interactions between partners, the curriculum design and implementation processes and the features that support this partnership approach. We discuss the challenge of partnership in teacher professional development that was identified by our partners as an area for improvement in this framework. This narrative has included the voices of our partners to bring deeper insights and perspectives into both policies and practice at the various stages of curriculum design and implementation. We hope that these insights and discussions would be useful to educators and curriculum developers who are involved in or exploring different approaches in curriculum design and the implementation process.

Keywords Science education • Primary science • Curriculum design • Curriculum implementation • Curriculum, pedagogy and assessment • Science curriculum framework • Science inquiry • Curriculum partnership framework • Teacher professional development • Teacher professional learning

---

T.-Y. Chin (✉)
Curriculum Planning and Development Division, Ministry of Education, 1 North Buona Vista Drive, Singapore 138675
e-mail: chin_tan_ying@moe.gov.sg

C.-L. Poon
Research and Evaluation, Planning Division, Ministry of Education, 285 Ghim Moh Road, Singapore 279622, Singapore
e-mail: poon_chew_leng@moe.gov.sg

A.-L. Tan et al. (eds.), *Inquiry into the Singapore Science Classroom*, Education Innovation Series, DOI 10.1007/978-981-4585-78-1_2,

Introduction

Curriculum can be viewed as 'an interrelated set of plans and experiences that a student undertakes under the guidance of the school' (Marsh & Willis, 2003, p. 13). Though stated simply, we like Marsh and Willis' definition of the curriculum as it encompasses the concepts of the planned, enacted and hidden curriculum that are relevant to the Singapore context. Singapore has a national curriculum and therefore a large part of the planned curriculum is very explicit to teachers, parents and even students. Marsh and Willis' inclusion of 'experiences' would embody the curriculum that is enacted in the classrooms, which may differ from the planned curriculum as teachers interpret the planned curriculum through their own beliefs and experiences and as they interact with students and with the curriculum materials. The culture of the school and classroom (e.g., whether teachers encourage or discourage students to speak their mind in class) and the interactions among teachers, students and the physical environment are all part of the hidden curriculum that shapes the 'experiences' of the students. What a student experiences in the classroom is often a result of a complex web of interactions and transactions between the actors (e.g., teachers and students) in the classroom, the physical environment, the materials (e.g., textbooks) and the values and social norms adopted by the different actors. Given this complexity, it is not surprising that the curriculum, or what goes into the curriculum, has been a subject of much debate and deliberations among educators through the years (e.g., Bruner, 1960; Dewey, 1916; Eisner, 1994; Tyler, 1949). The science curriculum is of no exception. What is to be taught in science, how to teach it, how students learn it and how much time should be allocated to it are some constant issues of interest and even tussles (e.g., Driver & Oldham, 1986; Hargreaves, 1994; Marsh, 2009; Schwab, 1962). Who makes these curriculum decisions and how is the curriculum negotiated and deliberated (McCutcheon, 1995) and at which points are also important areas of research that have an impact on what and how students learn.

Curriculum research is indeed a very broad field. Our interest in this chapter is focused specifically on the partnership roles of stakeholders in the various stages of the curricular design and implementation process. We regard any stakeholders who work alongside each other in the curriculum design and implementation process as 'partners'. Partnership refers to the various dynamic forms of working together. This is an important area of study particularly for countries or education systems that adopt a national curriculum, such as Singapore, as the entire body of curricular stakeholders (curricula developers, curricula resource developers, science educators, examinations board and teachers) has to understand the philosophy, intent and details of the curriculum to see through the successful implementation of the curriculum. Research in the field of curricula adoption and fidelity of implementation has shown that the processes of curricula implementation could impact learning (e.g., Adey, 2004; Lee & Chue, 2013). Knowledge on the approaches to engage teachers in curricula implementation is therefore critical.

Recent research on teachers' professional development, learning and change processes have also demonstrated the important roles teachers play, not just in curriculum implementation but also in curriculum design (Clarke & Hollingsworth, 2002; Darling-Hammond & Bransford, 2005). Teachers are viewed as active shapers of curriculum change to meet local school needs and to fill gaps in the design of the curriculum (Blumenfeld, Fishman, Krajcik, Marx, & Soloway, 2000). Hence, many recent studies focused on fostering collaboration between researchers and schools to encourage teacher ownership of curricular innovations (McMillan-Culp & Honey, 2000). This trend of devolving greater responsibility and ownership to teachers (e.g., in selection of content and methods of teaching) was also observed in an OECD (Organisation for Economic Co-operation and Development) project involving 13 participating countries (Atkin & Black, 2003). The teachers in the participating countries played different roles in curricular design and implementation. For instance, teachers in Germany initiated and developed the integrated science curriculum in consultation with academics from universities. Teachers in Japan trialled and influenced changes to the new elementary school curriculum in 'Environmental and Life Sciences' developed by their Ministry of Education. While these recent studies provided insights into how teachers were involved in developing and customising the curriculum and curricular materials, we felt that inadequate attention has been paid to understanding the dynamics of the partnership between stakeholders (e.g., curriculum planners, researchers and school practitioners) in the entire curriculum design and implementation processes.

This chapter, therefore, is an attempt to describe and reflect on the roles and dynamic relationships between these different stakeholders, explored specifically through the development of the Singapore primary science curriculum. We describe a partnership framework for curricular development and implementation that we used. Further, we elaborate on features of the framework and the challenges and opportunities afforded through the framework to initiate conversations among curriculum developers and school practitioners that could help improve the processes of curriculum design and implementation.

Method of Inquiry to Understand Partnership in Curriculum Design and Implementation

To better understand the curriculum partnership, we gathered information on the process of curriculum design and implementation across the various curriculum cycles since the 1960s. We examined curricular documents and other published materials about the Singapore primary science curriculum. Besides, we also invited 17 partners involved in the various cycles and stages of curriculum design and implementation to respond to a questionnaire. These partners include curriculum developers, school leaders, master teachers,[1] heads of science departments and partners from various science institutions. The questionnaire fielded open-ended questions such as:

1. Describe the role you played in the curriculum review. Describe how this contributed to the development of the curriculum.
2. In your opinion, what guided this primary science curriculum review/ implementation?
3. What is/are the focus in this primary science curriculum review/ implementation?
4. What process has helped ensure that the primary science curriculum stays connected and relevant to that of secondary schools and postsecondary education?
5. What do you think helped the curriculum planners, school leaders and teachers as well as the science community decide what is current and relevant to science and the Singapore society in the twenty-first century?
6. What is/are the challenge(s) you faced in implementing this primary science curriculum and how was it overcome?

Besides data from the questionnaire survey, we also conducted one focus group discussion with five respondents to gather deeper insights on the following key questions in the questionnaire:

1. What have been useful in guiding curriculum design and implementation over the curriculum cycles?
2. How have the curriculum design processes (e.g., scanning literature and practices, gathering feedback and consultations) (1) helped to ensure that the science curriculum in primary schools stay relevant and coherent with the curricula in secondary and post-secondary education and (2) helped the curriculum planners decide on what is current and relevant to science and the Singapore society in the twenty-first century?
3. What are the challenges in curriculum design and implementation and what minimises gaps between the intended, implemented and attained curricula?
4. How are partnerships in curriculum design and implementation important in encouraging greater curriculum ownership and higher levels of fidelity in curriculum implementation (encouraging teachers to take greater ownership as designers and facilitators of student learning as inquiry)? Why?

The interview session was audiotaped and later transcribed verbatim. Responses to the questionnaire and the interview transcript were systematically examined to fill in gaps in our knowledge about the different roles of the stakeholders in the

[1] School leaders, including the Principals, Vice Principals and heads of the science department provide instructional leadership for science teaching and learning in the schools. Master teachers are identified expert teachers who support teachers at the national level. They foster pedagogical leadership focused on teacher collaboration in learning communities within and beyond schools in professional networks. These networks serve to strengthen the culture of teaching excellence and raise the standards of practice in the classroom and across Singapore's education system.

curriculum design and implementation process. We also noted elements of partnership in the respondents' description of the curriculum design process, the partnerships which were enacted at each stage and the challenges and opportunities that arose from the partnership.

In writing this chapter, we drew on much of our institutional knowledge to describe the curriculum design processes, but acknowledge that the lens through which we narrate the curriculum partnership are limited by our own experiences and beliefs shaped by about 20 years (each) of involvement in science education. To augment this, we have therefore included our partners' voices liberally so that their perspectives and views will add to the richness of this narrative. This, in a way, is also consistent with the central philosophy of the chapter – that curriculum design is very much a concerted effort of the different stakeholders.

We begin the chapter with an overview of science education in Singapore and the evolving emphases in the primary science curriculum to provide the background and context to understand the curriculum design and implementation processes. This sets the stage for us to launch into the discussion on our curriculum partnership framework.

Overview of Science Education in Singapore

Structure of Science Education in Singapore

Singapore's national science curriculum spans from the primary levels, where it is mandatory, to the pre-university levels where science subjects are electives. Figure 2.1 shows an overview of the broad structure of science subjects offered across the educational levels.[2]

The responsibility of developing the curriculum for each of these science subjects lies with the Curriculum Planning and Development Division (CPDD) at the Singapore Ministry of Education (MOE). The division reviews the science curriculum on a 6-year cycle, with an intermediate third-year review to ensure currency of the curriculum. A science curriculum framework (Ministry of Education, Singapore [MOE], 2008), as shown in Fig. 2.2, guides curriculum design and ensures that a common set of philosophy threads through all of the science curricula across the primary, secondary and pre-university levels. The framework was conceptualised

[2] At the primary and lower secondary levels, students are offered general science. At the upper secondary levels, students can be offered pure science subjects or combined science subjects (comprising a combination of two science disciplines). At the pre-university levels, students can be offered science subjects at H1, H2 or H3 levels (with H3 level subject of the highest content and demand).

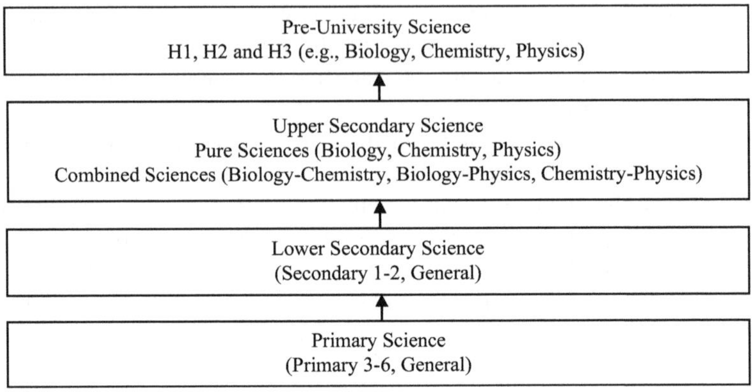

Fig. 2.1 An overview of the science subjects offered across educational levels in Singapore

from theories and practice of inquiry science, guided by a set of desired outcomes of education and vision of science education in Singapore. Inquiry is central to the curriculum framework and developed through the three integral domains of knowledge, skills and ethics. The meaningful pursuit of scientific inquiry is set in the contexts of science in daily life, science in society and science in the environment.

Evolution of the Emphases of the Primary Science Curriculum

As documented in Chap. 1 'Five decades of science education in Singapore', science first became a part of the formal primary school curriculum in Singapore in 1959. The primary science curriculum then and through the 1960s focused mainly on the teaching of plants and the environment, although much broader aims were stated in the curriculum, namely, '(a) to create an interest in nature and its working; (b) to encourage the natural curiosity of children and to inculcate a spirit of inquiry; and (c) to train the children to observe, to experiment and to seek further knowledge' (Yeow, 1982, p. 160). Although 'inquiry' and 'experimentation' were mentioned in the aims of the curriculum, Yeow inferred a more book-centric delivery of the curriculum.

In the 1970s, the primary science curriculum was broadened beyond the teaching of plants and the environment. Physical science topics were included to provide a more balanced learning experience and also a foundation in physical science concepts at the secondary level. This curriculum effort was fronted by the Ministry of Education – Science Teachers Association of Singapore (MESTAS), which comprised teachers from the Science Curriculum Branch of the Ministry of Education (MOE) and the Science Teachers Association of Singapore (STAS).

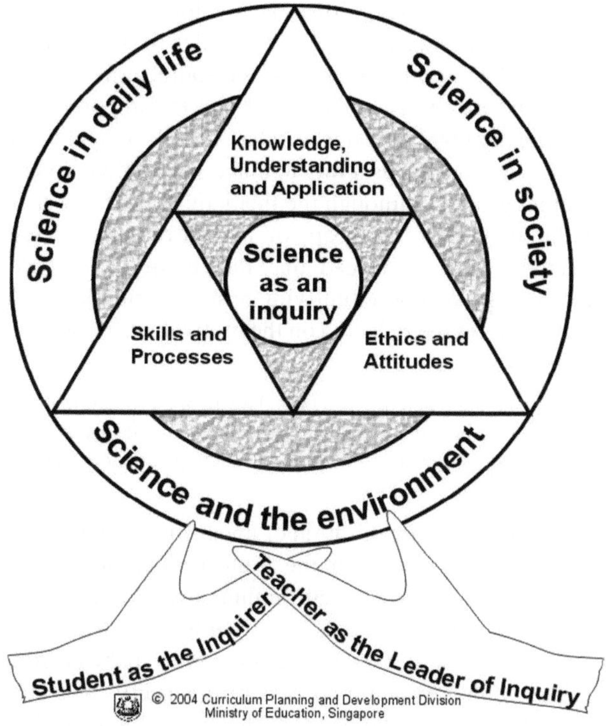

Fig. 2.2 Science curriculum framework in Singapore (Used with permission from the Ministry of Education)

In the 1980s, the primary science curriculum was further revised to include process skills. This curriculum revision was undertaken by a team of teachers and curriculum officers who made up the Primary Science Project (PSP) Team at the Curriculum Development Institute of Singapore (CDIS). A member of the PSP team who was involved in the review and implementation of the primary science curriculum explained that the decision to include process skills was partly inspired by 'discovery-centred approaches' observed in study trips to the United States. Singham (1987) described this first set of locally produced primary science curriculum materials (implemented from 1982) as giving more emphasis to the teaching and learning of science process skills such as observation skills, collection and interpretation of data and classification and measurement skills. The teaching and learning of science process skills continued to be emphasised in the 1990s curriculum.

At the turn of this century, the curriculum review committee initiated a thematic approach to organising the 2001 primary science curriculum (Ministry of Education, Singapore [MOE], 2001). Five integrated themes were developed – Diversity, Cycle, System, Interaction and Energy. The distinction in this curriculum was the effort to communicate a more coherent and integrated understanding of science that bridged the life science–physical science divide. The objective of this thematic organisation was to help students appreciate the links between concepts in different topics, thus progressing towards developing 'big ideas' in science (Harlen, 2004; Millar & Osborne, 1998). For example, in the theme of 'Diversity', a 'big idea' developed in the curriculum was the appreciation of the great variety of both living things and materials in the world and the importance of sustaining the diversity. Classification by observing similar and different characteristics and properties was a tool students learn as a way to make sense of the diversity and interaction between different groups. Although the thematic approach was not widely used internationally at that time, members of the curriculum review committee and teachers consulted were convinced that the themes will lead to better development of overarching concepts or unifying principles in science. A member of the curriculum review committee reflected on the motivation for organising the curriculum through themes:

> Thematic approach brings a more holistic understanding of the science concepts. I believe the themes do help (students) appreciate the concepts of diversity and organisation, cause and effect, systems, structures and functions, models and change. (Angie, survey, 21 Jul 2012)

This thematic organisation of the primary science (and lower secondary science) curriculum was retained in the 2008 curriculum. Beyond the thematic organisation, the 2008 primary science curriculum also introduced the curriculum 'white space', in line with the nationwide initiative to 'Teach Less, Learn More'. This involved reducing the number of learning outcomes in the curriculum (therefore having 'less to teach'), thus creating space and flexibility for teachers to better customise the curriculum to meet the needs of their students as well as to use more inquiry-based teaching and learning approaches (with the aim of students 'learning more').

The 2008 curriculum also gave explicit emphasis on inquiry in the teaching of science. This emphasis on inquiry took place against a backdrop of a number of key developments, both within and outside Singapore. One of these developments was the publication of the US *National Science Education Standards* by the National Research Council (NRC) in 1996. The inquiry-centric standards (National Research Council [NRC], 2000) generated a fair amount of discussion and interest among curriculum designers, educators and the science community in and beyond the USA. While students in Singapore have been doing fairly well in international benchmarking studies, such as the Trends in International Mathematics and Science study (TIMSS, e.g., Bybee & Kennedy, 2005), research into classroom practices surfaced areas for improvement. According to a study by Luke, Freebody, Lau, and Gopinathan (2005), Singapore's science classrooms displayed a largely 'didactic, traditional and rote reproductive character of pedagogy' (p. 11), which falls short of the MOE's call for a more student-centric active learning environment. Teaching and learning science through inquiry was seen as a way to help teachers better engage students in the learning of science.

Beyond the classrooms, Singapore was then gearing to meet the social, technological and economic challenges of the twenty-first century. Policymakers saw science education playing an important role in helping citizens in Singapore live, work and play in an era dominated by phenomenal advancements in computing technology, telecommunications, biotechnology, nanotechnology and alternative energy as well as critical environmental issues. Another consideration was the need for students to develop twenty-first century competencies that would enable them not only to be consumers of knowledge but also to be able to apply, transfer and create knowledge. These goals required a science education that supported flexibility of the mind, innovation and creativity. Then Minister for Education Mr. Tharman Shanmugaratnam, in his annual address to principals, teacher educators and education policy makers at the Ministry of Education (MOE) 2005 work plan seminar, reiterated this: 'We are progressively shifting the balance in education, from learning content to developing a habit of inquiry.... To engage them and prepare them for life than to prepare for tests and examinations' (Ministry of Education, Singapore [MOE], 2005).

Beyond curricular aspirations, this shift towards inquiry took time and resources to take root in practice. A science head of department (HOD) highlighted teachers' instrumental roles as well as their changing understanding and practices in supporting students in learning science as inquiry:

> Teachers need to invest quite a lot of time in inquiry lessons – questioning, assessing, linking what has happened in lessons towards explanations and learning outcomes...to make the most of the exploration phase, linking what children have discovered and allowing them to make use of that evidence to build explanations. Often, teachers are at the phase of carrying out hands-on activities and not inquiry based activities. But gradually, I see more teachers using questions that build understanding. Overall, teachers understand more about the benefits of using inquiry lessons. (Madeline, survey, 5 Jul 2012)

And while this chapter is being written, the new 2014 primary science curriculum is taking shape. 'Essential takeaways' would augment 'key inquiry questions' to better support teachers and students in uncovering the big ideas at the heart of each theme. The aspiration is towards a deeper and wider practice of inquiry, driven by classroom teachers.

Partnership in Curriculum Design and Implementation in Primary Science

Having provided an overview of the evolution of the primary science curriculum in Singapore, we now turn to a discussion of the partnership framework of curriculum design and implementation. Posner (1998) suggested that there are three common approaches to curriculum development. The 'procedural approach' or 'technical production approach' focuses on the procedures or steps of curriculum development (e.g., Schwab, 1970; Taba, 1962; Tyler, 1949). The 'descriptive approach' or 'events and decision-making approach' is about what curriculum planners actually do, including events that occur and decisions which are made (Walker, 1971). Finally, the 'conceptual approach' or 'levels of planning approach' examines elements of curriculum planning, implementation and evaluation and how the elements relate to one another (Goodlad & Richter, 1977; Johnson, 1977). Jackson

(1992) viewed this categorisation of curriculum approaches as not mutually exclusive but as a way of connecting thought and action based on the interest and experiences of people working on the curriculum.

Science curriculum development in Singapore has features of the three curriculum approaches highlighted by Posner (1998). This blended approach includes the adoption of a curriculum development process, working with different stakeholders to execute the various stages of the curriculum development process and making decisions based on the information and feedback gathered from a variety of sources and stakeholders. We will elaborate on the framework, features, challenges and possibilities of curriculum partnership in the following sections:

(a) Partnership framework of curriculum design and implementation
(b) Features supporting partnership framework
(c) Challenges and possibilities in partnership in teacher learning

Partnership Framework of Curriculum Design and Implementation

Central to Singapore's framework of curriculum design and implementation in primary science is the close partnership among curriculum developers, school leaders, science educators and practitioners. Figure 2.3 is our diagrammatic representation of this partnership process (inner concentric ring) interacting with the actors involved in the partnership (depicted in the outer circle where they can be involved in one or more stages of the curriculum design and implementation process). A brief description of the partnership framework follows.

- *Scanning literature and practices in primary science.* This process includes the scans of literature and practices locally and internationally by curriculum planning officers. Partners from the National Institute of Education (NIE) also contribute research knowledge of other education systems. Educators and practitioners from the various international schools in Singapore also share about their curriculum and how these were translated into practice. These scans of literature and practices provided useful insights on efforts in curriculum, pedagogy and assessment in the different education systems.

  A curriculum review committee member shared that the understanding of how the aims of primary science education were translated into curriculum learning outcomes and classroom practice in different education systems was valuable in informing the curriculum review:

  > I appreciate the rigour put into thinking through the aims, objectives and the conceptual framework of the curriculum during the review, the thoughts and thorough discussions in the currency, appropriateness of the knowledge and processes that are taught and also the study of existing curricula from the many countries and way the curriculum could support the practice and implementation in particularly. (Angie, survey, 21 Jul 2012)

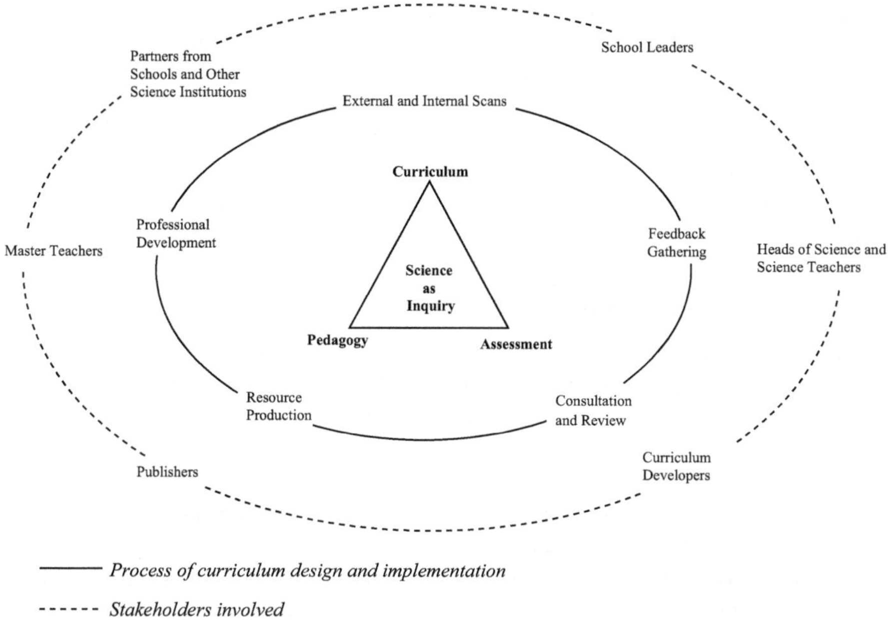

Fig. 2.3 A partnership framework of curriculum design and implementation of primary science in Singapore

- *School leaders and practitioners providing feedback on the existing curriculum.* In the feedback gathering process, at least one school leader (e.g., head of department) or teacher from each of the 178 primary schools participate in the pre-draft and/or post-draft focus group discussions. They provide useful feedback not only on the syllabus content but also on pedagogy, assessment and implementation issues. What is especially valuable is the sharing of personal experiences on curriculum implementation such as the following by a science HOD on how inquiry-based learning supported and motivated learning and thinking:

  > I see great interest in children for science – as it is one subject which is strongly linked to life and inquiry allows many skills and attitudes to develop. I found that children who are taught using inquiry always talk about turning into thinkers. They also become motivated to learn (this success spills over to learning other subjects) and they look forward to challenges. That interest can later on blossom into true passion – not necessarily in science but in learning. (Madeline, survey, 5 Jul 2012)

- *Consulting expert panel of curriculum developers, school leaders, science educators and practitioners.* This consultative process involves the convening of a curriculum review committee of experts and practitioners from the different institutions including the Academy of Singapore Teachers, the NIE, various departments (such as Educational Technology Division, Gifted Education Branch) and schools from MOE and the Singapore Examinations and

Assessment Board (SEAB is a government-funded independent Board that develops and administers national examinations). This close collaboration between the various stakeholders facilitates alignment of the principles, spirit and objectives of the curriculum with pedagogy and assessment (school-based and national assessments). A former curriculum specialist also highlighted the role of the panel in ensuring the coherence of curriculum development across levels of development, guided by the concept of a spiral curriculum where concepts are revisited in greater depth (Bruner, 1960):

> The concept of a spiral curriculum in which topics are revisited in greater depth at a later stage, even onto the secondary level. (Helen, survey, 12 Aug 2012)

- *Consulting senior educators and administrators on curriculum and implementation plans.* The proposed curriculum and implementation plans are discussed with a panel of senior educators and administrators from the Ministry of Education Headquarters. This consultative and approval process attempts to align the science curriculum to the overall goals of education and garner the necessary buy-in and support for resource allocation to implement changes.
- *Producing textbooks and resources, with reviews by school practitioners.* The resource development process involves the heads of science and science teachers in the textbook/resource review and approval process. This process gathers feedback on how the planned curriculum could be translated into learning experiences in everyday classroom contexts, in the textbooks (e.g., developed by publishers) and resources (e.g., developed by curriculum planning officers with teacher practitioners). Teachers provide reality checks on the feasibility of suggested teaching and assessment strategies and activities to support student learning.
- *Supporting professional development in collaboration with the Academy of Singapore Teachers, the NIE and schools.* The support for curriculum implementation includes working with the NIE in providing pre-service and in-service professional developments to support teachers in understanding and implementing the curriculum. The partnership with master teachers from the Singapore Academy of Teachers to foster pedagogical leadership focuses on teacher collaboration and leadership in learning communities within and beyond schools in professional networks. The long term goals of these collaborations are to raise the standards of practice in the classroom, thereby strengthening the culture of teaching excellence.

Besides the professional development opportunities provided by various partners, partnering schools in establishing ongoing communication channels and organising sharing platforms at various levels (national, zonal, cluster and school levels) also support teachers in curriculum implementation. Curriculum specialists from the Curriculum Planning and Development Division who developed the curriculum also join teachers in their lessons to observe and discuss issues and challenges during curriculum enactment.

Features Supporting Partnership Framework

From the responses in the questionnaire and in-depth interviews, we surface some features that our partners felt were important in supporting the partnership framework:

(a) Anchoring the review in the science curriculum framework
(b) Mechanisms that tap on multiple perspectives and experiences of various stakeholders
(c) Working with different partners in supporting different stages of the curriculum cycle
(d) Teachers as key curriculum partners in gathering local classroom-based evidence to inform curriculum review and implementation

Anchoring the Review in a Science Curriculum Framework

Many partners highlighted the importance of the science curriculum framework (Fig. 2.2) in guiding the curriculum review as it clarified the philosophy and broad goals of science education for the different partners. A member of the curriculum review committee who was a principal master teacher explained how the science curriculum framework served as a common compass that guided the curriculum design and review:

> I thought it is important to have a conceptual framework to curriculum, what needs to be learnt, what helps to facilitate the learning, what helps to understand the learning and most importantly, how the curriculum could build that spirit and processes of scientific mindset and literacy in the students so that they can function effectively as a science-literate person. (Ada, survey, 21 Jul 2012)

Marsh (2009) highlighted that one of the advantages in having a curriculum framework was to ensure coherence in the curriculum across levels and therefore the framework serves as a guiding post for partners. He cautioned against, however, having too much details in the framework which could hamper new ideas and flexibility at the point of curriculum design and also at the point of classroom implementation. It would be important to keep the curriculum framework sufficiently broad to provide a guide for partners in the curriculum review and implementation. At the same time, teachers should be supported in understanding and translating the understanding of the curriculum framework into the design of everyday learning experiences.

Mechanisms That Tap on Multiple Perspectives and Experiences of Stakeholders

Another essential feature of our partnership framework is the mechanisms that tapped on the multiple perspectives and experiences of stakeholders. In our case, we have broad-based mechanisms such as school questionnaires and feedback forms to tap on the broad swath of perspectives from all our primary schools. We also mount focus group discussions to tap deeper into teachers with varying years of teaching experiences and across the different grade levels. These focus group discussions provided not only insights on the science curriculum but also critical issues of enacting the curriculum. The importance of gathering multiple perspectives from teacher practitioners to better support curriculum implementation was highlighted by a primary school head of science:

> With the input from the teachers, the team looked at improving the syllabus for better clarity in communicating the intent to the teaching fraternity. There was deliberation on how the curriculum could be improved, for better classroom implementation. In this respect, input from the teachers are highly valued... as a teacher then, being involved in the review of the curriculum, I was able to appreciate the diversity of views and concerns expressed by different teacher representatives.(Wendy, survey, 12 Jul 2012)

The curriculum review committee, with partners from schools and the NIE, provided another layer of expertise and discipline knowledge and perspectives in curriculum, pedagogy and assessment. For instance, some gave input on the appropriateness of the depth and breadth of knowledge as well as the cognitive processes of learning. Others shared thoughts and views on how science inquiry could be integrated into the learning content and processes as well as how it could be facilitated and assessed in the classroom. Suggestions on the resources that could be used to facilitate the learning of science in the schools were also important inputs. A curriculum review committee member shared how she leveraged on her experiences in both primary and secondary school teaching to contribute to the curriculum review:

> Having taught science in secondary schools and supported science learning in both primary and secondary schools, I brought school-based science teaching and learning as well as assessment practices for the curriculum review committee to consider when we relook at currency, effectiveness and relevance of the curriculum ... I brought the practice into the curriculum review process, the science concept/knowledge (from science teacher perspectives) and the connections with assessment (feedback/inputs of learning) and resources (from a perspective of connecting learning to assessment and teaching strategies and resources).(Angie, survey, 21 Jul 2012)

Such a consultative approach in the curriculum review process in Singapore is supported by research on supporting change in educational practice (Fullan, 1982). Hart (1989) also reported that the analysis of the discrepancies between practice and policy resulted in changes in policy recommendations in Canada. This illustrates how the current ideas, research and practice in science education can be synthesised collaboratively.

Working with Different Partners in Supporting Different Stages of the Curriculum Cycle

In giving more agency to the different stakeholders of education, the curriculum review process values the contributions and inputs by various partners. However, we also recognise that in practice we could not involve *all* partners at *all* stages of the curriculum cycle. This would drain much of the resources and also tax too heavily on the time of our partners. Our partnership framework therefore acknowledges the need to strategically tap different partners at the different stages of the curriculum cycle. For example, in the recent 2008 curriculum cycle, the curriculum developers worked with teachers from schools and master teachers to design activities to support science teaching and learning as inquiry. They also worked with teachers and the Science Centre Singapore (SCS) to develop teachers' ideas into Science Teaching and Resource (STaR) kits to support the use of hands-on learning for students. The resources were used at both the NIE pre-service and in-service courses to support teachers in curriculum implementation.

In another example, schools partnered each other in curriculum implementation. Insights from a recent collaborative effort of a cluster of primary and secondary schools highlighted the value of teachers from the primary and secondary schools coming together as partners to observe and discuss student learning. Through pedagogical dialogue of classroom practices focused on student learning, teachers better understood the learning needs and misconceptions of students and how to better scaffold student learning across the primary and secondary science curricula. Such school-based partnership efforts not only supported teachers in understanding the progression of science learning beyond the primary level but built teachers' capacity as curriculum leaders.

Teachers as Key Curriculum Partners in Gathering Local Classroom-Based Evidence to Inform Curriculum Review and Implementation

Our partnership framework regards teachers as key partners in gathering local classroom-based evidence to inform their enactment of the curriculum. We recognise that teachers empowered to collect and make meaning of classroom data and information can make better decisions on how the curriculum can remain current and relevant for students and how learning can be more impactful (Cochran-Smith & Lytle, 1999; Crawford & Cornett, 2000; Hong & Lawrence, 2011). For example, data and information on how the curriculum impacted learning provided useful insights for teachers to design different teaching and assessment strategies that better supported students with different learning dispositions and aptitudes.

Teachers who played an active role in gathering classroom-based evidence of student learning had an added advantage in informing how teachers themselves could be better supported in resource provision and professional development. A master teacher shared how partnerships involving teachers in designing and field testing resources not only provided classroom-based evidence but also benefitted the curriculum developers, master teachers, heads of department and teachers who were involved in the process of resource design and curriculum implementation:

> The materials were planned with teachers and trialed in the classrooms by teachers and feedback given by a team of teacher-observers... Process was rigorous and thorough, which benefited all who were involved in it. (Ivy, survey, 12 Sep 2012)

Providing materials that are aligned to new curricula is generally recognised as important in supporting successful reforms. Adey (2004), however, cautioned that providing curriculum resources is not sufficient in themselves. A critical ingredient in designing the resources is the personal involvement of teachers who are the users of the resource. In their study of the use of CASE (Cognitive Acceleration in Science Education) materials to support science teaching, Shulman and Shulman (2004) found that teacher ownership of the resources was key to sustaining practice. Teachers, therefore, are important partners in resource development in the curriculum implementation process. These curricular resources can then play a key role in initiating and sustaining change because they are ‘concrete, tangible vehicles for embodying the essential ideas of a reform’ (Powell & Anderson, 2002, p.112).

Challenge and Possibilities in Partnership in Teacher Learning

While the partnership framework has enhanced the curriculum design and implementation processes, school leaders, teachers and partners from the various institutions have identified one process that is fundamental and requires more attention and further thinking. This is in the area of partnership to enhance teacher professional development that will in turn strengthen curriculum implementation. We have been exploring this issue and would like to share some thoughts on this.

In Singapore, the professional development effort draws on the partnership of many different groups of professionals to conduct formal workshops and sharing. These professional development activities have helped build teacher competencies in science teaching. However, there are limitations to ‘one-off’ workshops. Fullan and Stiegelbauer (1991) found that teachers often encountered difficulties in transferring experiences in one-off workshops to the realities of their own classrooms. Gusky (2000) also found that short one-off workshops often could not impact change at the level of belief and values and therefore were less effective in sustaining curricular changes.

Hence, educators and researchers are increasingly studying efforts in teacher learning opportunities beyond ‘one-off’ professional development (Clarke & Hollingsworth, 2002; Hoban, 2002) where teachers are active learners shaping their own professional growth as reflective practitioners (Schon, 1983). Shulman and Shulman (2004) have argued for a community-based approach to proliferate and sustain teacher learning. In their teacher learning model of Vision, Motivation, Understanding, Practice, and Reflection, they suggested that schools create environments that support, sustain and 'tune' the visions, understandings, performances, motivations and reflections of all its members to encourage learning. This goes beyond individual teachers' learning to a conception of teacher learning within a broader context of the community. We felt that this idea of community-based teacher learning is consistent with the partnership principle of curriculum design and implementation, where teachers are regarded as 'equal' partners creating the knowledge and environment to bring about change in the enactment of the curriculum. We would, therefore, have to evolve the current roles of the 'partners' in the teacher professional development process in our framework to reflect a stronger learning towards teacher learning within a community.

Spillane, Reiser, and Reimer (2002) argued that teachers are sense-making agents and that both formally structured social contexts and informally created social contexts therefore affect how teachers make sense of reforms, providing a shared base of beliefs and knowledge that teachers use to reason about implementation. Hoban (2002) and Adey's (2004) work has informed us that teachers providing curriculum leadership as part of a learning community could strengthen ownership and motivation to improve practices. We observe that some schools in Singapore have begun fostering the formation of professional learning communities to drive change in teaching, learning and assessment practices. Such ground-up curriculum leadership among teachers offers potential to address the challenge of effective professional development partnerships in curriculum implementation. We hope to further work on this aspect of teacher learning vis-à-vis the curriculum partnership framework. Such insights would contribute to the understanding of how teacher learning opportunities affect teaching practices and student outcomes as well as how teachers learn successful practices. This is an area which Darling-Hammond and Bransford (2005) reported to be lacking in the field of teacher learning and practice.

Looking Forward

This chapter documents an original primary science curriculum partnership framework which has features of three common approaches to curriculum development – 'procedural approach', 'descriptive approach' and 'conceptual approach' (Posner, 1998, pp. 79–100). We hope that insights gleaned on how different stakeholders were involved across various stages of the curriculum design and implementation process would be useful in driving stronger collaboration and ownership of curricular change in the classroom.

We also hope that this chapter will help science teachers reflect on their own changing and important roles as curriculum partners and leaders of inquiry in the classroom. The documentation of the framework provides a common set of terminology for conversations around partnership in curriculum design and implementation and we look forward to having rich dialogues with our partners in the science education community.

Acknowledgement We would like to thank the following colleagues, friends and many other partners from the Ministry of Education and Institutes of Higher Learning who have contributed their views and voices to this chapter: Aw Wai Lin, Chiang Wai Leng, Chien Mee Len, Chng Meng Seah, Gan Chee Hau, Goh Su Fen, Mohamed Azhar, Tan Hong Kim, Irene Tan, Tan Siok Cheng, Anne Wong, Betty Yeoh, Brenda Yeow and many others who have chosen not to be named.

References

Adey, P. (2004). *The professional development of teachers: Practice and theory*. Dordrecht, the Netherlands: Kluwer.

Atkin, J. M., & Black, P. (2003). *Inside science education reform: A history of curricular and policy change*. New York: Teachers College Press.

Blumenfeld, P., Fishman, B. J., Krajcik, J., Marx, R. W., & Soloway, E. (2000). Creating usable innovations in systemic reform: Scaling-up technology-embedded project-based science in urban schools. *Educational Psychologist, 35*, 149–164.

Bruner, J. (1960). *The process of education*. Cambridge, MA: Harvard University Press.

Bybee, R. W., & Kennedy, D. (2005). Math and science achievement. *Science, 307*, 481.

Clarke, D., & Hollingsworth, H. (2002). Elaborating a model of teacher professional growth. *Teaching and Teacher Education, 18*, 947–967.

Cochran-Smith, M., & Lytle, S. (1999). The teacher research movement: A decade later. *Educational Researcher, 28*(7), 15–25.

Crawford, P. A., & Cornett, J. (2000). Looking back to find a vision: Exploring the emancipatory potential of teacher research. *Childhood Education, 77*(1), 37–45.

Darling-Hammond, L., & Bransford, H. (2005). *Preparing teachers for a changing world. What teachers should learn and be able to do*. San Francisco: Jossey-Bass.

Dewey, J. (1916). *Democracy and education. An introduction to the philosophy of education*. New York: Free Press.

Driver, R., & Oldham, V. (1986). A constructivist approach to curriculum development in science. *Studies in Science Education, 13*, 105–122.

Eisner, E. W. (1994). *The educational imagination: On the design and evaluation of school programs* (3rd ed.). Upper Saddle River, NJ: Prentice Hall.

Fullan, M. (1982). *The meaning of educational change*. New York: Teachers College Press.

Fullan, M., & Stiegelbauer, S. (1991). *The new meaning of educational change* (2nd ed.). New York: Teachers College Press.

Goodlad, J. I., & Richter, M. N. (1977). Decisions and levels of decision making: Process and data sources. In A. A. Bellack & H. M. Kliebard (Eds.), *Curriculum and evaluation* (pp. 506–516). Berkeley, CA: McCutchan.

Gusky, T. R. (2000). *Evaluating professional development*. Thousand Oaks, CA: Corwin Press.

Hargreaves, A. (1994). *Changing teachers, changing times: Teachers' work and culture in the postmodern age*. London: Cassell.

Harlen, W. (2004). *Evaluating inquiry-based science developments: A paper commissioned by the National Research Council in preparation for a meeting on the status of evaluation of inquiry-based science education*. Washington, DC: National Academy of Sciences.

Hart, E. P. (1989). Toward renewal of science education: A case study of curriculum policy development. *Science Education, 73*, 607–634.

Hoban, G. (2002). *Teacher learning for educational change*. Buckingham, UK: Open University Press.
Hong, C. E., & Lawrence, S. A. (2011). Action research in teacher education: Classroom inquiry, reflection, and data-driven decision making. *Journal of Inquiry and Action in Education, 4*(2), 1–17.
Jackson, P. W. (Ed.). (1992). *Handbook of research on curriculum*. New York: Macmillan.
Johnson, M. (1977). *Intentionality in education: A conceptual model of curricular and instructional planning and evaluation*. Albany, NY: Center for Curriculum Research and Services.
Lee, Y. J., & Chue, S. (2013). The value of fidelity of implementation criteria to evaluate school-based science curriculum innovations. *International Journal of Science Education, 35*(15), 2508–2537.
Luke, A., Freebody, P., Lau, S., & Gopinathan, S. (2005). Towards research-based innovation and reform: Singapore schooling in transition. *Asia Pacific Journal of Education, 25*(1), 5–28.
Marsh, C. (2009). *Key concepts for understanding curriculum*. New York: Routledge.
Marsh, C., & Willis, G. (2003). *Curriculum: Alternative approaches, ongoing issues*. Englewood Cliffs, NJ: Pearson Education, Inc.
McCutcheon, G. (1995). *Developing the curriculum: Solo and group deliberation*. New York: Longman Publishers.
McMillan-Culp, K., & Honey, M. (2000). Scale and localization: The challenge of implementing what works. In M. Honey & C. Shookhoff (Eds.), *The Wingspread conference on technology's role in urban school reform: Achieving equity and quality* (pp. 41–46). Racine, WI: The Joyce Foundation/The Johnson Foundation/EDC Center for Children and Technology.
Millar, R., & Osborne, J. (1998). *Beyond 2000. Science education for the future. A report with ten recommendations*. London: King's College London.
Ministry of Education, Singapore [MOE]. (2001). *Science syllabus primary 2001*. Retrieved from http://www3.moe.edu.sg/cpdd/doc/Science_Pri.pdf
Ministry of Education, Singapore [MOE]. (2005, September 22). *Speech by Mr Tharman Shanmugaratnam*, Minister for Education, at the MOE work plan seminar 2005, on Thursday, 22 September 2005. Retrieved November 20, 2007, from http://www.moe.gov.sg/speeches/2005/sp20050922.htm
Ministry of Education, Singapore [MOE]. (2008). *Science syllabus primary 2008*. Retrieved from http://www.moe.gov.sg/education/syllabuses/sciences/files/science-primary-2008.pdf
National Research Council [NRC]. (1996). *National science education standards*. Washington, DC: National Academy Press.
National Research Council [NRC]. (2000). *Inquiry and the national science education standards*. Washington, DC: National Academy Press.
Posner, G. J. (1998). Models of curriculum planning. In L. E. Beyer & M. W. Apple (Eds.), *The curriculum: Problems, politics and possibilities* (2nd ed., pp. 79–100). Albany, NY: State University of New York Press.
Powell, J. C., & Anderson, R. D. (2002). Changing teachers' practice: Curriculum materials and science education reform in the USA. *Studies in Science Education, 37*, 107–136.
Schon, D. A. (1983). *The reflective practitioner*. New York: Basic Books.
Schwab, J. (1962). *The teaching of science as enquiry*. Cambridge, MA: Harvard University Press.
Schwab, J. (1970). *The practical: A language for curriculum*. Washington, DC: National Education Association.
Shulman, L. S., & Shulman, J. (2004). How and what teachers learn: A shifting perspective. *Journal of Curriculum Studies, 36*, 257–271.
Singham, J. K. (1987). *An investigation of the science process skills in the intended and implemented PSP of Singapore*. Unpublished Ph.D. thesis, University of Liverpool, UK.
Spillane, J. P., Reiser, B. J., & Reimer, T. (2002). Policy implementation and cognition: Reframing and refocusing implementation research. *Review of Educational Research, 72*, 387–431.
Taba, H. (1962). *Curriculum development: Theory and practice*. New York: Harcourt, Brace, & World.

Tyler, R. W. (1949). *Basic principles of curriculum and instruction*. Chicago: University of Chicago Press.

Walker, D. F. (1971). A naturalistic model for curriculum development. *School Reviews, 80*(1), 51–65.

Yeow, B. C. C. (1982). Development of primary science education in Singapore. *Hong Kong Science Teachers Journal, 10*, 160–165.

Chapter 3
Transforming Science Education by Expanding Teacher and Student Collaboration

Kenneth Tobin

Abstract Frequently policymakers and administrators inform teachers and students about what and how to teach and learn science and assume that curricula are enacted as stipulated. Not surprisingly, enacted curricula are resilient, do not change easily, and are radically dependent on continuously changing structures associated with material, symbolic, and social resources. In this chapter I advocate collaborative approaches to teaching and learning in which teachers and students undertake research on enacted curricula, reflect on what they learn from research, and, as desirable, design changes to teaching and learning roles and contexts. Dialogic inquiry and mindfulness are presented as specific examples of heuristics being used as interventions to improve learning environments to accord with stipulated values, policies, research, and theories.

Keywords Collaboration • Dialogic inquiry • Practitioner research • Mindfulness • Reflexive inquiry • Heuristics

Science Education in and for Society

It seems that there is one thing we can be sure about in science education – that reforms are always on the horizon. For as long as I have been involved in science education (since 1964 when I began science teaching), reforms have rained down on the field – changing the goals and recommended practices for teachers and learners. The methods for assessing learning also have been reformed, as have approaches to educating teachers, certifying them, and evaluating their competence. Despite the efforts of science educators, scientists, and policymakers, the problems persist, even increasing in number, taking on new forms while retaining a

K. Tobin (✉)
The Graduate Center, City University of New York, New York, NY, USA
e-mail: ktobin@gmail.com

A.-L. Tan et al. (eds.), *Inquiry into the Singapore Science Classroom*, Education Innovation Series, DOI 10.1007/978-981-4585-78-1_3,

family resemblance to those identified previously. Paul Hurd made it a hobby to count the number of national-level reports that advocated the reform of science education – pointing out the rapid rate of increase of such reports and arguing that the process of producing them was inherently flawed (Hurd, 1997). Now, as we approach the end of 2013, long after anybody is counting exhortations for reform, pervasive problems in science education still persist and calls for reform have a common ring to them. What is abundantly clear is that problems are not attributable only to science educators who have failed to heed the call for reform – would-be reformers appear to have little idea of how to enact reform so that what is intended has even the remotest chance of being implemented. There is a tendency to see the endgame in terms of changing macrostructures, specifying a vision for science education and then holding others, usually individuals, accountable for attaining that vision.

In this chapter, I make the case that the key stakeholders in any reform need to be involved in ways that are associated with understanding teaching, learning, and use of resources to improve the quality of learning environments. Any effort to successfully enact reform, I will argue, must involve deep understanding by all participants of what is intended and willingness to collaborate to attain the vision. Central to successful reform is an understanding on the part of policymakers, researchers, and educators that making sense of experience is inherently theoretical, involving frameworks that illuminate while at the same time they obscure what is happening and why it is happening (Tobin, 2008).

Purposes of This Chapter

In her chapter Chew-Leng Poon provides a succinct account of the development of science education in Singapore since it became an independent republic in 1965. She examined the history of Singapore in terms of phases in which progress was enhanced through industrialisation, efficiency, and ability. Although different labels might have been used to characterise five decades of independence, the author provided glimpses into ways in which national priorities shape curricula, especially in areas such as science, mathematics, and technology that relate so closely to economic competitiveness and advancement of industrialisation, technological edge, and human capacity.

There is no doubt that in a relatively short period of time, Singapore has accomplished a great deal – evidence being a reduction in class size, responsiveness of educational policy and curricula to national goals and people's needs, and an emphasis on the value of creative problem solving as an outcome of science education. Changes in the science curriculum began with trends that mirrored what was happening worldwide and then gradually addressed features of life in Singapore. Notable among these was an emphasis on understanding science and a need to concurrently build competence in home languages and English. Resolve and flexibility were shown in relation to when science would be taught, the qualifications of science teachers, and what would be taught. It was commendable, for example, that modern science could be added to the curriculum – but only if less suitable topics were eliminated. Similarly, an emphasis on inquiry and creativity recognised the importance of producing functional citizenry, not only being ranked consistently at the top of international comparisons of science achievement. History suggests that for five decades, science education evolved in parallel with developments in Singapore, with a keen eye on individual and collective interests. A question that arises is whether other lenses might have been used to represent the history of Singapore, thereby affording alternative parallels with developments in science education. For example, the foci used to represent the history of Singapore did not illuminate issues of equity and social justice, which have been prominent features of the science education landscape in many countries in the world. Equity issues associated with social categories such as gender, race, and official language proficiency are prominent in science education in many countries and have been for decades. Not surprisingly the equity issues that arise in science education are reflective of issues distributed throughout social life – for example, in countries such as the United States and Great Britain (e.g., Ryu, 2013). It is interesting to ponder whether an equity lens on social life in Singapore would have been generative in raising salient issues for curriculum reform in science education. I make this point about frameworks because it is salient to much of what I present in the remainder of this chapter.

Tan-Ying Chin and Chew-Leng Poon provide a rich account of ways in which collaborative partnerships can contribute to different aspects of curriculum review and reform. The authors highlight the salience of professional development and teacher learning as areas that warrant further thinking and attention. They also suggest there is merit in rethinking the roles of partners in professional development – making the point that they view teachers as equal partners. In the sections that follow, I take up these points and address the potential of including and expanding the roles of students in ongoing activities related to science curricular enhancement. My approach is to accept Chin and Poon's rich narratives of science education, contained in the two chapters, as springboards for expanding a conversation about changing curricula in rapidly changing times. My narrative weaves in some of what we have learned from ongoing research about the quality of science education, its purposes, and the enormous potential of collaborative contributions of participants who are usually excluded from substantive involvement in changing the faces of science education.

Getting Beyond the Facts

Interpretive studies of science education suggest that teachers, textbooks, and other media set the default switch for enacted curricula to emphasise the learning of science facts. There is a flawed assumption that productive changes to teaching and learning can be facilitated by policy changes, professional development, and assessment practices alone. Accordingly, a shift towards inquiry, which was a feature of the Sputnik-inspired curriculum revolutions of the 1960s, had spotty examples of success. There was an absence of widespread changes in teaching and learning practices. Despite the production and use of new curricula resources, the nature of science education did not change to attain the visions of those who planned and produced the new materials. Even though there was a tendency to try to teacher-proof the curriculum, classroom research highlights the central roles of teachers and learners in structuring and maintaining learning environments (Atkin & Black, 2003). Instead of teaching and learning as curriculum designers intended, most teachers and students continued to enact roles in much the same way they had done previously.

A possible shortcoming of most curriculum reform projects is failure to monitor continuously the fidelity or appropriateness of enacted curricula. It is self-evident that each classroom in which a curriculum is enacted is uniquely structured. Accordingly, enacted curricula vary from place to place and time to time. Furthermore, what are considered appropriate teacher and student practices depend on a dynamically changing structural flux. Even when a teacher teaches the same lessons to two different classes, there will be characteristics of the lesson that seem the same/similar and other characteristics that are distinctive/different. What is an appropriate form of enactment depends on continuously changing, interdependent macro-, meso-, and microstructures. To seek to reform curricula by only changing macrostructures is folly since changes at any level of social life contribute to what is experienced as a science curriculum (i.e., micro-, meso-, and macrostructures are dialectically related, constituents of a whole and interdependent). Just as changes in macrostructures can afford changes in meso- and micro-level actions (i.e., top-down transformations), changes in microstructures can transform meso- and macrostructures (i.e., bottom-up transformations; Collins, 1981). Similarly, changes in mesostructures can transform macro- and microstructures (i.e., middle-level changes transform the top and the bottom). Efforts could be made to move beyond stipulating teacher and student roles and assuming that they are enacted as intended. Furthermore, since so much of social life occurs under the radar of awareness (what Ann Swidler refers to as settled times), there seems to be some merit in participants such as teachers, students, and school leaders having roles in assessing the extent to which particular criteria are enacted appropriately, reviewing what happened, what changes are needed, and what seemed to work well (Swidler, 1986). I explore these ideas further in the following sections.

Interpretive Inquiry in Science Classrooms

When Fred Erickson first wrote about interpretive research, he described the term as an umbrella for all forms of participant observation research (Erickson, 1986). Interpretive inquiry seeks to find answers to two broad questions – what is happening and why is it happening? The approach embraces hermeneutic–

phenomenology, seeking to answer both questions in terms of the participants' voices. In the case of science classrooms, teachers and students are important stakeholders. When researchers use a hermeneutic approach to research, it is necessary for them to listen and learn from participants – an approach that requires more than just paying attention to whether the participants agree with the researchers' interpretations. Instead, there is a conscious effort to learn from all participants and arrive at outcomes that reflect what different participants agree and disagree on. Learning from sameness and difference is the essence of interpretive inquiry – embracing a methodology that focuses on understanding others' perspectives while respecting their rights to remain different.

When we began a program of interpretive research in urban classrooms (e.g., Tobin, 2000), we decided to incorporate student researchers who could be involved in all aspects of the research process (e.g., deciding what questions to answer; selecting and preparing data sources such as video segments and transcripts; interpreting data; and identifying and enacting changes to classroom resources and processes). In parallel to these studies, we also involved two to three students as teacher educators, assisting prospective teachers to answer the question: "How can you better teach students like us?" (Elmesky & Tobin, 2005). There were some highly beneficial elements of these activities, and others were deleterious. For example, a significant positive outcome was that students learned to speak to adults and they appreciated being listened to, taking it as a sign of respect. Similarly, teachers learned to listen attentively to youth and spoke to them respectfully and dialogically. It was apparent that both the teachers and students were producing adaptive culture that enabled them to be successful in the activities. However, a very obvious shortcoming was that many of the ideas on effective teaching and learning were behaviourist and somewhat draconian. For example, students advocated violence to subdue disruptive students and recommended removing them from the class. Even though there were some shortcomings, the power of giving voice to the teacher and student was very noticeable. Accordingly we decided to further develop collaborative activities.

Authentic Inquiry

In the science education research community, there has been a tendency to see the principal value of research in terms of producing new theory as distinct from improving practice. Consistent with a sociocultural framework, we regarded research as cultural enactment, and as such, theory is dialectically related to practice. Accordingly, to produce theory would simultaneously produce practices, which would be reproduced and transformed – just as the theory produced would be reproduced and transformed. This dialectical view of research is consistent with our ethical stance that participants in research should benefit from their participation and our acceptance of this tenet as an axiological commitment.

Part of our methodological bricolage included authentic inquiry, which consists of four components (each constituents of a whole). Ontological authenticity addresses the desirability of all participants changing their ontologies as they participate in a study. As a valued quality criterion, it is imperative that studies are designed and enacted to ascertain the extent to which ontological authenticity occurs. A second authenticity criterion is that all individuals develop an understanding of others' ontologies. The goal of this criterion is education, not imperialism. Research should be enacted to allow participants to build an understanding of one another's ontologies. There is no mandate that any participants change their understandings to become like the other.

The third authenticity criterion, catalytic authenticity, draws attention to the necessity that research should be beneficial to participating institutions and individuals. That is, the research should catalyse change in the individuals and institutions involved in the research, and the benefits of research should extend to all social clusters not just to researchers who may have planned the research and got it underway. A fourth criterion, tactical authenticity, also concerns beneficiaries – that benefits should be equitable and measures be adopted to ensure that all participants are assisted to benefit from research. To put it forcefully – research should be enacted to help participants who cannot easily help themselves.

Catalytic and tactical authenticity encouraged us to design interventions that could incorporate what was learned from a study to catalyse collective and individual improvement and promote equity through research interventions in which everybody improves. In the next section, I explain the nature of some of the interventions we have used to improve enacted curricula during a study.

The Development and Use of Cogenerative Dialogue

Because we adopted Egon Guba and Yvonna Lincoln's criteria for serial and contingent selection of participants in research, we endeavoured always to select participants based on their differences from those who were already involved (Guba & Lincoln, 1989). Accordingly, we experienced the strengths of having diverse participation in our research activities. We continued this practice when we created cogenerative dialogue (hereafter cogen) as an activity that could serve instructional improvement, teacher education, evaluation, and research (Tobin & Roth, 2006).

The original purpose of cogen was to actively involve teachers and students in the improvement of science education. A central structure is the necessity for each person to respect all participants by listening to what they have to say, build an understanding of what each person says, and encourage others to speak keeping in mind that over time all participants will have approximately equal talking time and similar numbers of talking turns. The rule is not to be enforced if people do not want to talk or feel that what they have to say is not yet ready to be shared. Furthermore, it is important to avoid a behaviourist stance (Kincheloe & Tobin, 2009) that only uttered words count as talk. Cogen recognises that individuals can enact culture internally as they interact with social artefacts associated with the given field (Vygotsky, 1962). For example, consider the following interaction:

Thomas (internal): Do you really think the volume will increase as the temperature rises? Why would you think that?

Irena (imagined by Thomas): Because objects expand when they are heated?

Thomas (internal): But what if it's a gas in a sealed container?

Irena (imagined by Thomas): You got me! Good one Thomas. (Thomas smiles outwardly and begins to speak.)

Thomas: Perhaps the idea only applies to solids and liquids. What do you think?

Irena: What are you thinking about gases?

Thomas: Well if you have a gas in a sealed container, the volume would remain the same but the pressure would increase.

Participation in cogen should be active to the extent possible, and participants should trust that others are active, even if they are silent. Of course, friendly overtures can be used by any of the participants to invite contributions from others in encouraging and nonthreatening ways. This could easily be justified with the understanding that talk benefits both individuals (speakers and listeners) and the collective.

The interactions in cogen should be dialogic in the sense that there is focus, synchrony, and entrainment during interaction chains (Bakhtin, 1986). Specifically, actors should be sure there is a connection between their actions and what came immediately before. Thus, focus is maintained until such time there is consensus on the need for change. Similarly, actors should make an effort to align their practices with previous verbal and nonverbal interactions.

When we first initiated cogen, it was designed to improve the quality of learning environments. Participants in cogen would identify an issue that was salient to the quality of the learning environments, and then subsequent dialogue would focus on the roles, resources, interactions, and changes that might be desirable and feasible. Sometimes the interactions led to invitations being sent to others, such as a student, a school administrator, or another teacher (Tobin, Roth, & Zimmermann, 2001). The rationale was that cogenerative dialogue among stakeholders could allow for understandings to grow about others' perspectives and for agreements to be negotiated on how to improve the quality of learning environments. We selected participants based on their differences from one another and a solid conviction that differences were resources for learning and building robust, equitable learning environments. It seemed obvious that an outcome of cogen would be consensus on what to do next. Accordingly, we developed a custom where one or more of the participants would call for a summary of "what did we agree to during this cogen?" This practice oriented participants' thinking towards the changes they would enact next time the class met. The practice also provided a basis for reviewing which agreements were enacted as agreed and the extent to which the intended changes actually occurred. That is, the focus on the outcomes of cogen resembled a form of action research. Even though the benefits of cogenerating ideas to improve the quality of science education were important and impressive, there may have been even more significant outcomes of participating in cogen. Participants dialogued respectfully, listened to one another, and learned how to enact successful interaction chains. Because the participants were selected based on their differences from one another, cogen enabled them to produce new forms of adaptive culture that afforded success in other fields – i.e., such as in the classroom, home, and workplace.

We began to view cogen as a seedbed for the growth of new culture. The implications of what we have learned from cogen, for teacher preparation and enhancement, are substantial. Preservice teachers can produce new culture to bolster their induction into teaching by participating in cogen with participants from the classes they are to teach or from classes like those they will be teaching. Similarly, if teachers are experiencing problems in initiating and sustaining productive learning environments, they can be involved in cogen with students from the class, professional development personnel, and school leaders. One caveat on this recommendation is that all participants in cogen also should have been active participants in one or more lessons. This ensures that all participants in cogen have experienced praxis, i.e., knowledge in action, as the curricula to be improved are enacted. Arguably, the experience of observing a lesson is significantly different than the knowledge experienced as a coteacher or a co-learner in a class. For this reason, we discourage cogen participants from watching from the side, making judgements about quality, and recording strengths and areas in need of improvement.

Resources to Support Cogen

Cogen can be structured in numerous ways and can have a vast array of resources to support the activity. Before looking at some of these resources, it is useful to ask what must be present in order for the activity to be considered cogenerative dialogue? The first essential requisite is that the social interactions that occur should be dialogic, featuring focus, synchrony, and entrainment. A second necessity is radical listening.

Radical Listening

Radical listening is grounded in hermeneutics, to learn from oral texts by listening attentively with a focus on making sense of what is being said from the perspectives of all participants in dialogue. As such, radical listening is an active process in which interactions focus on expansive understanding. During radical listening, it is important not to interrupt the conversational flow, but that does not mean a radical listener sits in absolute silence. First, there is internal self-talk that can raise questions and commentate on emerging understandings. Second, there are questions and comments that can be directed to others to clarify and expand what is being said and ascertain the affordances of what is being proposed. Radical listening will lead to understandings that include the logics underpinning others' oral contributions. To the extent possible, alternative propositions should not be introduced until after the affordances of a particular proposition have been identified and reviewed. When this stage is reached, an evaluative phase that includes comparisons with alternatives is beneficial.

Of course the most appropriate actions are contingent and emergent. If interactions are creating social and/or symbolic violence for others or they are in some other way violating an individual's ethical principles, then it is essential that the individual exercises right speech – speaking out if, when, and as necessary to disrupt interactions and redirect the interaction flow. Individuals practising radical listening should not tolerate dysfunctional practices if there is a moral imperative to intervene to protect participants (including self) from social violence.

Video Analysis

Participants in cogen (and others in the class) can have access to video recordings of the class. Initially we asked participants to watch the video and select events that included noteworthy examples of good or bad teaching and learning. The clips were not to exceed 2 min duration, and they would be used as objects for discussion during cogen. The use of video files provided a concrete focus for dialogue, a resource that could be viewed and reviewed at regular speed, slowed down, examined frame by frame, and sped up. The evidence from video recordings complemented knowledge of the class, gleaned from being in the class. Use of the video allowed all participants to undertake microanalyses of teaching and learning and evaluate teaching and learning through an expanded set of resources, including theories about expressed emotions, gestures and other body movements, and prosody. Teachers and students were reflexive in regard to what happened in the class – they became aware about actions and practices about which they were previously unaware (Bourdieu, 1992). Once their awareness had been heightened, all participants had opportunities to focus dialogue on whether changes could be made to improve the quality of the learning environments.

Multilogicality

A theoretical bricolage is like a patchwork quilt consisting of frameworks that can illuminate particular life experiences (in this case an enacted science curriculum). All participants have a bricolage that reflects different lived experiences – including education. As we soon found out when we asked students how to better teach kids like them and how to improve the curriculum – students have lots of bad ideas and an occasional good one. What was important was opening up the conversation, making frameworks visible, and being receptive to learning from others by being with them in cogen. Interactions with “different” others expanded the possibilities for learning and allowed new ways of thinking about teaching and learning to be introduced and learned. These opportunities were greatly expanded by the use of video clips that made visible expressed emotions and other aspects of enacted culture. Participants in cogen could identify patterns and contradictions and through dialogue become aware of new interpretive frameworks. To illustrate some of the possibilities we experienced in cogen, I present in the next several sections examples associated with interaction chains, expressed emotions, and dialogic inquiry.

Interaction Chains

Randall Collins presented a model that addresses chains of interactions and the production of emotions associated with success and failure (Collins, 2004). The model begins with a collective, such as participants in cogen, being in close proximity and having a mutual focus on a social artefact – such as a video clip, an issue to be resolved, or examples of good or bad teaching and learning. Because dialogue involves enacting consecutive actions in synchrony, Collins posited synchrony and asynchrony as salient parts of his model. He also addressed the tendency for culture to spread throughout a collective – synchrony across space and time and throughout a network of social artefacts, including people. He referred to this characteristic as entrainment, which is analogous to resonance. When entrainment occurs, there is a possibility for a build-up of emotions as participants within a field share emotions – Collins referred to this as shared mood. As the emotion builds up throughout a network, there is a chance of collective effervescence occurring. Examples include clapping, laughing, roaring with approval, sobbing uncontrollably, and booing a decision regarded as bad. When each of these characteristics occurs, there is heightened possibility that solidarity also will occur as individuals identify with one another and other social artefacts in the field. This model developed by Collins draws attention to interactions as a basic unit of analysis as culture is enacted in the field by a collective consisting of individuals and other resources.

Collins’ framework was one of the initial ones we taught participants in our research so that they could answer both broad questions associated with interpretive inquiry (i.e., what is happening and why is it happening?), thereby expanding the possible ways in which science classes and associated learning environments might be described, reviewed, and changed. Furthermore, the model is not complete in the sense that more theory is needed to ascertain, for example, whether there was evidence of synchrony. Collins’ framework was complemented with a multilogical bricolage that included prosody, proxemics, conversation analysis, and expressed emotions. There were many benefits derived from educating participants about Collins’ framework. One of these was that it provided them with new lenses for making sense of their experience as teachers and learners. If we were to ask students how they might improve their learning, instead of answering "try harder, or copy more accurately from the board, or do my homework", they could respond in terms of Collins' model. That is, the model expanded the possibilities for thinking about teaching and learning practices and changes that might produce more success, not just for an individual actor but also for a collective.

Expressed Emotions

Emotions were salient in numerous aspects of our research. In Collins' model, collective emotions were theorised as valenced (positive or negative) and varying in intensity. As well as being produced in the moment, it was clear that shared emotions, which we began to think of as emotional climate, also were historically constituted. We regarded it as important to dialogue about emotional climate and also to avoid making assumptions that positive was good and negative was bad. Our research showed that structures such as laughter was polysemic – being interpreted by different actors in different ways. It also was evident that laughter could be conducive to a build-up of positive emotional energy and at other times it could be a signal for disruptive practices that derailed the focus on learning science.

For example, we observed in our research that a teacher made a humorous remark and began to laugh. Two students laughed soon after and much more loudly than the teacher. Later, in a research meeting, the teacher interpreted the laughter as an example of synchrony, illustrative of the students supporting his teaching. Other members of the research team disagreed, arguing that the students were laughing at the teacher rather than laughing with him (Shady, 2013). The theoretical framework for analysing and improving teaching thereafter included the characteristics laughing at and laughing with. During cogen, teachers and students demonstrated their awareness that actors could interpret laughter in different ways and the issue of mindfulness in science then became salient – how to acknowledge emotions as they are produced, but not to allow them to break focus or divert the activity flow in undesirable directions. Being aware of the potential for emotions to "stick" to actors' enactment was to become an important focus of our research – mindfulness in education.

We elaborated our frameworks from Collins' work on interaction chains to include Jonathan Turner's primary emotions (happiness, fear, anger, sadness) and myriad secondary and tertiary emotions that were expressed in social life (Turner, 2002). Since we employed these frameworks during studies that embraced interpretive and authentic inquiry, and in cogen, we related emotions to different modes of expression, including the face, body movements and orientations, the voice, and in social artefacts such as language and the semiotics of the social spaces associated with our activities. Accordingly, there were contexts in which facial expression of emotion was central, in which case we added Paul Ekman's frameworks to our bricolage (Ekman, 2003). At other times, we focused on emotional climates and most recently we expanded our work to study physiological expression of emotions and neuroplasticity and its relationships to emotional styles (Davidson & Begley, 2012). The study of physiological expression of emotions included variables such as pulse rate, oxygenation of the blood, and blood pressure. The inclusion of variables such as these heightened awareness of the relationship of emotions to physiological expression and the wellness of teachers and students. These interests changed the foci of our research, the nature of discussions about teaching and learning in cogen, and heightened teachers' and students' awareness of wellness as a constituent of curricular enactment.

Dialogic Inquiry

A goal of cogen is to produce adaptive forms of culture that capitalise on the differences included among participants in cogen. At a later time, these new forms of culture are enacted in the regular science class to benefit enacted curricula through heightened awareness of how to adapt cultural resources to improve the learning of self and other. When the structures of cogen are enacted in a science class, we refer to the activity as dialogic inquiry. What this entails is adherence to the rules and other structures that characterise cogen as science curricula are enacted.

Since I began my career as an educational researcher in 1973, there has been contention about the nature of inquiry. The post-Sputnik curriculum reforms of the 1960s embraced science inquiry and to a marked extent what was meant by inquiry was grounded in the scientific method – though there were characteristic differences that reflected specific goals of particular projects and theories of learning that underpinned the development of curricula resources. For example, *Science: A Process Approach* embraced a hierarchical learning theory of Robert Gagné. The process approach, as it was referred to, included basic and integrated process skills that were predominantly derived from an analysis of the scientific method through the lenses of learning hierarchies. In contrast, the *Science Curriculum Improvement Study* was developed in accordance with a Piagetian oriented theory of learning under the direction of Robert Karplus, a physicist. The inquiry model, known as the Learning Cycle, involved three phases: exploration, invention, and discovery.

Mary Budd Rowe's research is especially salient to all curriculum projects developed in the post-Sputnik era and since. Rowe (1969) showed that the particular curriculum project enacted in the classroom did not make a difference to the incidence of inquiry evident in the verbal utterances of teachers and students. Instead, the length of the pauses between utterances, referred to as wait time, was related to the incidence of a number of inquiry characteristics that were indicative of opportunities to learn. This detail is most important because it reflects a difference between intended and enacted curricula and also to the idea of treatment fidelity. Just because inquiry is a goal of an adopted curriculum project, it does not mean that the students or the teacher will emphasise inquiry when the curriculum is enacted. Indeed, it is highly likely that teachers and students need to think about and enact other characteristics if inquiry is actually to occur. For at least a decade, an approach that was adapted from the *Science Curriculum Improvement Study*, referred to as the 5 E Learning Cycle model, has been advocated in much the same way that the Learning Cycle was advocated back in the 1960s. Insufficient attention has been given to the structures associated with enacted culture that would be indicative of inquiry. Based on our research, a 5 E approach that was augmented by the type of culture produced in cogen would be dialogic and collaborative and would heighten the value of a 5 E approach (Bybee et al., 1989). Accordingly, we developed a heuristic, for the construct dialogic inquiry, to allow teachers and students to understand dialogic inquiry in terms of salient characteristics and review the extent to which those characteristics were enacted in the class. The heuristic serves to heighten awareness about the characteristics of dialogic inquiry to provide a framework for describing what is happening in the classroom and why it is happening and to tweak classroom environments so as to increase the amount and quality of dialogic inquiry. The characteristics included in the dialogic inquiry heuristic are provided in the following list. Participants read each characteristic and indicate the extent to which the practice relates to them. A 5-point Likert scale is provided for this purpose (e.g., **5**, *Always/Very often*; **4**, *Often*; **3**, *Sometimes*; **2**, *Seldom*; **1**, *Never/Rarely*). After each characteristic, there is an open response space in which participants can write their thoughts about the characteristic and its salience to them.

1. I am respectful to others.
2. I try to get others to contribute to discussions.
3. I try to make sense of what others are saying.
4. Others are respectful to me.
5. Others have opportunities to speak as much as I do.
6. Others try to get me to contribute during discussions.
7. Others try to make sense of what I am saying.
8. There are opportunities for me to speak as much as others.
9. When I talk others listen to what I have to say.
10. When others talk I listen to what they have to say.
11. I collaborate with others.
12. I maintain focus.
13. I value others' perspectives.
14. Others maintain focus.
15. Others value my perspectives.
16. When I talk I build on what others have said.
17. When I talk others build on what I said.
18. There is a shared mood.
19. Participation is timely.
20. Participation is appropriate.
21. I test the potential of others' contributions.
22. I try not to judge the quality of others' contributions until I understand them.
23. My contributions are thoughtful.

24. Others do not judge the quality of my contributions until they understand them.
25. Others test the potential of my contributions.
26. Others' contributions are thoughtful.
27. Participation is anticipatory.
28. I try to find consensus.
29. I try to find contradictions for claims.

Interventions in Research in Science Education

Breathing Meditation

Our research has focused recently on the role of emotions in teaching and learning science (Tobin & Ritchie, 2011). A deep concern we had as a result of doing this research is that the intensity of emotions in classrooms was deleterious to the health of teachers and students. We wanted to intervene in such a way that all participants had heightened awareness about ways in which emotions were expressed and had knowledge of how their own bodies changed in different classroom contexts. However, we wanted participants to have tools to intervene if and when they decided that the intensity of emotions had reached a level that changes were necessary. Based on numerous studies, but especially the work of Richard Davidson (Davidson & Begley, 2012), a social neuroscientist, we developed a breathing meditation activity that we enacted with the whole class for 3–5 min at the beginning of a lesson. The interventions asked students to sit upright, place their hands on their abdomens, and slowly breathe in, pushing their belly out as they breathed in. They would then slowly breathe out through their noses, pushing the belly back towards the spine. To the extent possible they should focus on the in and out breaths, slowly breathing in and breathing out. As they had a thought or experienced an emotion, they were urged to let it go, to recognise it as a thought or emotion and return the focus to the in and out breaths, making an effort to match the duration of the in to the out breath.

The students we have done breathing meditation with have been college level – involved in graduate and undergraduate classes. The students typically look forward to the activity and report they have used breathing meditation in their own education contexts (i.e., as teachers), sport, commuting, faculty meetings, and home. The benefits include quieting the mind, increasing focus, and reducing the intensity and incidence of thoughts and emotions.

Emotional Climate

We asked students to register the emotional climate they experienced in a 3-min time interval. As the lesson proceeded, a soft bell signalled an invitation to record emotional climate using a Likert scale of 5, very positive; 4, positive; 3, neutral; 2, negative; and 1, very negative. Participants recorded their preference using a clicker that communicated with a computer via Bluetooth. We then undertook descriptive statistical analyses to calculate measures of central tendency and dispersion, identify regions of time in which particular trends were apparent, and to cluster students who tended to experience emotional climate similarly to one another but different from students in other clusters. Interestingly, we felt that our methods were somewhat intrusive and suggested that after a few lessons, we desist from measuring emotional climate. The students strenuously opposed the suggestion because they found it useful to be reminded of emotional climate during the class. They continued to measure emotional climate for the entire course, insisting that it did not disrupt their participation. Based on our interpretations of the video of the classes, we concur with their insistence – there appeared to be no evidence of disruption due to them regularly assessing emotional climate. On the other hand, their heightened awareness of emotions and the way emotions saturated social artefacts may have had desirable effects as far as ameliorating emotional climates were concerned. From a reflexive perspective, becoming aware of the unaware creates conditions necessary to change in ways that are deemed desirable.

Physiological Expression of Emotion

When we learned about the relationship between breathing patterns and emotions (Philippot, Chapelle, & Blairy, 2002), we decided to use a finger pulse oximeter to provide in-the-moment measures, three measures per second, of pulse rate, oxygenation of the blood, and blood pressure. The students wearing the finger pulse oximeter as they co-taught the class were able to monitor these three variables and on occasion the results shocked them. Some of the more striking examples included teaching with pulse rates close to 170 beats per minute and levels of oxygenation that dropped below 70 %. Unfortunately these were not isolated cases and it was interesting to see the participants taking proactive actions to remediate what they considered to be unacceptable levels of physiological expression of emotion. The most common approach was to use breathing meditation to bring the measures back to an acceptable level – from the perspective of the participant. Becoming aware of the relationship between physiological aspects of the body and emotions heightened awareness of other variables as well – including body temperature, which we are now measuring using a bio-harness.

The examples I have provided in this section are representative of the ways in which theoretical frameworks and empirical data can be used to think about constructs such as emotional climate, learning environments, emotional expression, and wellness. Each of these is an aspect of enacted science education that has been seldom considered in relation to effectively teaching and learning. In terms of educating people for literate and functional citizenry, it seems imperative that they should understand ways in which emotions and thoughts can ameliorate opportunities to learn and the quality of learning. The design of reflexive activities such as those described in this section might well be considered as an important part of teacher education for prospective teachers, professional development for practising teachers, and evaluation of teaching, learning, and curriculum.

Mindfulness in Education

A heuristic is like a model whose characteristics are contingent on the context of the social spaces to which it is applied. In developing heuristics, we undertake an analysis of a construct that is the focus of our research – like dialogic inquiry, and we list a set of characteristics that are salient in the context of the research. In so doing we acknowledge that the characteristics that are salient in different social space might not include many of the constituents of the heuristic we use in a study. The purpose of including characteristics of a construct is to heighten awareness of all participants about the characteristics. We go further than this by including a Likert scale that participants use to rate the extent to which each characteristic applies to them (e.g., **5**, *Always/Very often*; **4**, *Often*; **3**, *Sometimes*; **2**, *Seldom*; **1**, *Never/Rarely*). Reading the characteristic and selecting a point on a rating scale creates a link between the participant and the characteristic and strengthens the possibility that characteristics will serve as objects for changes of conduct and associated cultural enactment.

In our research we accept Erving Goffman's theorising of social interaction as the fundamental unit of analysis in social inquiry (Goffman, 1983). In a context of science education, when individuals interact with social artefacts, including teachers and students, they produce culture that includes science and knowledge associated with myriad fields in which the teachers and students are simultaneously engaged (e.g., by virtue of their in-the-moment thoughts and emotions that relate to other aspects of their social lives). For example, continuously produced emotions might be related to science or to another field of the lifeworld. In either situation, it is important that such emotions enhance science learning and are not detrimental to one or more individuals.

The heuristic we developed as a low-grade intervention for manipulating the quality and incidence of mindfulness consisted of characteristics we wanted participants to have heightened awareness of and an attachment to, in the sense that they would enact social life in ways that are in synchrony with them. In so doing, participants would not attach their thoughts and emotions to their practices, thereby enabling them to retain focus and minimise forms of conduct that would reduce their focus, or others' foci, on the activity in which they are engaged.

The characteristics included in the mindfulness heuristic address an individual's awareness of what is happening in the moment, the extent to which thoughts and emotions are attached to conduct, awareness of the physiological modes of expressing emotion, and five dimensions of emotional style that are associated with neuroplasticity (Davidson & Begley, 2012): resilience, to what extent and how quickly does an individual recover from adversity; outlook, how long does a positive emotion persist following success; social intuition, to what extent does an individual accurately detect emotional cues expressed by others; context, to what extent are the emotions expressed by an individual sensitive to the context; self-awareness, to what extent is an individual aware of others' detection of his/her expressed emotions; and attention, to what extent does an individual initiate and sustain focus. The characteristics for the *Mindfulness in Education* heuristic are provided in the following list. A 5-point scale like the one listed above and several lines for comments about the contingencies associated with making a rating of this characteristic accompanies each of the listed characteristics. I remind readers that the characteristics in a heuristic are contingent on context and should be changed to be salient to the circumstances of use. Mindfulness in science would have different characteristics possibly than mindfulness in a drama class, and mindfulness in ballroom dancing would differ from mindfulness while driving a car.

1. I am curious about my emotions.
2. I find words to describe my emotions.
3. I allow thoughts to come and go without being distracted by them.
4. I notice my emotions without reacting to them.
5. I am kind to myself when things go wrong for me.
6. I recover quickly when things go wrong for me.
7. Even when I am focused I use my senses to remain aware.
8. When I am emotional, I notice my breathing.
9. When I am emotional, I notice my heartbeat.
10. I maintain a positive outlook.
11. The way in which I express my emotions depends on what is happening.
12. The way in which I express my emotions depends on who is present.
13. I can focus my attention on learning.
14. When I produce strong emotions, I can let them go.
15. When my emotions change, I notice changes in my body temperature.
16. The way I position and move my body changes my emotions.
17. I use breathing to manage my emotions.
18. I am kind to others when they are unsuccessful.
19. I can tell when something is bothering another person.
20. I am aware of others' emotions from the tone of their voices.
21. I recognise others' emotions by looking at their faces.
22. When I am with others, my emotions tend to become like their emotions.

The Nature of Science Curriculum Reform for the Future

When I first became a professional educator, a curriculum was considered to be something written, which contained goals and various resources for meeting those goals. I was surprised when I studied a graduate course in curriculum to learn that this idea of curriculum was very limited, and there were curriculum scholars who had theorised curriculum as enacted (Marsh, 2009). This idea had significant explanatory power, and to my way of thinking, it appropriately centred teaching and learning. As enacted, a curriculum was dynamic, unfolding, and always unique, contingent on what happens and myriad interactions that constituted the flux that moved through space and time. From the perspective of enactment, curriculum reform necessarily was local as well as macro (i.e., focused at state and district level). In the 40 years that have passed since that curriculum course was taught, I have learned from theory and empirical studies that a successful curriculum occurs when interactions are fluent, when teachers and students enact appropriate culture that is anticipatory and timely. Because many schools are exceptionally diverse, and teachers and their students often have life trajectories that are radically different from one another, it is imperative that participants with significantly different social histories learn how to successfully interact with one another. In order to do this, all participants need opportunities to get to know others in the class and how to adapt their reservoirs of culture to provide a foundation for success.

Cogen is an activity that allows individuals to build adaptive culture that can be used to support successful, fluent interaction chains to afford science learning of self and others. At its best cogen embraces authenticity criteria and serves as a methodology for teachers and learners to do collaborative research in their own classrooms.

One of the most important aspects of our research is that emotions are continuously produced and can mediate learning and the quality of learning environments. There is a growing body of research that suggests that high-magnitude, negative emotions can be deleterious to wellness. Accordingly, it is important for all participants, teachers, and students to have a toolkit of interventions to ameliorate emotions when and as necessary. It is a priority to promote wellness for teachers and students by ensuring they know how to use breathing meditation and a host of reflexive strategies to bring intense emotions into check should they judge it as desirable to do so. Such a toolkit has obvious applications in the social world writ large, and it is essential that educators – including school leaders, teachers, students, teacher educators, researchers, and policymakers – acknowledge that the participants in curriculum cannot be separated from other constituents; all constituents are part of a multilectical whole (Fellner, 2014). Any effort to improve curriculum through reform needs to consider the whole and efforts to reform by focusing on one or more parts is folly. Similarly, those who seek to reform necessarily become a part of the whole and are just as accountable for any success or failure as any other constituent of the whole. In the new wave of curriculum reform efforts, there is a necessity to reconceptualise assessment, accountability, learning, and teaching to consider different frameworks for individuals and collectives collaborating to learn by coparticipating in activities. Finally, curriculum is not usually associated with wellness of participants. However, there are numerous signs that learning environments often may be toxic and injurious to the health of participants. Ongoing research needs to document and continuously monitor indicators of participants' health and educate all to intervene when and as necessary. For this to happen, the roles of teachers and students will necessarily include research and learning new frameworks for teaching, learning, and other salient parts of the curriculum. As well as research, the roles of teachers and students should incorporate professional development and evaluation. Professional development focuses on learning new frameworks for teaching, learning, and salient constructs like dialogic inquiry and mindfulness. New frameworks should then be used to describe what is happening and why it is happening. Such descriptions are expansive and become objects for considering change and curricular improvement.

Will curriculum reform along the lines I have described it be considered as a viable path forward? In my curriculum class of the early 1970s, the answer is certainly yes. In my doctoral classes of 2013, the answer also is a resounding yes. However, as it was in the 1970s, so it remains today. Policymakers probably will continue to consider learning as individualistic, prescribe curricula in terms of macro considerations, and commodify achievement and collective accomplishments derived from the means of individual test scores. Those with the power to constitute a committee of experts, charge the committee to formulate recommendations for change, and then prescribe changes need to work with stakeholders within institutions to create wave after wave of local success. The ripple effects from repeated success are likely to compound and impact curriculum across boundaries of space and time. Collaborative approaches to reform that are multilevel and multilogical appeal as highly promising to improve learning environments while transforming not only curriculum but also the nature of science education, science, and what constitutes learning and knowledge. If changes along the lines I have suggested are adopted, then there is a strong possibility that the prevailing inadequacies of science education will gradually diminish and disappear – only to be replaced by new problems and new key players. A long-term solution probably will acknowledge there is no permanent fix. Solutions to curricular problems will be emergent, contingent, and impermanent. Knowing this can refocus science education to being in the moment and learning not only for the benefit of self but also for the collectives associated with the individuals' life worlds

References

Atkin, J. M., & Black, P. J. (2003). *Inside science education reform: A history of curricular and policy change*. New York: Teachers College Press.

Bakhtin, M. M. (1986). *Speech genres & other late essays* (C. Emerson & M. Holquist, Eds., V. W. McGee, Trans.). Austin, TX: University of Texas Press.

Bourdieu, P. (1992). The practice of reflexive sociology (The Paris workshop). In P. Bourdieu & L. J. D. Wacquant (Eds.), *An invitation to reflexive sociology* (pp. 216–260). Chicago: The University of Chicago Press.
Bybee, R. W., Buchwald, C. E., Crissman, S., Heil, D. R., Kuerbis, P. J., Matsumoto, C., et al. (1989). *Science and technology education for the elementary years: Frameworks for curriculum and instruction*. Washington, DC: The National Center for Improving Instruction.
Collins, R. (1981). On the microfoundations of macrosociology. *American Journal of Sociology, 86*, 984–1014. doi:10.1086/227351.
Collins, R. (2004). *Interaction ritual chains*. Princeton, NJ: Princeton University Press.
Davidson, R. J., & Begley, S. (2012). *The emotional life of your brain*. London: Hodder & Stoughton.
Ekman, P. (2003). *Emotions revealed: Recognizing faces and feelings to improve communication and emotional life*. New York: St. Martin's Griffin.
Elmesky, R., & Tobin, K. (2005). Expanding our understandings of urban science education by expanding the roles of students as researchers. *Journal of Research in Science Teaching, 42*, 807–828.
Erickson, F. (1986). Qualitative research on teaching. In M. C. Wittrock (Ed.), *Handbook of research on teaching* (3rd ed., pp. 119–161). New York: Macmillan.
Fellner, G. (2014). Broadening our lenses of perception to advance learning: An introduction to multilectics. *Teaching and Teacher Education, 37*, 169–182. doi:10.1016/j.tate.2013.04.015.
Goffman, E. (1983). The interaction order: American Sociological Association, 1982 Presidential Address. *American Sociological Review, 48*, 1–17.
Guba, E. G., & Lincoln, Y. (1989). *Fourth generation evaluation*. Thousand Oaks, CA: Sage.
Hurd, P. D. (1997). *Inventing science education for the new millennium*. New York: Teachers College Press.
Kincheloe, J. L., & Tobin, K. (2009). The much exaggerated death of positivism. *Cultural Studies of Science Education, 4*, 513–528. doi:10.1007/s11422-009-9178-5.
Marsh, C. J. (2009). *Key concepts for understanding curriculum* (4th ed.). New York: Routledge.
Philippot, P., Chapelle, G., & Blairy, S. (2002). Respiratory feedback in the generation of emotion. *Cognition & Emotion, 16*, 605–627.
Rowe, M. B. (1969). Science, silence, and sanctions. *Science and Children, 6*(6), 11–13.
Ryu, M. (2013). "But at school . . . I became a bit shy": Korean immigrant adolescents' discursive participation in science classrooms. *Cultural Studies of Science Education, 8*, 649–671. doi:10.1007/s11422-012-9406-2.
Shady, A. (2013). Negotiating cultural differences in urban science education: An overview of teacher's first-hand experience reflection of Cogen journey. *Cultural Studies of Science Education*. doi:10.1007/s11422-013-9486-7.
Swidler, A. (1986). Culture in action: Symbols and strategies. *American Sociological Review, 51*, 273–286.
Tobin, K. (2000). Becoming an urban science educator. *Research in Science Education, 30*, 89–106.
Tobin, K. G. (2008). In search of new lights: Getting the most from competing perspectives. *Cultural Studies of Science Education, 3*, 227–230.
Tobin, K. G., & Ritchie, S. M. (2011). Multi-method, multi-theoretical, multi-level research in the learning sciences. *The Asia-Pacific Education Researcher, 20*(3), 117–129.
Tobin, K., & Roth, W.-M. (2006). *Teaching to learn: A view from the field*. Rotterdam, the Netherlands: Sense.
Tobin, K., Roth, W.-M., & Zimmermann, A. (2001). Learning to teach in urban schools. *Journal of Research in Science Teaching, 38*, 941–964.
Turner, J. H. (2002). *Face to face: Toward a sociological theory of interpersonal behavior*. Palo Alto, CA: Stanford University Press.
Vygotsky, L. (1962). *Thought and language*. Cambridge, MA: MIT Press.

Chapter 4
Teachers' Ideas and Concerns with Assessment Practices in Inquiry Science

Poh-Hiang Tan and Aik-Ling Tan

Abstract Issues with assessment have traditionally been viewed with great importance as it is used (1) for placement of students in different tracks, (2) as a proxy for allocation of funds to educational institutions, and (3) for assessing students' learning. In Singapore, assessment practices have generally been summative in nature and aimed at evaluating what students have learnt and gained from school. The idea of formative assessment is relatively new for most teachers in science classrooms. Using task probes in a questionnaire format, we examined the factors that 39 grade four science teachers consider as they decide on inquiry tasks for their students. Based on their considerations, we infer their ideas of formative assessment as they engaged in science as inquiry and highlight concerns and dilemma that emerge as they report on their practices of science inquiry and formative assessment. We raised the concerns of the low level of teachers' understanding of the principles related to formative assessment and the perceived conflicts with summative assessment requirements and their practices of science as inquiry in the classrooms.

Keywords Formative assessment • Primary school science • Task design • Teachers' ideas • Assessment practices • Students' abilities • Dilemma • Assessment requirement • Teaching and learning • Inquiry-based science

Introduction

Assessment is part and parcel of school experience and is integral to teaching and learning. Broadly, it refers to any procedures in which information about students' learning are collected to facilitate decision making. Consequently, assessment

P.-H. Tan • A.-L. Tan (✉)
National Institute of Education, Nanyang Technological University, Nanyang Walk 1, Singapore 637616, Singapore
e-mail: aikling.tan@nie.edu.sg

A.-L. Tan et al. (eds.), *Inquiry into the Singapore Science Classroom*, Education Innovation Series, DOI 10.1007/978-981-4585-78-1_4,

processes can be termed formative or summative depending on their purposes and associated actions. Summative assessment refers to assessment used for the purpose of review of programmes, selection, and certification or simply for accountability to stakeholders. As such, summative assessments are conducted usually at certain milestone stages during the students' learning experience such as the end of the course or the end of the year (Linn & Millar, 2005; Nitko, 2004). Over the past two decades, with new understanding of how students learn, there are many criticisms about the negative impact of summative assessment on classroom teaching and learning. Often, it is blamed for narrowing the curriculum that is enacted in the classroom (Gipps, 1994; Linn, 2000; Popham, 2003).

Formative assessment, on the other hand, serves a different purpose compared with summative assessment. It is intended to aid learning and is closely associated with curriculum and pedagogy and is said to be central to good teaching practices (Black, 1993). In their highly acclaimed review paper, Black and Wiliam (1998) used findings from 250 studies to show that appropriate formative assessment could help to increase learning gains of students, thus bringing attention to the value of using assessment *as a means* to support teaching and learning. The work by Black and Wiliam has since garnered much support from academics across the world, many embracing the idea of using formative assessment to enhance teaching and learning. According to Black, 'formative assessment' is not a commonly used term until 1986. Popham (2008b) calls it 'instructionally oriented classroom assessment' and advocated it as a means to improve students' learning. Formative assessment also provides students with a platform to know their weaknesses and strengths, thus allowing them to be active in their own learning (Harlen & James, 1997).

With the promise that formative assessment holds to improving teaching and learning, it is timely to ask how formative assessment practices can be aligned with reforms in science classrooms where science as inquiry is advocated. This chapter examines the relationship between teachers' ideas about science as inquiry and formative assessment from the lens of teachers who are involved in the daily business of teaching and learning science in the classroom. The research questions that guide this study are:

1. What factors do primary science teachers take into consideration when they select tasks for science learning?
2. What aspects of formative assessment do teachers consider when selecting tasks for science learning?

Review of Literature

In this section, we present the current debates in the area of formative assessment and science learning as teachers engage in teaching. While we acknowledge that teaching and learning exist in a complex ecology of multiple factors such as societal norms, school culture, administrative demands, students' profiles, and many others, we have chosen to focus specifically on the relationship between the teachers' ideas about science teaching and formative assessment for this particular chapter. This focus will enable us to develop a better understanding of the factors influencing teachers' pedagogical decision making as they navigate the messy terrain of formative assessment. Science as inquiry is the context in which the study is situated.

Formative Assessment

The term formative assessment is still a diffused and fuzzy concept. It has been defined and used in different ways by researchers, practitioners, and policy makers. For example, Black, Harrison, Lee, Marshall, and Wiliam (2003, p. 2) described it as an assessment activity from which the 'evidence is used to adapt teaching work to meet learning needs'. Stiggins and DuFour (2009) advocate that formative assessment is not about what teachers do to their students but rather how teachers partner students to track their progress towards learning goals. It is also about interpreting results and making decisions on how to help them successfully attain their goals. While responsiveness to students' learning is crucial, some viewed formative assessment as any assessment activity that provide information about how and what the students learnt (Stobart, 2008) and would include the use of summative assessment information to make adjustments to instructional process as formative assessment practices. Response to the information elicited could range from seconds to months (Wiliam & Thompson, 2007). Formative assessment is applied and used as a tool in Singapore classrooms to 'provide richer information that can be used by the teacher, pupils and parents to improve in specific areas. For example, for English compositions, teachers can use rubrics to provide feedback to pupils on specific areas such as presentation of ideas, grammar and vocabulary. Rubrics provide a structured way of giving qualitative feedback beyond just an overall mark' (Heng, 2010).

As can be seen from the various definitions of formative assessment, there is still a lack of a unified agreement of the characteristics of formative assessment. This makes understanding and practising formative assessment a challenge for teachers as well as researchers. Despite the different understanding, it is commonly agreed that formative assessment is aimed at enhancing teaching and learning. The process requires the teacher to elicit evidence about what students have learnt or have not learnt in order to inform subsequent instructional moves. As such, formative assessment can be summarised by three central questions for students and teachers, based on works such as Atkin, Black, and Coffey (2001), Sadler (1989), and Wiliam (2011) – (1) where are you going, (2) where are you now, (3) how are you going to get there. These three central questions are crucial for teachers to consider as they guide their students towards the learning goals identified. While these three central questions may appear simple to answer, we argue that their actual role and nature amidst the dynamic interactions in the classroom is one that is difficult to fathom. This is because formative assessment fundamentally needs to be integrated seamlessly into instructional practices (Abell & Volkmann, 2006) and that it is reflexive and relational in nature (Buck, Trauth-Nare, & Kaftan, 2010). Further, as highlighted by Coffey, Hammer, Levin, and Grant (2011), authentic and effective formative assessment is discipline dependent and hence requires the teacher to attend to the knowledge and nature of the discipline in question. For example, in science, the formative assessment practices need to be anchored in science contents as well as how students come to understand these contents.

The discussion of the characteristics of formative assessment leads us to conclude that formative assessment is a complex and skilled task and its enactment in the classroom is highly dependent on teachers' sound content knowledge, proficient pedagogical content knowledge, their understanding of context and learners, and their belief systems. The nature of formative assessment and the demands that are made of teachers for its enactment suggest that to enable enactment of formative assessment, we need to first understand teachers' beliefs and ideas about formative assessment and its relationship to teaching and learning science. This forms the focus of this study.

Science Education and Science as Inquiry

Building on the fact that formative assessment is discipline dependent, we review the literature in the area of science learning, particularly on learning science as inquiry. In this study, science as inquiry forms the context from which we distil teachers' ideas and beliefs about formative assessment.

In the past few decades, curriculum reforms in science advocate that science learning should move from transmission of scientific contents to one that focusses on both scientific knowledge and the process of developing and using scientific knowledge. In fact, learning science is seen as a means to develop 'the capacity to use scientific knowledge, to identify questions and to draw evidence-based conclusions in order to understand and help make decisions about the natural world and the changes made to it through human activity' (Organisation for Economic Co-operation and Development [OECD], 2003, p. 133). The National Science Foundation (NSF) described the process of inquiry-based teaching and learning and the nature of science as follows:

> . . .*inquiry teaching* leads students to build their understanding of fundamental scientific ideas through direct experience with materials, by consulting books, other resources, and experts, and through argument and debate among themselves. All this takes place under the leadership of the classroom teacher. This process of inquiry is modeled on the scientist's method of discovery.
>
> It views science as a constructed set of theories and ideas based on the physical world, rather than as a collection of irrefutable, disconnected facts. It focuses on asking questions, considering alternative explanations, and weighing evidence. It includes high expectations for students to acquire factual knowledge, but it expects more from them than the mere storage and retrieval of information. (NSF, 1997, p. 7)

One of the key objectives of the inquiry science classroom is to develop students as active learners to work with their peers in constructing scientific knowledge. When practised in classroom, this is often done through activities that give students opportunities to work together towards a common goal. It is possible that such student-centred activities could provide opportunities for the teachers, who may take the role of observer or participant of the discussion, to understand what their students learnt or have not learnt. More important than hands-on opportunities, students should also have minds-on experiences in which inquiry is central to their learning of science. In short, if students learn science through inquiry, they will know what they know, how they know what they know, and why they believe what they know (Duschl, 2003).

One key difference between the traditional classroom and an inquiry-based classroom is the role played by the students. In traditional classrooms, students played a passive role as recipients of knowledge. An inquiry-based classroom aims to develop students as active learners who take ownership of their learning. Such autonomy may allow them to set the direction of how the lesson progresses. Despite the promises offered by exposing students to learning in an inquiry context, teaching science as inquiry is potentially challenging for teachers, particularly in the arena of formative assessment. In order for the teacher to orchestrate inquiry-based learning, they must know where their students are located in the learning journey. This requires the teacher to actively and continuously elicit information about students' learning – what they already know and can do and also what they have yet to know – and then adapt and change their teaching plans to cater to their students' needs (Black & Wiliam, 1998). This iterative process helps to shape students' learning and is important and fundamental to teaching and learning – the essence is formative assessment. Using the argument by Coffey et al. (2011) and Bennett (2011), it means formative assessment in science must thus be conceptualised to take into account this shift towards inquiry-based teaching and learning. Teachers will not be able to make the adaptation towards effective formative assessment practices if they do not have sufficient knowledge on the subject matter of science and its practice as a form of inquiry (Duschl & Gitomer, 1997; Harlen, Brand, & Brown, 2003; Harlen & James, 1997).

Regardless of the forms of science as inquiry, teachers need to develop their abilities to listen to the inquiry, to monitor continuously the current understanding of students so that instructions and activities can be modified and enhanced based on the learners' understanding and engagement (Ruiz-Primo, 2011). Compared with a traditional teacher-fronted classroom whereby the purpose of teacher question is to evaluate students' learning of the science content, in an inquiry classroom, the purpose of questioning by the teacher is to elicit students' ideas to enable them to reflect and consequently to craft coherent explanations and argumentation. This shift on the purpose of questioning in the inquiry classroom demands a change in the roles and responsibilities of both the teachers and the students (Kawalkar & Vijapurkar, 2011). This change from a didactic classroom to an inquiry classroom also implies a change in ways of assessment of students' learning from one that focusses on the correct answers to one that privileges students' ideas 'in the making'. As such, it is reasonable to infer that teachers' understanding of the process and nature of formative assessment can influence their abilities to capitalise on opportunities to carry out formative assessment in the context of science as inquiry and hence impact the effectiveness of inquiry teaching.

Mirroring international trends in science education reforms, the science educators and policymakers in Singapore are also examining ways to infuse the inquiry approach to develop skills and knowledge deemed as necessary in the twenty-first century as well as formative assessment that is concerned about improving and supporting students' learning of science. Despite the fact that formative assessment and science inquiry are complex concepts, with careful and deliberate planning, it may be possible to link inquiry and formative assessment as both primarily share the same objective of improving students' learning. This commonality between the two may provide opportunities for assessment to be integrated with activities that encourage scientific inquiry in the classroom and moves a step closer towards realising Popham's (2008b) idea of instructionally oriented classroom assessment or Shepard's (2000) assessment in a learning culture.

While it seems possible to integrate the two initiatives, it should be noted that the key to the success of any educational reforms lies with the teacher in the classroom. As highlighted in some earlier research, teachers' enactment of new reforms may not always be consistent with the intent and the expectations of the reforms (Lefstein, 2008; Tobin & McRobbie, 1996; Yerrick, Parke, & Nugent, 1997). In this study, we focus on how teachers conceptualise formative assessment in the light of the emphasis on inquiry in science classrooms. Earlier studies on science inquiry in Singapore classrooms by Tan and Wong (2012) and Kim, Tan, and Talaue (2013) suggested that the primary science teachers' classroom instruction remained predominantly didactic with only a handful exhibiting science as inquiry practices as described in the curriculum documents. This provides the current context of practices in the Singapore primary science classroom.

As can be seen from the discussion, one of the key contributing factors to successful science teaching is an alignment of the inquiry process to assessment practices. Accurate alignment is dependent upon teachers' knowledge and ideas of science as inquiry and the formative assessment process. As such, this study attempts to understand the factors that teachers take into consideration when they make decisions about tasks used to enable students to learn science concepts as well as to be engaged in scientific inquiry. Based on these factors, we examine if teachers give explicit attention to the principles of formative assessment as they go about their daily business of teaching science in the classrooms. This study models after the study by Tomanek, Talanquer, and Novodvorksy (2008) in terms of design, but we sampled elementary science teachers (as compared with secondary science teachers in the Tomanek et al. study) and the probes we used were for the topic of electricity. Primary science teachers' ideas form the focus of this study as they are usually trained as generalist teachers with no specialised academic training in science. As such, we hypothesise that their narrow content knowledge of the discipline might potentially contribute to their uncertainty about science as inquiry as well as formative assessment. In this chapter, we aim to distil the views of elementary science teachers pertaining to formative assessment as they plan for science inquiry. The findings from this study will contribute thus towards the body of knowledge on classroom instruction and assessment and increase our understanding of elementary science teachers' formative assessment practices in the science classroom with an orientation towards science inquiry.

Context of Study

This study is carried out in Singapore where there are efforts to move towards a more holistic system of assessment so that assessment practices can enhance learning. At the primary school level, students are exposed to various continual summative assessments throughout the school year. The results of these summative assessments are aggregated at the end of the year as a summative score of students' achievement for the year (Ministry of Education, Singapore [MOE], 1998). The final scores could be used to place students in different classes according to their achievement, but at the end of grade four, each student's achievement in individual subject is used to ascertain if he or she should proceed to take the subject at foundation or standard level in grade five. At the end of primary six, there is a national placement examination, known as the Primary School Leaving Examination (PSLE), which will determine the placement of students into different secondary schools. The impact of testing and examination on Singapore elementary schools has been highlighted in a number of reviews on elementary education. In the review on Singapore education in 1998, the external review committee, comprising academics from tertiary institutions and private sectors, noted that there was a lack of positive learning culture and attributed this to the overemphasis on class tests as well as terminal examinations in elementary schools. Pen and paper test is the predominant assessment mode used with the intent of finding out what is learnt rather than how it is learnt. It was also noted that within the elementary classroom, the pace of teaching is dictated by the imperative to complete all the workbooks that accompany each textbook and instructions are focussed on drill and practice to train students to produce the correct answers.

The science curriculum in Singapore undergoes a 6-year review cycle. In 2008, a new science curriculum was rolled out. The overarching idea that guides the new science curriculum is teaching and learning science as inquiry. Training workshops were conducted by curriculum specialists from MOE to equip teachers with the pedagogical skills to conduct inquiry-based lessons. One of the more frequently used inquiry-based strategies in Singapore classrooms is based on the 5E model developed by the Biological Sciences Curriculum Study (BSCS). The instructional model known as the 5E model includes five stages in the teaching and learning process, namely, Engage, Explore, Explain, Elaborate, and Evaluate (Bybee et al., 2006). This model, widely used in science classrooms in the United States and also in Singapore (Bybee et al.), provides a framework to enable teachers to plan lessons that have inquiry components infused in them. In the Singapore context, this is of particular importance, since many elementary school teachers do not have tertiary education in science.

Two years into the new science curriculum, a new reform was introduced into the primary classrooms for all subjects. In 2008, another review on education known as Primary Education Review and Implementation (PERI) was carried out. The committee, headed by the Minister of State for Education, comprises representatives from MOE, the private sector, and government agencies such as institutes of higher learning. Schools were recommended to 's(S)hift assessment practices away from an over-emphasis on assessment of learning as an end-outcome . . . and shape mindsets to view assessment as an integral part of ongoing processes to support learning' (Ministry of Education, Singapore [MOE] , 2009, p. 30). The review committee also recommended that there should be more assessment of students in academic and nonacademic areas to provide stakeholders with a holistic picture of the students' development in all aspects. It envisioned that students' learning could be enhanced if teachers were equipped with effective facilitation, questioning, and assessment techniques that included the use and development of rubrics to assess students. To equip teachers with the necessary skills in pedagogy and assessment, the committee recommended that S$100 million be allocated solely for teacher development in these areas. This recommendation was subsequently accepted by the government. With these recommendations, the focus is now shifted to formative assessment. A website dedicated to this initiative was set up to provide teachers with the resources for teaching. Training workshops focussing on classroom assessment strategies as well as alternative modes of assessment for specific subjects are currently being conducted.

Method

This study aims to provide perspectives into teachers' ideas about formative assessment as they plan for teaching science as inquiry. These perspectives include nature of tasks, students' abilities, resource availability, and constraints and curricular demands that teachers potentially consider or experience as they make pedagogical decisions in their planning. Research has suggested that teachers are often not conscious of their assessment practices during instruction (Bell & Cowie, 2001). As such, we use a task probe (in the form of a questionnaire) to elicit the tacit perspectives that teachers have as they go about planning for science as inquiry and formative assessment. We assumed that task/activity selection is part of lesson planning and is what Wiliam and Thompson (2007) referred to as 'upstream planning'. Upstream planning refers to classroom activities planned in view of learning outcome before the lesson and such activities would naturally include formative assessment events. Shavelson et al. (2008) noted that since formative assessment has a salutary impact on learning and achievement, it is essential that what counts as formative assessment should be aligned to the national curriculum on which national assessment is based. The alignment would not only enhance learning but also better inform students on the expectations of the national curriculum. However, under normal circumstances, the teachers' choice of activity would depend on what they see as opportunities for learning and how learning could be enhanced through the activity. Whether or not formative assessment forms part of the factors affecting decision making would depend on how they view the role of assessment in teaching and learning. This baseline study is designed to enable us to find out teachers' views about task selection and whether formative assessment is featured in their decision making in task selection and planning.

The sample consists of 39 primary school teachers teaching science. The six schools represent schools that have responded favourably to an email sent to primary schools in Singapore. The teachers are all teaching science in the schools, and they have had experiences teaching primary five science. They had either volunteered to be part of the study or have been nominated by their heads of department to participate. The teachers had teaching experiences ranging from 1 to 40 years. A questionnaire consisting of three tasks (labelled T1, T2, and T3) on electricity was used (see Appendix 1). The three tasks were aimed at assessing the following three learning outcomes found in the primary science syllabus:

1. Recognise that an electric circuit consisting of an energy source (battery) and other circuit components (wire, bulb, switch) forms an electrical system.
2. Show an understanding that a current can only flow through a closed circuit.
3. Identify electrical conductors and insulators (Curriculum Planning and Development Division [CPDD], 2008).

To develop the tasks, we analysed some questions from the PSLE in previous years and noted that there was an emphasis on reasoning. Generally, the questions aimed to assess higher-order thinking skills which included the ability to analyse, interpret and evaluate data, and make informed decisions or prediction. Based on this understanding, task T1 was designed in the format similar to the questions in PSLE, requiring the student to give a written response. Task T2 was similar to the activities found in science resource books, requiring the students to work in pairs. It followed the predict-observe-explain sequence with which teachers are familiar. The apparatus needed in this activity are commonly found in the science laboratory. Task T3 was a group activity that required each group of four to five students to create a puzzle for another group of students to solve. At the end of the activity, the students were required to share with the owner of the puzzle how they arrived at the answers. The set-up for this activity required the use of an empty shoebox, which is easily available and inexpensive. It also required a switch, two light bulbs, and a set of batteries to be mounted on the box cover. The task was developed in alignment with the new science inquiry curriculum and required students to pose questions, investigate, and solve problems through student collaboration.

The tasks were crafted with the intent to understand how teachers view the possibility of finding out about students' learning through written assessment versus hands-on activity as well as an activity with predictable outcomes versus an activity which allows a range of responses. A summary of the three tasks is found in Table 4.1.

Table 4.1 Features of the three tasks

Task	Format of response	Deliverables	Process of working
T1	Individual written response to questions	Written answers	Individual analyzes and solves problem based on diagram of circuit
T2	Individual written report based on instructions in the task	Written report	Pair predicts and then constructs actual circuit to test predictions
T3	Group, presentation	Actual shoe-box puzzle Presentation	Group designs, constructs puzzle. Group analyzes and solves problem

The responses in the teacher questionnaire after being shown the three tasks form the data corpus for this study. As suggested by Saldaña (2009), two cycles of coding were adopted. The first cycle marked the beginning of analysis in which the data was split into individual coded segments using NVivo 9. Ideas frequently expressed by the teachers were examined for similarities and differences. Subsequently, axial coding was carried out in the second cycle of coding where the coded segments were compared and reorganised into categories to show the relationships among the frequent ideas expressed by the teachers. At the start of each round of coding, the two coders discussed to establish a common understanding of each category and coding resulted in an inter-rater reliability of 85 %. Four themes (implementability, learning outcomes, nature of task, and students) were thus developed based on the responses of the teachers. To further ensure reliability of the themes that emerged, the details and context of the responses in the questionnaires were re-examined to finalise the themes among the responses so that the factors guiding the teachers' decision making for selecting tasks for formative assessment could be elucidated.

Findings and Discussions

Factors Affecting Task Selection

Related to the first research question on the factors that teachers take into consideration when they select learning tasks, we found that there were four key factors: (1) nature of task, (2) students, (3) learning outcomes, and (4) implementability of task. As shown in Table 4.2, the nature of the task (18.9 %) is the most important consideration when teachers select tasks for formative assessment. Teachers are also cognisant of students' ability and motivation (18.3 %) when they are selecting the task as can be seen by the similar percentage coverage of the codes.

While we can see quantitatively from the coding of teachers' responses that the nature of task and consideration of students' abilities influence their selection, we argue that the four factors interact with one another to collectively influence a teacher's decision. We describe the interaction between the four factors next.

Table 4.2 Factors affecting teachers' decision making for formative assessment task selection

Categories	Sub-category	Percentage coverage (%)
Nature of task (18.9 %)	Task is good for enrichment, to prepare for the PSLE, for tuning in or as an extension	5.6
	It is challenging / simple enough	8.4
	To assess students' understanding or for summative assessment	3.4
	To identify misconceptions or students' understanding of conductor and non-conductors	1.5
Students (18.3 %)	Have or do not have the ability to do the task, for high (er) ability, for average ability or for low(er) ability	10.2
	Students will like or enjoy or find interesting	5.3
	Collaboration, group work or work as a team	2.8
Learning outcomes (14.7 %)	Understand concept of electrical circuit/electricity/conductors/electrical components	8.9
	Hands-on/process skills/predict/analyze	5.8
Implementability (12.8 %)	Will be able to handle, manageable or too complex to carry out	4.4
	Investment of time	3.8
	Teachers' guidance, help/need to monitor or scaffolding	2.5
	Easy to get materials, time-consuming or apparatus is available	2.1

Approximately 8.4 % of the responses described the tasks as 'simple' or 'challenging' without clarifications on its meaning. Tasks could be seen to be simple or challenging for teachers in terms of the quantity or type of materials that they have to prepare ahead of time or it could also refer to its complexity and time required for marking. In contrast, the task could also be simple or challenging for students (as such, reference is to students' abilities) in terms of its cognitive demand on students or the complexity and time required for completing the task. These concerns were expressed by the following teachers:

> I think the middle ability and high ability classes will be able to do this activity although I think the high ability classes may find Task 3 more interesting and challenging. (Tr 24)
>
> The preparation of the materials used in the task is too time consuming. (Tr 22)

Some of the teacher participants (4.4 %) also referred to the manageability of the task for students, that is, whether the students are able to carry out the task. While we have categorised 'simple' or 'challenging' under the category of nature of task and manageability under the category of implementability of the task, we are aware that the participants could be referring to the same thing. In our analysis, nature of task describes the intrinsic property of the task and the cognitive demand required from the students to work on the task. Hence, when teachers describe the tasks as 'simple' or 'challenging', the assumption we make is that they are referring to the intrinsic cognitive demands of the task on students. Alternatively, implementability refers to aspects such as resources required for the task and does not change the nature of the task.

Our analysis of the teachers' responses about their considerations during task selection reveals that only 2.5 % were deliberations on the roles that teachers and students have to play. This group of teachers discussed their role as one of monitoring, checking, guiding, or scaffolding students' learning. There was no further elaboration on the teachers' role. Considerations of students' involvement was completely absent although there was some acknowledgement (2.8 %) that collaboration was a good thing to have in classroom activity. The absence of mention of students' role in task selection is likely to be indicative of the lack of consideration for students' active participation in the assessment process. This reiterated the concern by Stiggins and DuFour (2009) and Shepard (2000) about the active roles that students need to take up during classroom assessment.

In most primary schools, between 1.5 and 2.5 h per week is devoted to science lessons. Teachers indicated that the allotted time for science was short and they usually have insufficient time to complete their teaching plans. The availability and sufficiency of curriculum time is hence an important consideration for task selection and explains why 3.8 % of the responses indicated that it was a factor for consideration. Factors such as time and ease of administration can also potentially overshadow other educational benefits during task selection. As expressed by Tr 1: 'I prefer T1 as it is a neat way to get an idea of whether the learning outcomes have been attained in the shortest time. T2 and T3 provide the pupils with lots of hands-on opportunities… t(T)he pupils will enjoy T2 and T3 much more. They may be more engaged and if given the right guidance may learn much more about electricity (beyond syllabus coverage)'. Here we sense that the teacher is tempted by the efficiency of task 1 in helping her achieve the intended learning outcomes within the shortest possible time and hence her choice of task. The teacher implied that hands-on and problem-solving tasks will require more time for her to get an idea of whether her students have attained the intended learning outcomes.

Understanding of Formative Assessment Revealed by Teachers

Related to the second research question about the aspects of formative assessment that teachers revealed from their involvement in task selection, we infer that the teachers show a lack of understanding about the use of formative assessment in assessing the working conceptions of students rather than the achievement or mastery of content knowledge. When assessment is identified as the purpose that the task can fulfil, the participating teachers focus their attention on whether the task is able to help them differentiate and prepare the students for summative assessment, often giving little attention to how information about students' learning from engaging in the task would inform their subsequent pedagogical move. For instance, Tr 4 said 'I prefer this task because it can be used to differentiate the higher ability pupils from the middle/low ability' and similarly, Tr 38 said '…*T1 helps students to prepare for PSLE*'. The participating teachers' concern with placement of students as well as the national examination is not surprising as previous research (Gipps, 1994; Linn & Millar, 2005; Stobart, 2008; Popham, 2008a) has established the strong influence of the national examination on classroom practices. For the 1.5 % of the responses who expressed possibilities of assessing students' understanding while they do the tasks, it was not obvious how the information would be used to help students' progress in their subsequent learning. In a nutshell, even if the task were intended for formative purposes, the focus was on elicitation of evidences and it was unclear how evidences would be used to improve students' learning.

Secondly, with respect to the characteristic or nature of formative assessment, the teachers showed that they lacked the concept of integrating assessment into the daily ritual of teaching and learning. Their ideas about the developmental purpose of formative assessment are somewhat diffused and adulterated with their deeply ingrained ideas of summative assessment. When planning for science-related activities to engage students in class, teachers primarily considered the nature of the task. They tend to perceive each task as serving distinct functions – either for teaching and learning or for assessment. As Tr 16 said, 'The other two tasks are more direct and can be used immediately after the basic lessons are taught. I think they are not very suitable for formative assessments'. The participant teacher seemed to suggest that if a task is designed for teaching and learning, then it is not suitable as formative assessment. Further, when the participating teachers examined the three tasks, they related the task to the purpose with which it can help them fulfil in the classroom – as enrichment, as an extension activity, to diagnose students' misconceptions or to prepare for examination (refer to Table 4.2). This is consistent with the notion that tasks chosen for lessons must be designed with a specific purpose in mind and it is likely that they are also implemented to fulfil that specific purpose only. This somewhat superficial perspective can be limiting in teachers infusing formative assessment practices in their classrooms.

Summary of Findings

To answer the first research question, it is evident that teachers consider the intrinsic nature of the task and the cognitive demands required from the students to work on the tasks when they are making a decision. They relate levels of difficulty of the task to the abilities of the students and try to establish a match between the task and the learner ability. With respect to formative assessment, our findings suggest that teachers use prior evidences of students' abilities to help them select tasks for their students. If they believe that students have lower abilities in science, they will select tasks that are less demanding. Only when the students have exhibited that they can master less demanding tasks will more demanding tasks be selected. Teachers also examine the tasks for the learning outcomes that will be fulfilled when the task is carried out. They consider learning outcomes on the two aspects of scientific concepts as well as science process skills. Finally, the teachers are also cognisant of the logistical constraints and availability of resources when they are selecting tasks for students to work on. Our findings revealed that considerations such as students' roles and involvement, student naïve conceptions, and systematic documentation of student learning and interactions are not considerations which are privileged by teachers during task selection to enhance science learning.

The findings from this study reveal very similar findings from Tomanek et al.'s (2008) study although the participating teachers teach at different grade levels. Many of the teachers consider students' ability as an important factor in selecting tasks, and consequently, students from the lower ability group are given more routine and familiar tasks. While students' learning is an important consideration, other practical considerations, such as time constraint in terms of time needed for preparation as well as curriculum time needed to complete the tasks, may result in teachers forsaking any perceivably less familiar or complex tasks for the more routine ones.

Pertaining to the second research question, our findings suggest a rather disturbing lack of systematic and conscious thought about enactment of formative assessment in the classroom. With regard to the process of formative assessment, the participant teachers showed a lack of understanding about the use of formative assessment as a *process* to develop students' working conceptions of science rather than assessing their content mastery and attainment. Related to the characteristic of formative assessment, the participating teachers showed that they need to develop the concept of integrating formative assessment seamlessly into the teaching and learning process and that they have to develop a clearer understanding of the purposes of formative assessment (as opposed to summative assessment). Teachers' concerns about task and syllabus completion to prepare the students for final placement examinations appear to overwhelmingly influence their decision making. This is not surprising since the results of summative final assessment are more tangible than that of formative assessment. The issue is further compounded by the fact that there is still a lack of consensus on what constitutes formative assessment (Bennett, 2011; Black & Wiliam, 1998; Popham, 2008b) – is it an instrument or a process? Bennett cautioned that there is a lack of clearly defined artefacts or practices associated with formative assessment and specific to science and how the contents of science can be integrated into generic formative assessment ideas. The other finding from this study shows a dichotomy of purposes into summative and formative. Even within formative purposes, the responses from teachers were silent about how students may be actively involved. This further raises questions on the nature of formative assessment in the classroom.

Implications and Conclusions

The results from this study suggest that a set of descriptors about formative assessment for science as inquiry is necessary. These descriptors will assist teachers to incorporate principles of formative assessment in their lesson planning and task selection so that enactment of formative assessment is systematic and deliberate rather than accidental. Teachers will also benefit from having exemplars and narratives of formative assessment practices within science inquiry classrooms. These exemplars can potentially provide teachers with a mental image of the forms of interactions that will take place between teachers and students when formative assessment is used in the classroom. Further, these exemplars will be useful for the professional development of teachers as well as to aid professional dialogue among communities of teachers. The infusion of formative assessment ideas in upstream planning will lead teachers to be more cognisant about how to use information about students' learning to decide on the next pedagogical move. Explicit knowledge of formative assessment practices will also move teachers' attention away from logistical concerns about tasks to a more student-centred decision-making process. This will enable them to customise their lessons better to meet the diverse learning needs of different students.

There are a few limitations in this study that we like to highlight. Firstly, the results of this study relied only on teachers' responses to the questionnaire. Analysis of their written responses is hence interpretative in nature, and there was no in-depth interview carried out to clarify any of the responses. To achieve a more robust idea of their views, follow-up interviews can be carried out with selected teachers. Secondly, classroom observation data could have been included to provide a more accurate picture of what actually goes on within the classrooms where these teachers interact with the students and the task selected.

Appendix 1

Survey: Electrical Circuits

Objective of the Survey

To study teacher's reasoning when they select tasks for use in formative assessment.

Introduction

Thank you for your participation in this study.

This questionnaire is based on the three formative assessment tasks given in Appendix A.

The formative assessment tasks were set based on the following learning outcomes:

1. Recognise that an electric circuit consisting of an energy source (battery) and other circuit components (wire, bulb, switch) forms an electrical system.
2. Show an understanding that a current can only flow through a closed circuit.
3. Identify electrical conductors and insulators.

Please help to complete the survey questions in pages 2–3. Your feedback should be based on your own recent experience when you taught the topic 'electricity' to one of your classes.

Thank you for your help.

Information about you:
Name (optional): ________________
Race: ________________ Gender: ________
Number of years of experience in teaching science: ________
Highest qualifications (academic): ________________
Professional qualifications: ________________

Background of the Lesson

You have introduced to your students ideas of electrical circuits, complete circuit, and electrical conductors.

(a) If you were now to formatively assess your student when teaching a lesson with the above objectives, what assessment tasks, T1, T2, or T3, would you choose?

Possible Considerations

1. Suitability of task
2. Students
3. Other reasons

(b) Why did you not choose the other two assessment tasks? In what situations do you think these two tasks could be used?

Possible Considerations

4. Suitability of task
5. Students
6. Other reasons

Task 1: Direct Questioning/Process Skill Textbook: Individual Work (T1)

The diagram below shows an incomplete circuit with gaps at X, Y, and Z and identical light bulbs L1 and L2.

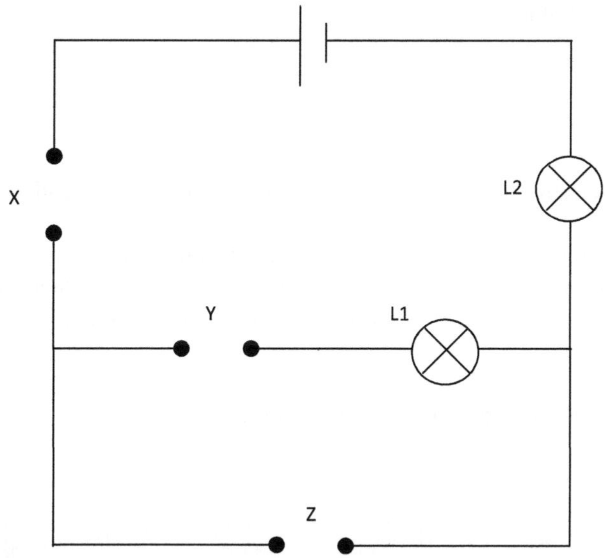

John placed three rods, one made of glass, one made of iron, and another made of copper on X, Y, and Z.

Only L2 lights up.

What can you conclude about the objects at X, Y, and Z? Explain your answer.

Are there other possible answers? If yes, what are they?

Task 2: Inquiry-Based Activity: Pair Work (T2)

Question: Can we tell if an object is a good conductor of electricity by its appearance?

Predict if the following items are good conductors of electricity:

Item	Good conductor of electricity	
	Yes	No
1. Aluminium foil ([a]rolled as a stick)		
2. Rubber stick		
3. Thick copper wire ([a]there is a layer of transparent insulation that needs to be removed for it to conduct electricity)		
4. Strip from a soft drink ([a]colourful can, e.g., coca cola)		

[a]( ) information for teachers

You are given the following apparatus that need or need not be used:

1. Battery in holder
2. Connecting wires with crocodile clips
3. Light bulb in holder
4. Switch
5. Sandpaper
6. A4 paper
7. Scotch tape

(a) Design a circuit to test whether the above items are good conductors of electricity.
(b) Use your circuit to confirm if your predictions are correct.
(c) Is/are there any item(s) that did not behave as predicted? What could be the reason? You may conduct other tests to check on your reasons.

Task 3: Guess My Connection: Group Work (T3)

Shoebox Puzzle

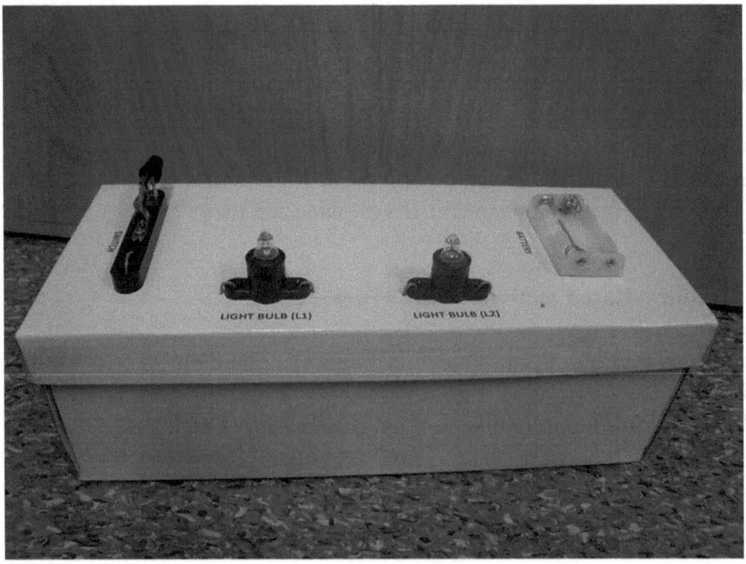

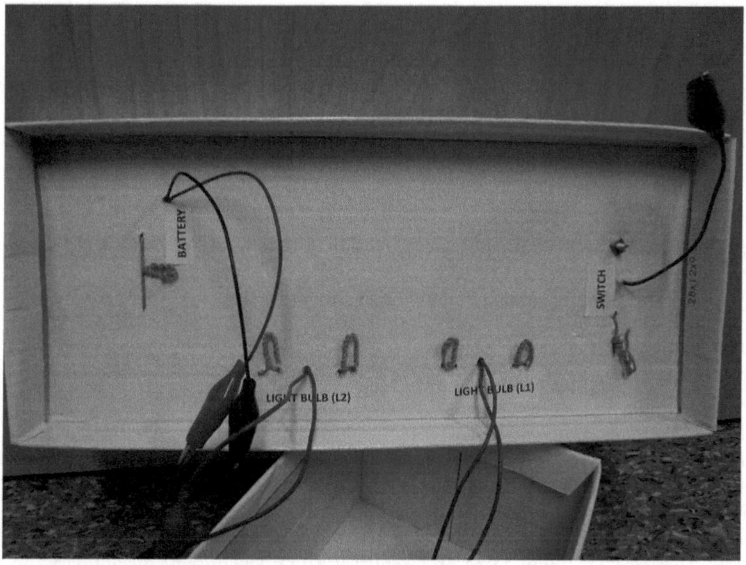

Your group is given the following apparatus:

1. A shoebox with a switch, a battery, and two light bulbs (L1 and L2) on the cover of the box.
2. A glass rod
3. A aluminium foil
4. An iron rod
5. Six connecting wires
6. A piece of A3 paper

Make a shoebox puzzle using the following steps:

1. On the given paper, design a possible circuit which contains a battery, a switch, two light bulbs, and three rods (one made of glass, one made of aluminium foil, and another made of glass).
2. Predict what happens when you switch on the circuit.
3. Connect this circuit using the given apparatus. The connections must not be seen and must be made under the cover of the box.
4. Is your prediction correct? If not, why?
5. Repeat steps 1–4 until you are satisfied with your design.

After you have completed your shoebox puzzle, exchange your shoebox puzzle with another group. While the other group will try to solve your puzzle, your group must also try to solve their puzzle in as short a time as possible by drawing out the possible connections.

Present your solution to the owners of the puzzle.

References

Abell, S. K., & Volkmann, M. J. (2006). *Seamless assessment in science: A guide for elementary and middle school teachers*. Portsmouth, NH: Heinemann.

Atkin, J. M., Black, P., & Coffey, J. E. (2001). *Classroom assessment and the National Science Education Standards*. Washington, DC: National Academy Press.

Bell, B., & Cowie, B. (2001). The characteristics of formative assessment in science education. *Science Education, 85*, 536–555.

Bennett, R. E. (2011). Formative assessment: A critical review. *Assessment in Education: Principles, Policy & Practice, 18*(1), 5–25.

Black, P. J. (1993). Formative and summative assessment by teachers. *Studies in Science Education, 21*(1), 49–97.

Black, P. J., Harrison, C., Lee, C., Marshall, B., & Wiliam, D. (2003). *Assessment for learning: Putting it into practice*. Maidenhead, UK: Open University Press.

Black, P., & Wiliam, D. (1998). Assessment and classroom learning. *Assessment in Education: Principles, Policy & Practice, 5*(1), 7–71.

Buck, G. A., Trauth-Nare, A., & Kaftan, J. (2010). Making formative assessment discernible to pre-service teachers of science. *Journal of Research in Science Teaching, 47*, 402–421.

Bybee, R. W., Taylor, J. A., Gardner, A., Scotter, P. V., Powell, J. C., Westbrook, A., et al. (2006). *The BSCS 5E instructional model: Origins and effectiveness*. Colorado Springs, CO: BSCS.

Coffey, J. E., Hammer, D., Levin, D. M., & Grant, T. (2011). The missing disciplinary substance of formative assessment. *Journal of Research in Science Teaching, 48*, 1109–1136.

Curriculum Planning Development Division [CPDD]. (2008). *Science syllabus – Primary 2008*. Singapore: Curriculum Planning and Development Division, Ministry of Education.

Duschl, R. A. (2003). Assessment of inquiry. In J. M. Atkin & J. Coffey (Eds.), *Everyday assessment in the science classroom* (pp. 41–59). Washington, DC: National Science Teachers' Association Press.

Duschl, R. A., & Gitomer, D. H. (1997). Strategies and challenges to changing the focus of assessment and instruction in science classrooms. *Educational Assessment, 4*(1), 37–73.

Gipps, C. (1994). *Beyond testing: Towards a theory of educational assessment*. London: The Falmer Press.

Harlen, W., Brand, J., & Brown, R. (2003). *Enhancing inquiry through formative assessment*. San Francisco: Exploratorium.

Harlen, W., & James, M. (1997). Assessment and learning: Differences and relationships. *Assessment in Education: Principles, Policy & Practice, 4*, 365–379.

Heng, S. K. (2010, September 16). *Parliamentary replies: Primary one assessment*. Accessed May 9, 2013, from http://www.moe.gov.sg/media/parliamentary-replies/2010/09/primary-one-assessment.php

Kawalkar, A., & Vijapurkar, J. (2011). Scaffolding science talk: The role of teachers' questions in the inquiry classroom. *International Journal of Science Education*. doi:10.1080/09500693.2011.604684.

Kim, M., Tan, A.-L., & Talaue, F. (2013). New vision and challenges in inquiry-based curriculum change in Singapore. *International Journal of Science Education, 35*, 289–311.

Lefstein, A. (2008). Changing classroom practice through the English National Literacy Strategy: A micro-interactional perspective. *American Educational Research Journal, 45*, 701–737.

Linn, R. (2000). Assessments and accountability. *Educational Researcher, 29*(2), 4–16.

Linn, R. L., & Millar, M. D. (2005). *Measurement and assessment in teaching* (9th ed.). Upper Saddle River, NJ: Pearson Education.

Ministry of Education, Singapore [MOE]. (1998). *Learning, creating, communicating: A curriculum review – Report by External Review Team*. Accessed October 6, 2012 from http://www.moe.gov.sg/media/press/1998/980321.htm

Ministry of Education, Singapore [MOE]. (2009). *Report of the primary education and implementation committee.* Accessed on October 6, 2012 from http://www.moe.gov.sg/media/press/files/2009/04/peri-report.pdf

National Science Foundation [NSF]. (1997). *The challenge and promise of K–8 science education reform* (Foundations: A monograph for professionals in science, mathematics, and technology education, Vol. 1). Washington, DC: National Science Foundation.

Nitko, A. J. (2004). *Educational assessment of students* (4th ed.). Upper Saddle River, NJ: Pearson Education.

Organisation for Economic Co-operation and Development [OECD]. (2003). *The PISA 2003 assessment framework – Mathematics, reading, science and problem solving knowledge and skills.* Retrieved September 4, 2012, from http://www.oecd.org/edu/preschoolandschool/programmeforinternationalstudentassessmentpisa/33694881.pdf

Popham, J. (2003). *Teach better, test better: The instructional role of assessment.* Alexandria, VA: Association for Supervision and Curriculum Development.

Popham, W. J. (2008a). *Classroom assessment: What teachers need to know* (5th ed.). Boston: Pearson Education.

Popham, W. J. (2008b). *Transformative assessment.* Alexandria, VA: Association for Supervision and Curriculum Development.

Ruiz-Primo, M. A. (2011). Informal formative assessment: The role of instructional dialogues in assessing students' learning. *Studies in Educational Evaluation, 37*(1), 15–24.

Sadler, D. R. (1989). Formative assessment and the design of instructional systems. *Instructional Science, 18*(2), 119–144.

Saldaña, J. (2009). *The coding manual for qualitative researchers.* Cornwall, UK: Sage.

Shavelson, R. J., Yue, Y., Furtak, E. M., Araceli Ruiz-Primo, M., Ayala, C. C., Young, D. B., et al. (2008). On the role and impact of formative assessment on science inquiry teaching and learning. In J. Coffey, R. Douglas, & C. Stearns (Eds.), *Assessing science learning: Perspectives from research and practice* (pp. 21–36). Washington, DC: NSTA Press.

Shepard, L. A. (2000). The role of assessment in a learning culture. *Educational Researcher, 29*(7), 4–14.

Stiggins, R., & DuFour, R. (2009). Maximizing the power of formative assessments. *Phi Delta Kappa, 90*, 640–644.

Stobart, G. (2008). *Testing times: The uses and abuses of assessment.* New York: Routledge.

Tan, A.-L., & Wong, H.-M. (2012). "Didn't get expected answer, rectify it.": Teaching science content in an elementary science classroom using hands-on activities. *International Journal of Science Education, 34*, 197–222.

Tobin, K., & McRobbie, C. J. (1996). Cultural myths as constraints to the enacted science curriculum. *Science Education, 80*, 223–241.

Tomanek, D., Talanquer, V., & Novodvorsky, I. (2008). What do science teachers consider when selecting formative assessment tasks? *Journal of Research in Science Teaching, 45*, 1113–1130.

Wiliam, D. (2011). What is assessment for learning? *Studies in Educational Evaluation, 37*(1), 3–14.

Wiliam, D., & Thompson, M. (2007). The future of assessment. In C. A. Dwyer (Ed.), *The future of assessment: Shaping teaching and learning* (pp. 53–82). New York: Taylor & Francis Group, LLC.

Yerrick, R., Parke, H., & Nugent, J. (1997). Struggling to promote deeply rooted change: The 'filtering effect' of teachers' beliefs on understanding transformational views of teaching science. *Science Education, 81*, 137–159.

Chapter 5
The Development and Implementation of a Guided-Inquiry Curriculum for Secondary School Physics

Darren Wong and Chor Yam Lau

Abstract The Physics by Inquiry (PbI) curriculum developed by the Physics Education Group at the University of Washington has been introduced in Singapore since 2005 to help teachers teach physics in a way that emphasises the development of fundamental concepts and reasoning skills through first-hand laboratory-based experiences. A key question for us is whether the success of the PbI curriculum and approach for teacher preparation could be transferred to effective learning of physics for secondary students in the Singapore context. In particular, we are interested to find out about students' attitudes and perceptions and their conceptual learning through the inquiry approach. This chapter describes our journey in the development and implementation of a guided-inquiry curriculum using the PbI 'research-curriculum development-instruction' iterative approach. The 3-year project was conducted in a secondary school which involved eight teachers and over 1,000 students from three cohorts of secondary one (grade 7) classes. The data sources for this study included notes from interactions with teachers, classroom observations, pretests, posttests, surveys and interviews. We report on the effectiveness of the curriculum package and the perceptions of teachers and students towards the guided-inquiry instructional approach and discuss the main areas of modification in adapting the PbI curriculum to the school context. Lessons learned in this study may provide insights to other schools, teachers and educators who wish to embark on a similar journey to facilitate inquiry-based curriculum and instruction in the learning of school science.

Keywords Guided-inquiry science education • Inquiry-based science • Secondary school science • Physics education research • Teacher professional development • Students' attitudes towards science • Student alternative science conceptions

D. Wong (✉)
Curriculum Planning and Development Division, Ministry of Education, Singapore, Singapore
e-mail: darren_wong@moe.gov.sg

C.Y. Lau
National Institute of Education, Nanyang Technological University, Singapore, Singapore

A.-L. Tan et al. (eds.), *Inquiry into the Singapore Science Classroom*,
Education Innovation Series, DOI 10.1007/978-981-4585-78-1_5,

Introduction

The inquiry-based approach to science instruction has been advocated in science education documents of many countries for well over a decade. One of the most widely cited descriptions of 'inquiry' is offered in the National Science Education Standards (National Research Council, 1996, 2000). The inquiry approach aligns closely with the activities and processes scientists engage in: focussing on scientific theories and models, asking investigable questions, forming hypotheses, gathering evidence, generating explanations from evidence and justifying claims and explanations. The inquiry approach is also consistent with constructivist beliefs, where students are challenged to form deep understandings about natural phenomena by engaging in the construction and critique of scientific knowledge and evidence through an active process of investigation. The term 'inquiry' refers to both what scientists do and how students learn (and the corollary of this being how teachers teach).

The primary and lower secondary science curricula in Singapore has placed 'science as an inquiry' as the core emphasis in guiding curriculum design and the way science should be taught in schools since the turn of the century in 2000 (for the latest copy of the science syllabuses, refer to http://www.moe.gov.sg/education/syllabuses/sciences/). A large part of the professional development effort since then had been to provide teachers with good models of inquiry-based science instruction and to equip them with the knowledge and skills to facilitate inquiry-based learning in schools. In particular, the PbI curriculum (McDermott & The Physics Education Group, 1996), developed by the Physics Education Group at the University of Washington, was introduced through a series of workshops for primary science, secondary and pre-university physics teachers since 2005. The PbI curriculum is primarily designed for in-service teachers to strengthen their subject matter knowledge in physics/physical science topics, with an emphasis on the development of fundamental concepts and reasoning skills through first-hand laboratory-based experience. There is a strong emphasis on the development of important scientific skills, such as distinguishing between observations and inferences, controlling variables, proportional reasoning and deductive and inductive reasoning. PbI fosters the simultaneous development of physical concepts, reasoning ability and representational skills within a coherent body of content. The teachers go through the reasoning in depth and are guided in synthesising what they have learned into a coherent conceptual framework (McDermott, Heron, & Shaffer, 2005). An extensive list of publications gives strong evidence of the effectiveness of the PbI approach in developing deep conceptual understanding in various topics for undergraduates as well as prospective and practising teachers at the University of Washington (see http://www.phys.washington.edu/groups/peg/pubs.html).

Can the success of the PbI curriculum and instructional approach for teacher professional development be also transferred to effective learning of physics for secondary students in the Singapore context? This was the key question in our minds and we hope that this research can contribute towards the relatively scant literature on the implementation and effectiveness of the PbI approach at the secondary school level, where 'effectiveness' is evaluated based on both cognitive and affective aspects.

The research team was mindful that the early years of secondary education are most critical in shaping students' attitudes towards science (Osborne, Simon, & Collins, 2003). Recent studies have indicated that students may hold dichotomous attitudes towards science. Data from the Relevance of Science Education (ROSE) project (Sjøberg & Schreiner, 2005) reveal that the issue of dichotomous attitudes towards science is especially prevalent among youths from developed countries. In particular, this is seen in England (Jenkins & Nelson, 2005) where students feel that 'science is important (to society) but not for me'. In a recent study, Singapore teachers felt that physics in particular seemed to be difficult and abstract to students, and they emphasised the importance of co-curricular activities and laboratory work to enhance interest of students in learning physics (Oon & Subramaniam, 2011). Results from our own baseline study (Yeung et al., 2010) showed the cognitive and affective components of self-concept in learning physics (i.e., competence and interest in physics, respectively) to be highly correlated, with competence in physics (cognitive component of self-concept) having an effect only on engagement in learning physics (a short-term outcome) but not on the aspiration to learn physics in the future (a long-term outcome). This means that a student who shows a sense of competence in learning physics may not necessarily aspire to take up physics in the longer term. However, interest in physics (affective component of self-concept) had an effect on both engagement and aspiration and is a key predictor of whether the student will likely aspire to learn physics in future.

Hence the goal of this study is to develop an inquiry curriculum that not only advances students' understanding of scientific concepts and procedures, but that also personally engages them and shapes positive perceptions and attitudes towards science (physics), particularly at the lower secondary level. The key research questions guiding this study are:

1. What are the attitudes and perceptions of teachers and students on teaching and learning physics by inquiry in the classroom?
2. What are the differences in the conceptual learning outcomes when students learn physics through the inquiry approach as compared to the standard classroom practices?

Background of Study

While all the teachers who attended the workshop found the PbI workshops relevant and enriching for their personal development as teachers, there remained challenges to systemic implementation of the inquiry approach in school. Through our interaction with teachers and schools, we found four common challenges in using the PbI curriculum. Firstly, teachers found it difficult and time-consuming to adapt the PbI instructional materials by pitching it at an age-appropriate level and in a coherent manner to match their students' needs and abilities. Secondly, additional time and effort were needed to prepare and test laboratory equipment to ensure good experimental data can be collected, since the PbI approach requires students to craft their own evidence-based conclusions and explanations based on their experimental results. Thirdly, another major challenge in implementing PbI was the large class size of typically 40 students. Instead of pair work, students were usually grouped in fours in order to meet the demand for laboratory apparatus while still allowing for active participation by all students. Even so, teachers found it difficult to juggle the facilitation of the PbI processes of checkout interviews and questioning for ten groups of students and to monitor all their students' progress in learning. Lastly, teachers needed to make adjustments to their current teaching practices. Teachers could no longer contend with only teaching by telling, where having a well-rehearsed line of reasoning might seem sufficient to get the job done. With the guided-inquiry approach of PbI, teachers must now learn to guide their students' learning and thinking, often coming from various angles, through skilful questioning, and to manage the different learning needs of each group.

Recognising the challenges teachers faced in using the PbI approach in schools, the research team embarked on the PbI@School project in an effort to better support schools in adopting the inquiry approach to science instruction. The main goal of the project was to develop, implement and evaluate a guided-inquiry curriculum using the PbI instructional approach for lower secondary physics through the iterative cycle of research, curriculum development and instruction. The research-curriculum development-instruction process was carried out in three phases: Phase 1, baseline study (2009); Phase 2, pilot intervention (2010); and Phase 3, refined intervention (2011).

Methodology

Participants

The 3-year project was conducted in a secondary school and involved a total of eight teachers and three successive cohorts of secondary one students in the express stream. The students were of above average ability, with a typical Primary School Leaving Examination (PSLE)[1] T-score of about 240. The number of students who participated in the baseline study was 230 (6 classes) and 430 in each of the subsequent 2 years (11 classes each).

[1] The Primary School Leaving Examination (PSLE) is the first key stage national examination taken at the end of primary education. The aggregate score from PSLE will determine students' placement in one of three streams in secondary school, with an aggregate cut-off score of around 188 for express stream students.

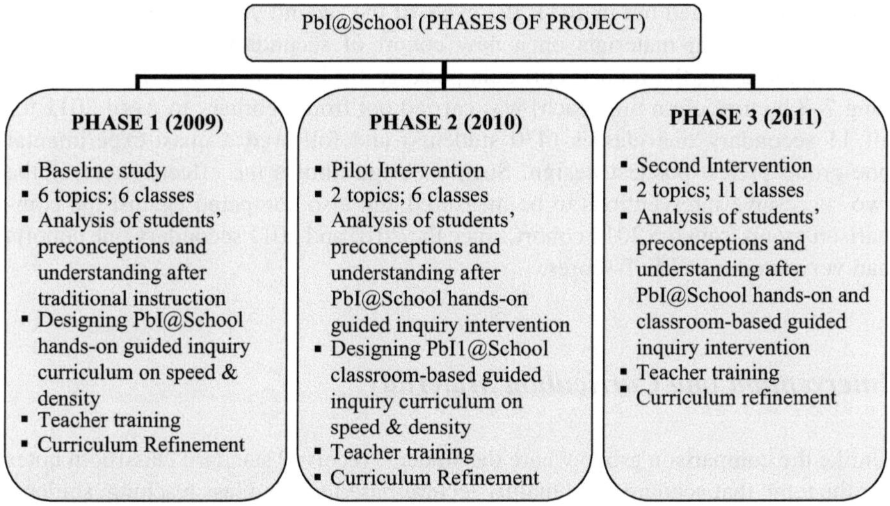

Fig. 5.1 Phases of the PbI@School project (2009–2011)

Research Design

This study adopted the PbI 'research-curriculum development-instruction' research design approach in three phases, as outlined in Fig. 5.1.

The first year of the project (2009) was a baseline study of the learning and teaching of lower secondary science carried out on about 230 students from six secondary one classes. There was no intervention for this cohort of students. The main objective was to study the kinds of difficulties students encountered for lower secondary physics topics such as *speed and density*, *force and pressure* and *moments and work done* before and after traditional instruction. A preliminary study was also done to obtain the students' views of science and their reasons for liking or disliking science (Kuppan, Munirah, Foong, & Yeung, 2010). The baseline study served as a research base to guide the development of adapted PbI materials which served to match the needs and abilities of the students.

The guided-inquiry curriculum was subsequently designed based on the findings of the baseline study, with the focus on two topics: *speed and density*. The laboratory-based inquiry curricula materials for the two identified topics (spanning about 10 h of curriculum time each) were developed and piloted for six secondary one classes (experimental group) in January to March 2010. The rest of the other five classes acted as the comparison group, where instruction was done by another group of teachers with science lessons conducted for the same topics mainly through lecture-based, teacher-centred instruction spanning about 7 h of curriculum time for each topic. The Phase 2 pilot study in 2010 followed a quasi-experimental pretest-posttest comparison group design, where the samples were chosen conveniently based on the classes assigned by the school to the teachers.

In Phase 3 carried out in 2011, the cycle of the second year was repeated using refined curriculum materials on a new cohort of secondary one students. This intervention using the refined curricula package for both *speed and density* (spanning 7–8 h curriculum time each) was carried out from February to April 2011 for all 11 secondary one classes (430 students) and followed a quasi-experimental one-group pretest-posttest design. Such a design allows the effectiveness of the two successive interventions to be measured and also compared against the comparison group from the 2010 cohort, since the 2010 and 2011 secondary one cohorts had very similar PSLE T-scores.

Intervention and Curriculum Materials

Unlike the comparison group where the students received standard classroom notes for the topic that accompanied mainly lecture-based whole-class teaching, students in the experimental group were given carefully sequenced PbI worksheets that helped them construct the key concepts in the topic as they examined their experiences with the laboratory materials. The PbI curricula are designed around a learning model of *elicit-confront-resolve*. The activities deliberately elicit student ideas and preconceptions and trigger common learning difficulties and then guide students to confront and resolve them directly. Students work in groups while the teacher's role in PbI is to use questions to probe, clarify or extend student thinking and understanding. The students are active participants in the learning and are required to go through the reasoning involved in the development and application of concepts.

For example, for the topic of *density*, students first learn the more basic concepts of mass, volume and volume displacement and then proceed to learn about the significance of the ratio of mass to volume, which gives density. They learn that density is a characteristic property of a material, which affects whether an object made from that material will sink or float in a particular liquid. Eventually, building on the logical flow of ideas based on a series of activities, students devise a rule that would help determine if an object would sink or float in a given liquid (i.e., the concept of relative density of the object and the liquid it is placed in will finally determine the buoyancy of an object in the liquid). An excerpt of the PbI worksheet is shown in Fig. 5.2.

In contrast, the comparison group had limited opportunity to construct their ideas and to experience with phenomena. The typical lesson flow comprised the introduction of the concept for the topic, with the relevant formulae, followed by quantitative worked examples and practice questions that were mainly 'plug-and-chug' questions. Any experiments done were usually confirmatory-type experiments, with the purpose of verifying a formula or reinforcing the theory taught. An excerpt of the class notes on the same topic of *density* is shown in Fig. 5.3.

Teachers participated in workshops introducing the PbI approach to prepare them for each intervention (Table 5.1). The teachers role-played as students to work

		a	b	c	d	e	f
Object	*Do both cylinders look alike?*	*My Predictions:*	Cylinders			Density of water = ...g/cm^3	Density of saltwater = ...g/cm^3
		Float/sink	mass g	volume cm^3	density g/cm^3	Float/Sink in water	Float/Sink in saltwater
Cylinder 1	☐ yes	*Floats/sinks in water *Floats/sinks in saltwater					
Cylinder 2	☐ no	*Floats/sinks in water *Floats/sinks in saltwater					

Fig. 5.2 Excerpt of activity worksheet for the PbI group

through the PbI worksheets and, in the process, familiarise themselves with the conceptual flow of the activities and the possible difficulties that students might face when carrying out the activities. Members in the research team role-played as facilitators, demonstrating the facilitation process in guiding student learning through inquiry. Teachers were also given opportunity to put on the 'teacher hat' to think of questions that they could ask at various points of the worksheets to probe, clarify or extend student thinking and understanding. Examples of questions include: 'What exactly do you mean by...?', 'How did you come to know that...?', 'What evidence do you have to support your idea?' and 'Based on previous activities, how can you explain...?'

Instruments, Coding and Analysis

The data sources for this study included field notes from interactions with teachers, classroom observations, pretests, posttests, surveys and interviews with students and teachers. The focus of the pretest questions was on qualitative understanding of concepts as demonstrated by students' ability to explain the reasoning for their answers and did not require application of formulae and mathematical calculations. The pretests were administered to students during curriculum time prior to formal instruction. They served a dual role of eliciting students' preconceptions as well as motivating their engagement in exploring the key concepts through the carefully sequenced activities. These tests were developed mainly by research team members responsible for the topics, but reviewed by the whole research team and the physics teachers of our research school on the suitability of the questions in terms of difficulty level, language and format.

For the topic of *density*, three pretests were designed to examine students' functional understanding of the key concepts for the topic. By functional understanding, we mean the reasoning necessary to construct and apply conceptual models to the

Floating and Sinking

Does a watermelon float or sink in water? Make a prediction. ___________________

Then place a watermelon in a basin of water to find out. Can you explain the outcome?

Reading a newspaper while floating comfortably in the middle of the ocean sounds like an outrageous idea! But you actually can – in the Dead Sea (east of Jordan, west of Israel). But why is this so?

Have you noticed how some substances float while others sink?

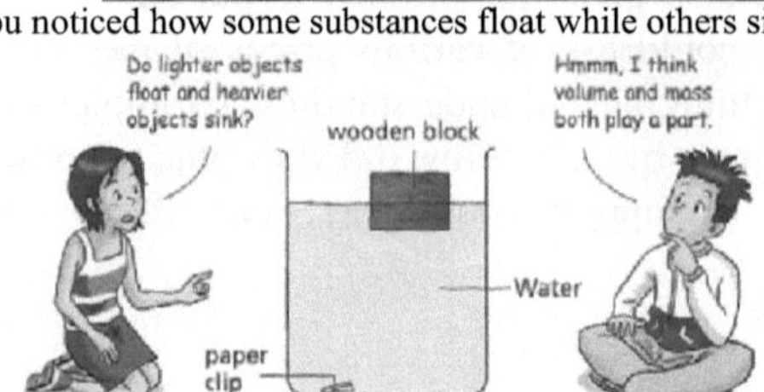

The wooden block floats on water while the paper clip sinks. Compare the densities of the 2 materials with the density of the liquid. What can you say about the densities?

In general, a denser substance will sink while a less dense substance will float. The wooden block, being less dense than water floats in water. The paper clip, on the other hand, being denser than water, sinks in water. Less dense substances rise when placed in denser fluids (liquids and gases).

Fig. 5.3 Excerpt of class notes for the comparison group

Table 5.1 Comparison between the PbI intervention and standard instruction

PbI intervention (experimental group)	Standard instruction (comparison group)
Activity worksheets to record findings and discussions, with no notes given	Standard notes given, with relevant definition, formulae, examples and practice questions
A series of experiments/activities in each topic to help construct key concepts	One experiment per topic to verify formulae and reinforce the theory taught
Students mainly work in groups	Instruction is mainly whole-class teaching
Teacher facilitates learning by asking questions	Instruction is mainly lecture based and teacher centred

interpretation of physical phenomena (McDermott, 2001). Pretest 1 probed for student's ability to distinguish between mass and volume and their understanding that the volume of an object can be measured using the water volume displacement method. Pretest 2 probed for student's understanding that the density of an object is a characteristic property of the material and that the density of a material is independent of its size. Pretest 3 probed for the idea of the relative densities of the object and the liquid into which it is placed to determine the buoyancy of the object in that liquid. The posttests that were administered to students after instruction also probed for the same corresponding concepts as in the pretests. See Fig. 5.4 for a description of the pretests.

The rubric used to analyse student responses to the questions in the pretests and posttests was based on the completeness of conceptual understanding as determined by the reasoning that the students wrote to support their answer rather than just looking simply at the 'correct' answer. Responses that showed sound reasoning based on scientific concepts and ideas were awarded a maximum score of 2. Incomplete responses which displayed partial conceptual understanding and reasoning were awarded a score of 1, and those with wrong reasoning, no reasoning or not attempted were given the minimum score of 0. The coding scheme for students' responses is summarised in Table 5.2.

In studying students' attitudes towards science, a survey instrument was developed that comprised factors such as self-efficacy, competence, interest, inquiry, engagement, awareness of job opportunity, career aspiration and educational aspiration (see ; Yeung, Kadir, Kuppan, & Foong, 2010; Yeung et al., 2010 for details). Besides the attitude survey, students were also asked about their perceptions of the features of the intervention. Of the 110 students who responded to the survey, 20 students were randomly selected for an interview session, where two students were interviewed by two researchers at one time. Pair interviews were conducted to yield best information from our students. The pair shared the same experience in their class and was able to prompt each other to recall with greater accuracy. Students also felt less stressed in pairs and were less hesitant to speak as they could take turns to respond to questions and add on to each other's comments. As for the teachers, a list of survey questions was emailed to them and they were given a month to return an email response. About 2 weeks after submitting their written response, the teachers were interviewed individually by two researchers at one time.

Findings

We first present students' and teachers' views and perceptions towards the inquiry-based physics curricula and instruction based on surveys and interviews that were conducted during each phase of the project. We then illustrate the effectiveness of the inquiry-based curriculum on students' conceptual and reasoning abilities for the topic of *density*. Findings for other topics describing students' common

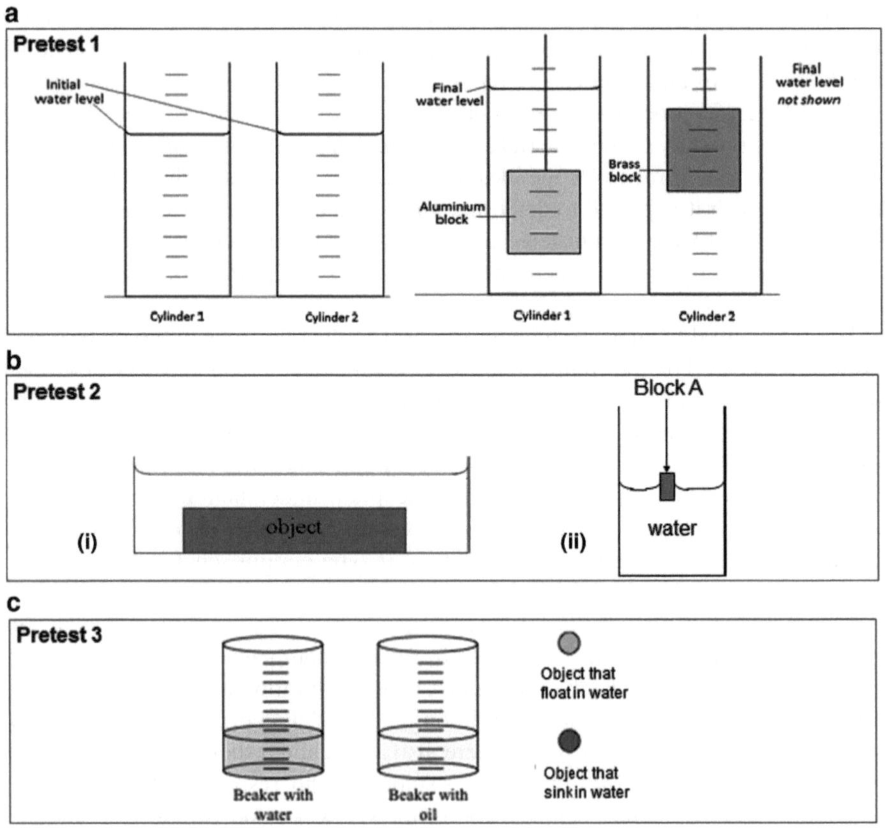

Fig. 5.4 The pretests that were administered for the topic of density. (**a**) Diagram used in Pretest 1. Respondents were asked to make a comparison of: (1) the displaced water levels where the brass block was suspended at a higher level than the aluminium block but still completely immersed. (2) the final displaced water levels when both the blocks were released and allowed to sink to the bottom of the cylinders. (**b**) Diagram used in Pretest 2 showing: (1) a large object (sinker) immersed in a tank of water, and respondents asked to make a prediction about the buoyancy of each of the 1,000 pieces that were broken off from the object. (2) a small block A (floater) floating on water, and respondents asked to make a prediction about the buoyancy of another block B, made from the same material, that was 1,000 times larger in size. (**c**) Diagram used in Pretest 3 showing: (1) a floater in water and respondents asked to make a prediction about its buoyancy when placed in oil. (2) a sinker in water and respondents asked to make a prediction about its buoyancy when it was placed in oil. The final question asked specifically if an object's buoyancy depended on the liquid it was placed in

Table 5.2 Coding scheme for students' responses

Coding criteria	Score
Correct answers displaying complete understanding and reasoning	2
Correct answers displaying partial understanding and reasoning	1
Incorrect reasoning or no reasoning	0

preconceptions and difficulties have been reported elsewhere and will not be repeated here – refer to Munirah, Foong, Wong, and Kuppan (2011) for the topic of *speed*; Munirah et al. (2009) and Munirah, Foong, and Lee (2010) for the topic of *work done and moments*; and Wong, Lim, Foong, and Kadir (2009) for the topic of *force and pressure*. The lessons learned from the curriculum adaption process will be discussed in the next section.

Students' and Teachers' Perceptions of the PbI Curricula and Instruction

The survey responses of students at the end of the refined intervention in 2011 showed that they generally held favourable views of the PbI lessons over the more traditional science lessons. The students were asked to rate the items in a randomised order on a 4-point scale (1 = strongly disagree to 4 = strongly agree). Table 5.3 gives a summary of the descriptive statistics of students' views of the key features of the PbI@School intervention. These features include learning through hands-on activities and experiments, learning from peers and group discussions, exploring learning on their own through guided worksheets and consolidation checkout sessions (where teachers asked questions to check on student understanding and to ensure they were on the right track) and homework.

Hands-On Activities

The opportunity for learning through hands-on activities was the most notable positive feature cited by the students. Students recognised that the PbI lessons, unlike traditional lessons, had more activities that they needed to work through and to solve on their own. From the survey, 95 % of students either strongly agreed or agreed that they 'enjoyed the hands-on activities', 93 % felt that 'hands-on activities helped them to develop their understanding of the topics better' and 87 % felt that 'hands-on activities made them want to find out more about the topic'. Doing more hands-on activities helped students to better visualise the phenomena in real life and better understand the concepts: 'We did many experiments for the topics of speed and density, which gives us more understanding to the topics'; 'We get to see the effects of speed and density in front of our eyes'; 'There were a lot of hands-on activity that made me more easier to imagine the question scenario'. However, when asked to choose between learning from the teacher teaching in front of the class and learning through hands-on activities, close to 40 % of students still preferred the former way of learning. This could be because students are still used to having teachers explaining to them concepts in the class and feel more secure with having notes that highlight the important concepts to focus on. In our interviews with the teachers, they also observed that the experiential nature of the curriculum made the students more interested and more engaged and focused

Table 5.3 Descriptive statistics of survey items regarding students' views ($N = 110$) on the key features of the PbI@School intervention in 2011

Key feature	Item	Mean	SD
Hands-on activities	I enjoyed the hands-on activities	3.28	0.59
	Hands-on activities helped me to develop my understanding of the topics better	3.16	0.51
	Hands-on activities made me want to find out more about the topics	3.14	0.63
	I prefer learning from teachers teaching in front of the class rather than through hands-on activities (−)	2.74	0.99
Self-directed learning	I was able to clearly follow the instructions in the worksheets	3.08	0.48
	The questions in the worksheets guided me in my thinking about the topics	3.07	0.51
	I am confident that my results from the hands-on activities helped me to form good conclusions	3.07	0.59
	I prefer the teacher to question me until I arrive at the answers myself rather than for the teacher to give the answers directly	2.92	0.77
	I prefer analysing data given to me in the class worksheets rather than the data that my team collected during experiments (−)	2.39	0.89
Learning from peers	Discussing my work in groups helped me to develop my understanding of the topics	3.23	0.57
	I was able to contribute well to the group discussions	3.20	0.45
	I prefer to work on my own rather than in groups (−)	2.73	0.95
	I prefer the teacher to reveal the answers to the questions directly rather than discussing the findings of other teams in the class (−)	2.72	0.99
Checkouts and homework	During checkouts (discussion with teacher) in class, I paid attention to the discussions	3.19	0.51
	I find the checkouts (discussion with teacher) useful in developing my understanding of the topics	3.17	0.52
	The homework for each topic helped me to reflect and strengthen the ideas I learnt during lessons in school	3.08	0.60

Note: The items were randomised in the survey. Items were coded such that higher scores reflected more favourable perceptions
(−) = reverse coded item

during the PbI lessons: 'Experiencing hands-on is better for their learning; it keeps them awake and they understand better because they get to find out the concepts by themselves'.

Self-Directed Learning

The carefully designed guided-inquiry worksheets served their function well to guide student independent learning, where 92 % of students felt that they were able to 'clearly follow the instructions in the worksheets' and 91 % felt that 'the questions in the worksheets guided their thinking about the topics'. Instead of relying on the teacher as the sole authority for knowledge, the students were able to take greater ownership of their learning in making predictions, observing for patterns, looking out for sources of errors and checking for consistency of their ideas and reasoning. Rather than provide model answers, teachers were also encouraged to respond to students' queries with questions that clarify, probe and extend their thinking. While teachers were initially apprehensive about this method, for fear of frustrating the students, 73 % of students actually agreed with the statement: 'I prefer the teacher to question me until I arrive at the answers myself rather than for the teacher to give the answers directly'. During interviews, while some students admitted that they preferred model answers because they were 'lazy to think', most acknowledged that this 'self-learning' method of instruction was effective in making them think more carefully about the ideas and concepts and be able to reason out their own conclusions based on experimental evidence 'instead of the teacher telling them the conclusions straight'.

Opinions were split almost equally on preference between analysing given 'clean' data in the class worksheets versus 'messy' data that have been collected by the students themselves during the experiments. During the interview sessions, students explained that they felt the need for both types of data. They wanted to analyse their own data because it is more 'real' to them and they get to experience how they are being measured. On the other hand, some preferred 'clean' data given by the teacher. One student put it this way: 'If I do myself and got wrong data, I will get confused and get the wrong conclusions'. One of the fears the teachers had was the inaccuracy of the data collected by the students leading to erroneous conclusions and students developing the wrong concepts. For example, in one of the activities, students were asked to find the density of different volumes of water. If the mass and volume of water were measured accurately, students will find that the density of water is constant for all volumes. However, if they find the density of water decreases as the volume gets lower (due to the higher percentage of error when magnitudes of measure are small), they might wrongly conclude that density decreases as volume decreases.

In our interviews with the teachers, the teachers stated that for the pilot intervention, the students in the experimental group were observed to be more independent and confident after the PbI intervention. They were more task oriented and better prepared to do experiments on their own, with minimal supervision: 'Pupils are forced to reason and predict and not just wait for the suggested answer'; 'students no longer just take in whatever information the teacher gives, they are trained to question and find out the concept themselves'. Prior to participating in this PbI project, teachers reported that science lessons tended to be more teacher centred and theory focused, with students being relatively more passive in their learning. Experiments were carried out to confirm the theory taught in the class rather than as a means to help students make hypothesis and observations, identify patterns and develop and test out their ideas. The teachers' responses to their pre-intervention practices are in line with field observations made of the lessons. After the intervention, the lessons became more student centred, with the teacher acting as facilitators to guide students to learn independently and collaboratively through hands-on experimentation. A notable impact on teachers from participating in this project was the tendency for them to withhold directly giving answers to students but to ask more questions in class: 'I used to answer my own questions, but now I have to restrain'. They were also more aware of the different kinds of questions they could ask in order to clarify, probe and extend student thinking. Overall, teachers were positive about the guided-inquiry approach to science instruction and stated that they would continue teaching the *speed* and *density* curricula package to future cohorts of students even after the project was over. One teacher said that 'I have myself started to think about how to create activity-based lessons to enable students to learn concepts themselves'.

Learning from Peers

Learning from peers in groups was generally considered a positive feature of the instructional approach for the students. 95 % of students agreed that discussing their work in groups helped them to develop their understanding of the topics. Typical answers include 'I like it as we get to work in groups, discuss and share answers'; 'I like that we were in groups and got to discuss. This made learning and understanding more fun and easy'. When asked to choose between working on their own versus in groups, about 30 % of students indicated that they preferred working on their own. Group dynamics is an important consideration to the student. Some students were deterred from group work for fear of encountering uncooperative and disruptive team members. A teacher also observed the importance of good team dynamics: 'They have to learn to work as a team, and it can be rather challenging, as team dynamics will affect how well students can understand concept and carry out activities'. In general, students still preferred working in groups as they were able to get different perspectives and also to clarify their doubts and thinking with their team members. Interestingly, close to 40 % of students preferred the teacher revealing the answers to the questions directly as opposed to discussing the findings of other teams in class. While about 60 % of students valued the learning arising from discussion with other teams in the class, there were others who still preferred the security of the teacher's answers. As one teacher put it, 'Students want suggested answers, they are worried about phrasing as different groups may have different phrasing and they want a higher authority telling them that they are correct'. However, the teachers agreed that 'discussion in teams allows students to learn from each other, they question one another and clarify their doubts during the activities'.

Checkouts and Homework

Consolidation in the form of small group and whole-class checkouts (where the teacher asked questions to check on student understanding) and homework in the PbI intervention is another key feature of the PbI project. The checkout sessions played an important role in ascertaining if students were on the right track and have learned the requisite concepts as designed in the worksheets. Ninety four percent of students found the checkout sessions useful in developing their understanding of the topics and paid attention to the whole-class checkout discussions with the teacher. Eighty eight percent of students felt that the homework for each topic helped them to reflect and strengthen the ideas they learned during lessons in school.

Effect on Students' Conceptual Understanding (Example from Density)

Inability to Distinguish Mass, Volume and Density

Prior to formal instruction in secondary science, about half of the students were unable to distinguish between the concepts of mass, volume and density. A common preconception was that the mass (or weight) of the object and/or the depth at which an already completely immersed object is placed below the surface of the liquid affected the level of displaced liquid.

The pretest used (Pretest 1) showed two same-sized cylinders filled with water to the same level. It was stated in the pretest that two same-sized cubic blocks made from different materials were lowered into two separate measuring cylinders to completely immerse them in water (see Fig. 5.4a) and that the brass block was heavier than the aluminium block. Students were asked to mark out the level of displaced water for the cylinder with the heavier brass block suspended at a higher level than the aluminium block. Students were finally asked to make a comparison of the final displaced water levels when both the blocks were released and allowed to sink to the bottom of the cylinders.

Responses from students showed that many thought that the level of displaced liquid depended on the mass (and/or weight) of the immersed object: 'Brass is heavier than aluminium; hence the brass block puts more pressure on the water so the water level rises up more'. Statements such as 'Both the aluminium block and the brass block have equal mass and take up the same amount of space,' and 'The brass block has a higher density, which results in a higher water level in cylinder 2' indicate that students did not distinguish the concepts of mass, volume and density. Even when students used the term 'volume', they could be thinking of 'mass': 'the volume of the brass block is bigger than the aluminium block; thus the water level should be higher'. In addition, there were students who gave the correct response on the resulting water level but based on wrong reasoning such as 'the water level in cylinder 2 will be of the same level as cylinder 1 because even though the brass is heavier than the aluminium block, it is immersed at a higher level in the water'.

Student responses from Pretest 1 also revealed that some students thought that the depth at which an already completely immersed object was placed affected the level of displaced liquid. For example, when asked to compare the final displaced water levels when both the blocks were released and allowed to sink to the bottom of the cylinders, sample responses included 'the water level in cylinder 1 rise a little, but the water level in cylinder 2 is at the top of the cylinder as it had overflowed', 'water will flow out of the cylinder when they are released, and the brass block is dropped from a higher place, so more water will overflow' and 'the weight of the blocks does not affect the volume, but the brass block is higher up so it pushes less water up'. Admittedly, some students could have overlooked the given information and thought that the brass block was not completely immersed, even though the question clearly stated that it was completely covered by water: 'the brass block is not completely lowered into the water in cylinder 2, so it will occupy less space and hence, the water level will be lower', 'the weight of the blocks does not affect the volume, but the brass block is higher up so it pushes less water up' and 'the brass block is not completely put into the water, so the mass of it is not completely added to the water level'.

Heavy Object Sinks and Light Object Floats

Another common preconception is the idea that the mass (or weight) of the object determines its buoyancy (i.e., whether it will sink or float). Prior to instruction, about a third of students specifically stated that the buoyancy of an object depended on its size or mass: 'If its mass was small enough, it would float and if its mass was big enough, it would sink', and 'those which are lighter will float and those which are heavier will sink'. A further 20 % of students seemed to be able to explain why they thought an object sank or floated in terms of relative density. For example, a typical response for why they thought an object sank in water was 'it is denser than the water, so it sinks'. However, upon further probing (see Pretest 2 in Fig. 5.4b), they stated that smaller pieces from an initially large sinker may float, while a big object made from small floaters of the same material may sink. Evidently, students may have memorised stock answers without having a real functional understanding of the concepts. Some even reasoned that the density of the object changed with the size of the object: 'the density of every 1,000 smaller pieces decreases when it was broken, and smaller pieces have less density, hence may be able to float. If the size of Block B is 1,000 times the size of Block A, its density is also 1,000 times the density of Block A, hence can't float'. Students' think that density of objects can change with size and that of liquids can change with amount (or volume) highlights the need to emphasise the concept of density as a characteristic property of a substance during instruction.

Reasoning Difficulties in the Context of Two Liquids

Students also encountered reasoning difficulties when making comparisons about the sinking and floating behaviours or levels of liquid displaced in the context of multiple objects and liquids. Responses to Pretest 3 indicated that 25 % of students were not able to make correct deductions about an object's buoyancy in a second liquid, given the object's buoyancy in the first liquid and the relative densities of the two liquids. For example, 15 % of students stated that the numerical values of the densities of the object and liquids must be explicitly given before any conclusions can be made. They did not realise that if the object sank in the water, it would also sink in oil, since oil was less dense than water. The other 10 % of students simply stated the object floated or sank in oil based on sweeping assumptions made about its density relative to oil: 'the object is about the same density as oil, but it could possibly be denser, so it sinks in oil'. Responses to Pretest 3 revealed that 15 % of students thought that objects were more likely to float in liquids with lower density: 'as water is denser than oil and the object that floats in water can surely float in oil too'. Students think that objects float 'better' in liquids with lower density.

Effectiveness of Density Inquiry Package

The different student groups performed significantly better after instruction in all concepts except for the comparison group (2010) which did not show significant improvement for Pretest 2 (see Fig. 5.5). Results from the first intervention in 2010 showed that the experimental group using the inquiry curriculum achieved greater conceptual understanding, with an average normalised gain (see Hake, 1998)

$$\langle g \rangle = \frac{\langle \text{posttest score} \rangle - \langle \text{pretest score} \rangle}{< \text{maximum possible posttest score} > - \langle \text{pretest score} \rangle}$$

of 0.65 compared to 0.39 for the comparison group.

It is acknowledged that the intervention in 2010 did not control for curriculum time – the experimental group had in fact about three more hours of class time for each topic compared to the comparison group. While the greater gain in conceptual understanding for the experimental group in 2010 could be attributed to more class time, we would like to argue that this is probably not the case, as an equivalent additional time for the comparison group in a 'teaching-by-telling' and 'drill-and-practice' approach may not necessarily translate to better student conceptual understanding. In the PbI approach, students are actively engaged in their own learning through a series of hands-on activities that deliberately elicit students' ideas and preconceptions and guide them to construct their own understanding in a logical progression, supported by peer discussion and teacher questioning.

In our second intervention in 2011, we used a refined curriculum which had been streamlined by reducing the number of activities. This allowed us to compare the learning gains of the students using the refined curriculum in 2011 that matched the curriculum time experienced by the comparison group in 2010. It was found that consistent good learning gains were achieved similar to that of the pilot intervention in 2010, with an average normalised gain of 0.68. Table 5.4 summarises the learning gains for the different groups of students for the respective pretests.

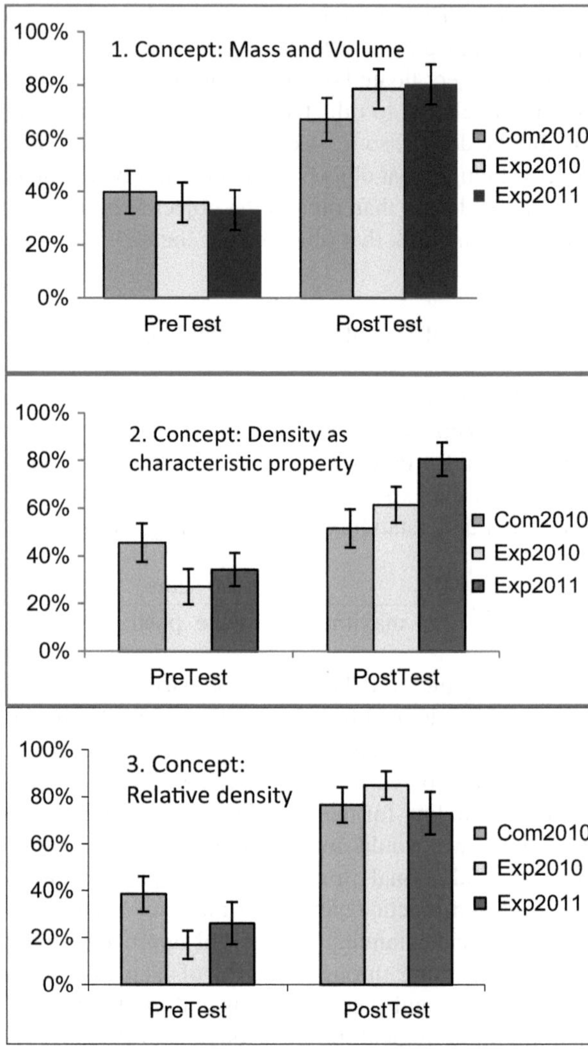

Fig. 5.5 Comparison of students giving correct answers (with correct reasoning) for the respective concepts tested

Table 5.4 Comparison of learning gains

	Com2010	Exp2010	Exp2011
$\langle g\rangle_{\text{Pretest 1}}$	0.45	0.67	0.71
$\langle g\rangle_{\text{Pretest 2}}$	0.11	0.47	0.70
$\langle g\rangle_{\text{Pretest 3}}$	0.62	0.82	0.64
$\langle g\rangle_{\text{All}}$	0.39	0.65	0.68

Discussion

We now devote this section to a discussion on the lessons learned from working with the research school in the curriculum development, implementation and adaptation process. Refinements to the curriculum and instruction were made iteratively over time based on on-going feedback by teachers and students, and evaluation by team members in the light of classroom observations and research data. For example, the initial PbI curriculum was developed based on the findings pertaining to students' preconceptions, difficulties and interest in science during the Phase 1 baseline study as well as teachers' inputs on the learning outcomes and concepts that may be explored through inquiry. Prior to the pilot intervention, we trialled the draft worksheets with ten secondary one students to check the suitability of the worksheets in terms of the nature of the activities as well as the language used. The team members also modified and sourced for laboratory materials and kits to improve the effectiveness of the experimental results of the inquiry activities. At the end of the pilot intervention, we analysed the effectiveness of the PbI curriculum from student data collected as well as from teacher feedback and field notes from lessons. For the intervention in the second year, the activities were better pitched to an appropriate level of challenge and streamlined by reducing the number of activities while still maintaining the coherence and flow of the curriculum.

We next discuss the main areas of modification in adapting the curriculum to the school context. In scaling up the implementation of the project to the whole cohort of secondary one students, there was a need to ensure efficient use of the limited science laboratories available. This constraint meant that each class can spend only about one-third of their science curriculum time in the laboratory. This necessitated an important change in the way the curriculum was designed. Unlike the pilot curricula which included only laboratory-based activities, the refined package included both laboratory follow-up and classroom-based exercises. The laboratory follow-up session, done after the laboratory activity, served as a platform for students to share their data and findings with the class. Questions were included in the laboratory follow-up worksheet to reinforce students' construction of their conceptual understanding by providing scenarios and problems similar to the laboratory activities that may include some pre-given data. Classroom-based materials were designed to be flexible in that they can be conducted either before or after the laboratory activities, depending on the class timetable. They served to provide opportunities for students to discuss and debate in groups of three or four students about their differing conceptual ideas, share strategies to solve given problems and analyse data. There were also computational exercises for students on the concepts introduced in the topic to allow the students to practise solving quantitative problems as well.

Over the 3-year period, the three main areas of modification that are evident in the final adapted PbI@School package are:

1. Guided format of inquiry activity worksheets
   Secondary one students require a lot of guidance through the use of simple diagrams or tables in the activity worksheets in order to be able to carry out the required inquiry activities. They were also more comfortable in giving short sentence responses and explanations as opposed to writing their explanations in long open-structured sentences. Hence, the format of the activity worksheets was made more guided and requiring less unstructured written responses.
2. Integrated use of laboratory-based and classroom-based materials
   There was an integrated use of lab-based and lab follow-up activities with classroom-based materials. A balance was sought between students having to physically carry out the activities to collect their own 'messy data' (so as to have an idea of the associated experimental constraints and measurement errors) versus giving students 'clean data' so that the focus was on the analysis and interpretation of the data, without the fear and distraction of the noise and reliability of the data affecting their conclusions.
3. Use of classroom checkouts and embedded probing questions
   The PbI@School package was designed to be conducted by one teacher in a typical class of 40 students. A necessary modification to the process of guiding and extending students' understanding by questioning was to use whole-class checkouts in addition to group checkouts. This allowed teachers to be able to hold class discussions when there was insufficient time for individual group checkouts during a lesson. In addition, probing questions were included in the inquiry worksheets to facilitate student discussion and thinking in place of having the teacher moving from one group to another to facilitate the discussions.

Conclusion

There needs to be continued effort in nurturing students' interest in science early and through the lower secondary school years, given that students' interest in science is a key predictor of future aspiration to learn the subject. Inviting overseas experts to conduct workshops on innovative inquiry-based curricula and instruction has been a useful starting point in supporting teachers to implement 'science as an inquiry' in our national curriculum. This study has shown that inquiry-based teaching can be supported by research-validated curriculum materials that have been developed through a successful adaptation of existing overseas instructional approaches to suit our local secondary school context. Hence, more support could be given to teachers to actively participate in the process of curriculum making and adaptation for other parts of the lower secondary science curriculum.

Through our interactions with teachers, we noticed that as teachers overcome the confidence barrier, they are capable of effectively facilitating inquiry-based learning. This was done through an iterative, collaborative process, with colleagues in the school and with the help of a team of researchers to provide the crucial initial support for curriculum adaptation and the development of teaching kits. This process takes time and with a small step involving a few classes at the start. When teachers gain the necessary mastery experiences to conduct inquiry, the implementation can then be scaled up to all classes, taking into account the existing school context. An interesting follow-up study would be to examine how teacher efficacy affects student learning by inquiry.

It would be good to expose prospective and practising teachers early to good models of inquiry-based science teaching so that they could draw on the necessary experience and repertoire of strategies. Having inquiry curriculum materials that have been adapted for use with students in our local classrooms is useful in helping to reduce the barrier and shorten the learning curve for other teachers who might plan to embark on their own inquiry-based teaching in science. Whether the same materials will work as effectively in another school context is of course something that cannot be guaranteed, since different schools have different school practices, organisations and learning cultures. Hence, there is still the need for these teachers to go through a similar learning process of adapting and implementing the existing PbI materials that are contextualised for their own students. It is important that teachers work through the curriculum materials in a manner that is consistent with how their students learn so as to be able to better facilitate the learning process. Teachers need to be cognisant of the conceptual flow of the curriculum materials that have been specially designed to support the students' own construction of ideas and concepts. Without this deep understanding, teachers might inadvertently short-circuit the students' learning process and their construction of a coherent conceptual framework necessary for a functional understanding of the science concepts.

This study has provided us with useful insights into how to support schools in implementing inquiry-based teaching. We acknowledge the different challenges and tensions that teachers face when implementing inquiry-based teaching. But with proper support and guidance, inquiry-based teaching need not be carried out at the expense of test scores. Indeed, the benefits of PbI lie not as much in the improvement of standardised test scores as in the enhancement of student conceptual understanding, reasoning abilities and interest towards science learning.

References

Hake, R. R. (1998). Interactive-engagement versus traditional methods: A six-thousand-student survey of mechanics test data for introductory physics courses. *American Journal of Physics, 66*, 64–74.

Jenkins, E., & Nelson, N. W. (2005). Important but not for me: Students' attitudes toward secondary school science in England. *Research in Science and Technological Education, 23*, 41–57.

Kuppan, L., Munirah, S. K., Foong, S. K., & Yeung, A. S. (2010). On the attitude of secondary 1 students towards science. *AIP Conference Proceedings: International Conference in Physics Education 2009, 1263*, 118–121.

McDermott, L. C. (2001). Oersted Medal Lecture 2001: Physics education research – The key to student learning. *American Journal of Physics, 69*, 1127–1137.
McDermott, L. C., Heron, P. R. L., & Shaffer, P. S. (2005). *Physics by inquiry: A research-based approach to preparing K-12 teachers of physics and physical science*. Forum on Education of The American Physical Society, Summer 2005 Newsletter.
McDermott, L. C., & The Physics Education Group. (1996). *Physics by inquiry* (Vols. I and II). New York: University of Washington/Wiley.
Munirah, S. K., Foong, S. K., & Lee, P. (2010). On the conceptual understanding of 'Work Done' for secondary one students in Singapore. *AIP Conference Proceedings: International Conference in Physics Education 2009, 1263*, 187–190.
Munirah, S. K., Foong, S. K., Wong, D., & Kuppan, L. (2011, June). *PbI1@School: On secondary one students' understanding of speed.* Paper presented at the 4th Redesigning Pedagogy International Conference 2011. Singapore: OER/NIE.
Munirah, S. K., Lim, J., Foong, S. K., Prasanthee, R., Lee, P., & Yau, C. M. (2009, June). *PbI1@School: On Singapore secondary one students' perception and understanding of work done and moment of force.* Paper presented at the 3rd Redesigning Pedagogy International Conference 2009. Singapore: OER/NIE.
National Research Council. (1996). *National science education standards.* Washington, DC: National Academy Press.
National Research Council. (2000). *Inquiry and the national science education standards.* Washington, DC: National Academy Press.
Oon, P.-T., & Subramaniam, R. (2011). On the declining interest in physics among students – From the perspective of teachers. *International Journal of Science Education, 33*(5), 727.
Osborne, J., Simon, S., & Collins, S. (2003). Attitudes towards science: A review of the literature and its implications. *International Journal of Science Education, 25*(9), 1049–1079.
Sjøberg, S., & Schreiner, C. (2005). How do learners in different cultures relate to science and technology? Results and perspectives from the project ROSE (the Relevance of Science Education). *APFSLT Asia Pacific Forum on Science Learning and Teaching, 6*(2), 1–17.
Wong, D., Lim, C. C., Foong, S. K., & Kadir, M. S. (2009, November). *Secondary 1 students' understanding of force and pressure.* Paper presented at the International Science Education Conference 2009, Singapore.
Yeung, A. S., Kadir, M. S., Kuppan, L., & Foong, S. K. (2010). Boys' and girls' self-beliefs, engagement, and aspirations in physics. *International Journal of Learning, 17*, 397–417.
Yeung, A. S., Kuppan, L., Foong, S. K., Wong, D. J. S., Kadir, M. S., Lee, P., et al. (2010). Domain-specificity of self-concept and parent expectation influences on short-term and long-term learning of physics. *New Horizons in Education, 58*, 55–73.

Chapter 6
From Transmission to Inquiry: Influence of Curriculum Demands on In-Service Teachers' Perception of Science as Inquiry

Aik-Ling Tan, Frederick Talaue, and Mijung Kim

Abstract A new primary science syllabus with science as inquiry as its foundation and guiding philosophy was implemented in Singapore since 2008. In this study, we present perspectives of science as inquiry as experienced by teachers teaching science under the current educational landscape that is routinised and highly teacher-fronted. We invited 41 in-service teachers to participate in (1) questionnaires and narratives, (2) reflective writings and (3) group discussions related to science inquiry. Data analysis in the form of thematic coding using NVivo8, with over 80 % inter-coder coding agreement level, was carried out. Three key interrelated aspects of in-service teachers' perceptions of science as inquiry were revealed: (1) pressure of assessment demands, (2) perceived need for strong teacher intervention in inquiry and (3) favouring content knowledge rather than process skills. These considerations surface conflicts of teaching science as inquiry that are driven by assessment demands and resulted in a mode of science as inquiry that has tight teacher control with a strong focus on learning the canonical science content. We also explored the underlying reasons behind the perceptions of inquiry practices held by these in-service teachers.

Keywords In-service teachers • Curriculum • Nvivo • Canonical science • Teachers' beliefs • Teacher control • Science as inquiry • Fidelity • Teachers' perception • Teacher professional development

A.-L. Tan (✉) • F. Talaue
National Institute of Education, Nanyang Technological University, Nanyang Walk 1, Singapore 637616, Singapore
e-mail: aikling.tan@nie.edu.sg

M. Kim
Department of Curriculum and Instruction, University of Victoria, Victoria, BC, Canada

A.-L. Tan et al. (eds.), *Inquiry into the Singapore Science Classroom*, Education Innovation Series, DOI 10.1007/978-981-4585-78-1_6,

Introduction

The advent of science as inquiry in school science curriculum began with Dewey in 1910 based on his observation that there was too much emphasis on facts and insufficient attention paid to scientific thinking and attitude. This science as inquiry movement has endured many societal changes, debates and demands of current society and managed to retain its position of importance in many science curricula worldwide. The twenty-first-century society of today is once again reminded that scientific knowledge should be perceived as *process* (science in the making) more than scientific knowledge as *product* (ready-made science). In line with this view, the National Research Council (NRC, 1996) initiated the science education reform movement in the United States more than a decade ago, urging teachers and schools to engage in forms of science learning that are authentic. Science as inquiry was promoted as a means to present more accurate ideas about science as a discipline and more so as a way of knowing. To solve problems in the present and future society, it is more important for students to learn how to creatively seek knowledge in order to make decisions about everyday situations, rather than merely recapitulating knowledge obtained from textbooks. Recognising this, science as inquiry is then envisioned as a means as well as an end in science curriculum change. Science as inquiry has hence been widely emphasised among science educators in order to develop students' scientific literacy and has become a central part of science curriculum for the twenty-first century in many countries (Abd-El-Khalick et al., 2004). In summary, science as inquiry is regarded as a platform for good and successful teaching of science.

Science as inquiry suggests that science teaching is no longer only about students' acquisition of content knowledge. Attention needs to be paid to skills, strategies and habits of mind associated with decision-making and problem-solving processes that students face in their everyday lives (Hurd, 1998; Kolsto, 2000). In practising science as inquiry, it is widely established that students should be provided with opportunities to gather evidence, decide on the value of the evidence and craft coherent scientific explanations based on the available evidence. The central role that scientific explanations play in the learning of scientific knowledge and understanding the process of science is indisputable. Recent research has focussed on working towards a better conceptualisation of the components of scientific explanation and the role that it plays in science learning (Braaten & Windschitl, 2011; Osborne & Patterson, 2011). Some characteristics of scientific explanations include the incorporation of theories and theoretical models to account for the observations made in nature. In a classroom where science is practised as inquiry, students and teachers will hence be engaged in first-hand observations of evidence, discussion of ideas and generation of scientific explanations based on available evidence and theories.

With science as inquiry being the pervasive theme in many school science curricula, teachers are encouraged to practise inquiry as a goal as well as a way of teaching science. The prolific discourses and enactment of science as inquiry have also surfaced various difficulties and tensions associated with its enactment, teachers' professional development, curricular design and students' science learning experiences. Despite the clear goals and efforts initiated to ensure high fidelity and successful enactment of science as inquiry in classrooms, researches in international communities of science education show that the implementation of inquiry-based approaches has been a daily struggle for science teachers (Crawford, 2007). A significant number of studies have discussed the difficulties of practising science as inquiry in elementary science classrooms – external factors (constraints with time, curricular demands, students' abilities and classroom structure) and internal factors (lack of knowledge, beliefs and attitudes) hindering the use of inquiry approach (Chin, Goh, Chia, Lee, & Soh, 1994; Lee, Tan, Goh, Chia, & Chin, 2000; Yoon & Kim, 2010). Many elementary science teachers lack the necessary background knowledge and experiences to teach science and thus lack the confidence in teaching the subject (Appleton, 2002; Appleton & Kindt, 1999; Gustafson, Guilbert, & MacDonald, 2002; Howitt, 2007; Kielborn & Gilmer, 1999; Palmer, 2006). Even for science teachers with science background knowledge, the implementation of science as inquiry is challenging for reasons such as the lack of administrative support (Zion, Cohen, & Amir, 2007), conflicts between personal experiences and beliefs in scientific inquiry (Crawford, 2007; Trumbull, Scarano, & Bonney, 2006) and cultural and institutional expectations of the responsibilities of science teachers (Kim & Tan, 2011). With the lack of external support of time, materials and laboratory assistance, conducting inquiry-based learning can certainly be challenging. Another fundamental hurdle in enacting science as inquiry in elementary science classrooms is that the inquiry approach is perceived to be very demanding. As science is not a priority subject for elementary school teachers (Appleton & Kindt, 1999), they struggle to find a balance among various subject areas, and so from their perspective, the inquiry approach may not seem to be the most efficient way to teach science (Appleton, 2002; Mulholland & Wallace, 2003).

The lack of an agreement on the meaning of science as inquiry (Martin-Hauser, 2002) among science educators, science teachers and curriculum developers has divided the community. The differences in understanding and interpretation of science as inquiry have resulted in varied forms of enactment and learning experiences of science among different students. Barrow (2006, p. 265) made a call to science teacher educators to "reach consensus about what is inquiry, in both pre-service and in-service situation". While a common understanding of science as inquiry is fundamental, it is also important to be cognizant of social contexts and variations to science as inquiry learning experienced under different cultural conditions. Hence, it is crucial for science educators to firstly agree on the invariants about science as inquiry, understand the demands and conditions that need to be made available for inquiry to take place and subsequently proceed to understand the realities of science as inquiry as enacted in classrooms by teachers and students. Such knowledge will enable us to diagnose the misalignment between the ideals of science as inquiry and the realities of science as inquiry in the classrooms as experienced by teachers and students. This forms the motivation for the current study as we start to explore teachers' understanding and perception of science as inquiry and look into how these perceptions are intertwined with the difficulties of inquiry teaching in the current social contexts.

Science Teachers' Beliefs About Inquiry

Most studies exploring teacher beliefs about inquiry-based teaching began with the assumption that beliefs influence classroom actions. Beliefs have been defined as understandings, premises or propositions that people know and feel to be true or reasonable enough to hold on to (Richardson, 1996). They conceptualise something as existent or non-existent (existential presumption), reflect idealised situations (alternativity) and borne of feelings (affect and evaluative loading) and personal experiences (episodic nature) (Nespor, 1987). However, while people are conscious of some of their beliefs, other beliefs remain tacit, stable and unchanging unless intentionally challenged (Kagan, 1992). The durability of beliefs is partly attributed to the slow and gradual process of enculturation into a certain social group whose members share and guard particular systems of beliefs (Hubbard & Abell, 2005).

Teachers' difficulties in adopting inquiry-based approaches in the classroom have been partially linked to the difficulties in changing teachers' beliefs about teaching and learning science (Furtak & Alonzo, 2010). Beliefs serve both as lenses for their interpretation of the interplay between reform innovations and actual classroom events, and as filters for decision-making on instructional goals and classroom management (Luft & Roehrig, 2007; Pajares, 1992). For example, some teachers believe inquiry-based approaches are too complex and therefore impractical to implement due to time and material constraints (Britner & Finson, 2005; Smolleck, Zembal-Saul, & Yoder, 2006). Other teachers base their pedagogical choices on what they believe their students are capable of doing. Even in just considering the implementation of hands-on lessons, middle school teachers evaluate whether students can learn particular science concepts better if they get to "see" and "do" the science activities themselves (Eick & Reed, 2002). Teachers do have varying levels of expectations of students that are calibrated to their perception of students' knowledge and skills, as well as motivation and inquisitiveness. Expectedly, there is a general biased perception that science as inquiry can be implemented only for students with above-average abilities (Dietz & Davis, 2009; Moseley, Ramsey, & Ruff, 2004).

Notwithstanding the strong influence of beliefs on actions, changes in beliefs do not necessarily translate into changes towards more inquiry-oriented teaching practices. Traditional practices are often difficult to align with beliefs consistent with the reform agenda (Cohen & Ball, 1990a; Guskey, 1985). The fact that teachers begin to use reform jargon (e.g., teacher as facilitator, students constructing knowledge, hands-on) more frequently is not absolutely indicative of reformed practices in the classroom (Bryan, 2003; Yerrick, Parke, & Nugent, 1997). Change that simply modifies intrinsic variables, such as teachers' knowledge about and attitude towards inquiry teaching, may also not be sufficient by themselves to revamp classroom practice (Simmons et al., 1999). Even the inclusion of authentic inquiry experiences in pre-service training programmes, believed to be a critical component for achieving desired practice outcomes (Roth, McGinn, & Bowen, 1998), have fallen short of researchers' expectations (Brown & Melear, 2006). Furthermore, several studies that documented the influence of inquiry-oriented professional development programmes reported that what teachers professed to believe in is somehow disconnected from their actual classroom actions (e.g., Fang, 1996; Johnson, 2006; King, Shumow, & Lietz, 2001; Lee, Hart, Cuevas, & Enders, 2004). Interestingly, it has been found, following Luft's (2001) implementation of an inquiry professional development (PD) programme, that changes in beliefs are remarkable among beginning teachers, whereas changes in practice among veteran teachers are comparatively less obvious.

At the very least, what can be drawn from all these studies is that while beliefs and practices are evidently linked, there is quite a complex relationship. It has been argued that it might be more productive to conceive of beliefs as interactive with practice rather than to think of them as dictating modifications in behaviour (Richardson, 1996). This perspective implies that reflecting on classroom actions then becomes critical in aiding teachers to gradually appropriate changes in their beliefs (Blanchard, Southerland, & Granger, 2009). It also means that such a reflective process will make practice always look like "partial" change, which by no means should be construed as corruption of a reform effort because teachers are thoughtful facilitators in the classroom that take account of various influences in the environment (Cohen & Ball, 1990b; Leander & Osborne, 2008). The synchronous achievement of changes in beliefs and practices is indeed a challenging accomplishment because, as Volkmann and Zgagacz (2004) argued, changing teaching practices does not only entail scrutinising one's beliefs but also opening one's self to the possibilities of assuming new identities as a teacher and learner.

Science Curriculum in Singapore

In Singapore, science as inquiry received a renewed emphasis in 2008 when inquiry was identified as the guiding pedagogical framework for science education change. While the inquiry approach had always been informally encouraged, the new 2008 primary science syllabus was designed to be explicitly more inquiry-centric (Ministry of Education [MOE], 2007), stating that "central to the curriculum framework is the inculcation of the spirit of scientific inquiry" (p. 1). Teachers are expected to play the role of leaders in the inquiry process. They are to nurture students as inquirers so as to deepen their conceptual understanding and enhance their problem-solving skills. Correspondingly, they are to sustain interest in science and prepare them better for the knowledge-based economy in the twenty-first century. The new syllabus promotes the practice of inquiry in three integral domains of (a) knowledge, understanding and application; (b) skills and processes; and (c) ethics and attitudes. The science syllabus is designed around five themes:

cycle, systems, diversity, energy and interaction. Each theme includes activities and inquiry processes grounded in the knowledge, issues and questions that relate to the roles of science in daily life, society and the environment. An analysis of the learning outcomes spelt out in the syllabus document for Primary 3 to Primary 6 surfaced 86 outcomes in the knowledge, understanding and application category, 35 outcomes in the skills and processes category and 45 outcomes in the ethics and attitude category.

Comparing the science content in this reformed inquiry-centric curriculum with the earlier pre-2008 curriculum, we observed that the content learning outcomes remain largely similar. Hence, the recent curricular change primarily focussed on the pedagogical approach (from didactic to inquiry and from teacher-centred to student-centred) to achieve similar content learning outcomes.

To support the curriculum change, the Singapore MOE worked with publishers to write science textbooks that are theme-based. It also developed materials and resources for teachers to use in their classrooms (besides those stated in the official syllabus documents) and provided PD programmes. For instance, as part of the continual teacher PD to prepare primary science teachers for the new inquiry-based primary science curriculum, five customised courses were developed by the Natural Sciences and Science Education (NSSE) Academic Group at the National Institute of Education (NIE) and implemented since 2007. These PD efforts were important both to ensure accurate communication of the intent of the new science curriculum and also to make certain of the currency of teachers' pedagogical skills and knowledge. At the school level, teachers initiated and joined professional learning communities to discuss and support one another's inquiry efforts. They re-examined their current lesson plans and developed ways to incorporate inquiry activities into the two hours of science lessons per week.

The change in curricular focus to one that is inquiry-centric is not free from problems despite attention paid to teacher PD. Similar to their counterparts in other parts of the world (see Crawford, 2007), teachers in Singapore also encountered issues in their practice of science as inquiry in classroom situations. In this study, we attempted to understand ways of enhancing science as inquiry in the current science education climate in Singapore. To do so, it is necessary to explore how teachers experience inquiry in local classrooms and to delve into the possibilities and difficulties of inquiry teaching in school contexts. Since the pedagogical approach of science inquiry is relatively new in Singapore schools, it is important to listen closely to teachers' stories and expectations in order to ensure its meaningful enactment. The voices of teachers in classrooms will help us understand the challenges of the intended curriculum in the community.

Giving importance to the participatory actions of teachers in curricular change, we gathered teachers' stories and experiences of the new science syllabus and their interpretation of inquiry. The research questions guiding our study were:

1. What are in-service teachers' perceptions of inquiry-based science teaching?
2. What are the challenges of inquiry-focussed science curriculum from in-service teachers' perspectives?

Research Context

Participants

In this study, we focus on teachers' perceptions and discussions of science as inquiry in our attempt to comprehend how the inquiry-based curriculum has been understood, experienced and enacted in schools two years after the implementation of the inquiry-centric science curriculum. To understand the experience of science as inquiry teaching as practised by in-service elementary school teachers in Singapore, this study involved 41 in-service teachers from five public primary schools selected through convenience sampling methods. The backgrounds of these in-service teachers vary in terms of gender, education and years of teaching experience. Prior to teaching, all have earned either a bachelor's degree and/or diploma in teaching. For this group, the number of years of actual teaching experiences ranged from less than a year to 30 or more years. The teachers in this study taught science from Primary 3 to Primary 6. There were seven male and 34 female teachers.

Data Collection and Analysis

To probe the perceptions and conflicts of teachers' understanding of science as inquiry, we used narratives and survey questionnaires, reflective writings and discussions. We used the questions suggested and developed by Kang, Orgill and Crippen (2008) in the *Ideal Inquiry Lesson* narratives in an open questionnaire to survey teachers' perceptions of inquiry and inquiry lessons. This 10-item questionnaire was chosen as it allowed teachers to express their ideas about the ideal science as inquiry lesson in a semi-guided manner through prompts. Subsequently, we invited the participant teachers to review video materials that have been previously developed by pre-service teachers to facilitate inquiry for classroom teaching. The video materials were used as a tool to initiate teachers' discussions on their perceptions and experiences of inquiry teaching. They were asked to identify which aspects of the videos were "inquiry" and to state reasons for their choices. Finally, they reflected on personal experiences about the dilemmas of inquiry teaching in a reflection log. Based on their reflections, the in-service teachers were also engaged in group discussions to share their ideas, experiences and conflicts of inquiry practice. At this stage, the researchers played the role of a facilitator to ask clarification questions and also to encourage all the teachers to share their ideas.

For data analysis, NVivo8 software was employed for data coding. All videos and written data were uploaded and reviewed independently by the three researchers in this study. Responses to the survey questionnaire were coded using a scheme based on the five essential features of inquiry (NRC, 1996). The five essential features of inquiry spelt out by NRC was chosen for coding for two reasons: (1) it was adopted as a framework in the Singapore science curriculum, and (2) it was also used as a framework for data analysis by Kang et al. (2008) from which the questionnaire for this study is adopted. For the narratives and reflections, the coding schemes were developed as follows: to establish reliability and validity, we randomly selected five samples and individually coded them. We then discussed our initial individual coding schemes and coding results. A consensus coding scheme was agreed upon after discussions. Subsequently, we individually coded another ten samples to further refine the consensus coding scheme into its final form.

Results generated independently by applying the final coding scheme onto the entire data corpus were then compared by calculating for inter-coder agreement, which we set to a satisfactory 80 % level. In cases where the rate was below 80 %, we had discussions over the items to re-examine each other's interpretation and then recoded them separately. Several iterations of this coding agreement process were conducted until the satisfactory rate was reached or exceeded. Based on these coding results, we analysed the emergent themes of Singapore teachers' stories and concerns of inquiry teaching practice and discussed the visions and challenges of the new curriculum implementation.

Findings

Salient Features of Inquiry in Teachers' Responses

We coded teachers' survey responses on the reasons for their choice of an event as "inquiry" or "not inquiry". The coding was conducted based on the five essential features of inquiry: (1) learner engages in scientifically oriented questions [questions], (2) learner gives priority to evidence in responding to questions [priority to evidence], (3) learner formulates explanations from evidence [explanation from evidence], (4) learner connects explanations to scientific knowledge [connection to knowledge] and (5) learner communicates and justifies explanations [communication and justification] (NRC, 1996). We did not explain the five essential features to teachers before the survey took place. Coding results, reported as "% of references coded", revealed the categories of teachers' ideas about inquiry (see Table 6.1).

In-service teachers expressed ideas in all five elements of inquiry but with differing degrees of salience. It was noted that they gave greatest attention to *priority to evidence* (38.1 %) and the least to *explanations from evidence* (7.2 %). In their justifications, references to seeking and gathering evidence were more frequent than those about students providing explanations based on the evidence gathered. The most frequent justification for *priority to evidence* was that the student is directed to collect certain data through laboratory experiments or hands-on activities. There was hardly any mention of teachers providing students with the opportunity to design investigations and thus determine for themselves the

Table 6.1 In-service teachers' understandings on the five features of inquiry

Essential features of inquiry	% of references
Priority to evidence	38.1
Connect to knowledge	21.6
Questions	18.6
Communication and justification	14.4
Explanations from evidence	7.2
Total	100.0

method of data collection. In other words, teachers prioritised data gathering activities in inquiry learning and did not realise the importance of interpreting and explaining the gathered data.

The relatively low frequency of *explanations from evidence* could suggest that rather than putting emphasis on having students formulate explanations organically from the evidence gathered, teachers instead favour canonical science knowledge as the outcome of learning through inquiry. Indeed, the features of inquiry that relate to the acquisition of science content knowledge – *connection to knowledge and questions* – scored higher at 21.6 % and 18.6 %, respectively. One teacher's response is quite telling in this regard. She commented that students giving a presentation (explanation) cannot be considered inquiry "unless pupils are given a question to begin a process where they engage prior knowledge and determine the new knowledge they need to explain and answer a question".

The emphasis on acquiring the prescribed knowledge was similarly pervasive during the workshop discussions on inquiry teaching. In the following excerpt (Excerpt 6.1), a teacher used her experience on teaching about *insects* to emphasise the importance of school science knowledge. She stated that while students build knowledge on their own through the learning activities provided for them, a teacher must still guide students to the "right answer" through proper questioning in order to "correct misconceptions".

Excerpt 6.1. (Emphasis Added)

Teacher They make connections when they hear from others. All students are supposed to connect differently from each other because there are different learnings. So when the class comes back to share, **the teacher must provide input, as well as correct misconceptions**. Thus, as a teacher, you are actually just a facilitator. I do agree [with my colleague] that the questioning techniques are a big issue. What is important is **how you guide the students' learning** when they come back with their findings. This is **their input for the science knowledge** – information that they basically formed on their own. So let's say a child has mistakenly classified an item as an insect. The child, and also the whole class, will **know the right answer** because you would have **corrected their misconception**. But you didn't feed that information to them. So the children will make their own connections, and I think they will retain these... And I think as young kids, we want them to reach that kind of learning rather than we feed them

Similar to the *explanation from evidence*, the category *communication and justification* (14.4 %) was found infrequently among the teachers' responses. There were few references to students engaged in discussions to justify claims and present logical arguments. One teacher viewed discussion as just a "repeat of what they have learnt in class". Another teacher expressed the idea that inquiry "should not just be a discussion". Instead, she continued, "it should provide hands-on experience for them to connect their ideas to real-life experience". This result thus seemed to suggest that only few teachers hold the view that discussions are valuable in generating and sustaining student reasoning and argumentation. Alternatively, it may also suggest that the teachers viewed engagement in practical, hands-on activities as exemplars for inquiry teaching, more than those activities that involved explaining and justifying knowledge claims.

Teachers' Ideas on Requisites for Ideal Inquiry Lessons

In-service teachers featured in their narratives the importance of certain elements of inquiry science lessons (Table 6.2). The teacher participants opined that students' attitudes and knowledge of science (23.5 %) are fundamental for successful enactment of science as inquiry. They highlighted students' attitudes, such as being curious, interested and enthusiastically engaged in questions and activities, as cornerstones to making science as inquiry possible. For students' knowledge development, they mentioned the importance of knowing students' misconceptions, connection of knowledge to activity, achieving conceptual knowledge etc. as important prerequisites for science as inquiry.

Besides students' attitude and knowledge, the other important component of an ideal inquiry lesson was teachers' skills to facilitate inquiry (16.8 %). They recognised that teachers should be able to carry out purposeful questioning to guide children as they investigate and solve problems. Further, teachers were responsible for providing the necessary conditions and materials to allow the inquiry in the classroom to take place.

For some in-service teachers, a conducive classroom learning environment was also an ideal scene of inquiry classrooms (11 % of responses). They mentioned that students need to feel free and safe to explore their own interests and to share their ideas and questions without being afraid of being judged by teachers. Teachers also mentioned collaboration and interaction, questions and resources for inquiry teaching, student-centred approach and hands-on experiments as requisites for ideal inquiry lessons.

In contrast, teachers less frequently pertained to teaching models, students' process skills, teachers' attitudes and knowledge, laboratory equipment and open-ended questions. This suggests that teachers viewed students' knowledge and attitudes as more important than process skills for successful enactment of more open-ended inquiry.

Table 6.2 Elements of in-service teachers' perception of ideal inquiry lessons

Topic	%
Students' attitudes and knowledge	23.5
Teacher's skills to facilitate inquiry	16.8
Conducive environment	11.4
Collaboration and interactions	9.4
Questions	8.7
Resources in general	7.4
Student-centred orientation	6.7
Hands-on and experiments	6.0
Inquiry teaching models (5E, PBL, POE)	2.7
Students' process skills	2.7
Teacher's attitudes	1.3
Teacher's knowledge	1.3
Laboratory equipment	1.3
Open-ended answers (open inquiry)	0.7
Total	100.0

Teachers' recognition of students' knowledge and teachers' facilitation seemed to explain why teachers tend to focus on students' acquisition of conceptual knowledge through inquiry teaching rather than emphasising process skills, thinking or exploring through open-ended inquiry approach. To further probe in-service teachers' notion of facilitating inquiry, we also examined transcripts of the discussions on the narratives they wrote. We found that they indexed facilitating inquiry with good questioning, guiding students to knowledge, having the necessary content knowledge, doing enough preparation and creating an encouraging environment for students. The teachers also spoke of facilitating inquiry in relation to learners' characteristics and abilities, as exemplified in Excerpt 6.2. This particular teacher expressed that a guided inquiry approach benefits not only low-ability students but also those with high ability. However, she felt that most teachers need more experiences to build their confidence in using this approach.

Excerpt 6.2. (Emphasis Added)

Teacher For the inquiry approach, it is not necessary that we do it for every [textbook] chapter. We **look at one that really interests the children** then we pick that out. We **start slow then we do the guided inquiry**. We guide the so-called lower-ability pupils. We do a very guided one **so that the children can see**. I believe even the higher-ability pupils we cannot just throw in. So what we do is we start with a very guided one. And maybe our teachers are not confident enough. This where, in a way, we are also groping in the dark. If let's say there is some kind of model lesson that you can find, then you can actually try out. If it works, it gives us teachers a little bit confidence in carrying out the approach

In addition, during the discussion, it became clear that teacher quality and mindset are important factors in successful enactment of science as inquiry. The teachers emphasised the necessity of scaffolding and other guiding actions, complementing the common depiction in their narratives of teachers helping and guiding students in their work. The teacher's role in enacting successful inquiry is expressed quite candidly in a teacher's sharing during the discussion, as presented in Excerpt 6.3.

Excerpt 6.3.

Interviewer	What would be the roles of facilitators in inquiry teaching? She said questioning skills and what else?
Teacher	In a way, you also have to be **patient.** In my experience, I find myself so **eager to give** them the correct answer. But I have to pull back because I know that if I were to tell them the answers that defeat the whole inquiry approach. Sometimes, I don't know, they take a long time to finish a particular task we think is simple and easy to do. But to them it's harder. So we have to **change our mindset** and we have to be very – I mean we are, we are patient enough but we have to be **extra patient** when we do this. And we must also expect that there will be failure. There will not be like a perfect experiment that they will carry out.

Difficulties of Inquiry Teaching

The teachers' ideas about the difficulties of inquiry teaching were elicited through reflective writing and group discussions. They highlighted the tension between their current practices and the ideals of inquiry science and identified several factors that they believe create and sustain such tension. Coding results of their written reflections (Table 6.3) revealed that the predominant concern is the students' abilities and their readiness for inquiry instruction (16.8 %). This concern really seems to be important to teachers as it also emerged in their idealisation of inquiry lessons, as mentioned in the previous section. Many teachers held the view that students were not ready to participate in inquiry teaching because the approach was new to them. Having been accustomed to teaching that provided "model answers" in preparations for examinations, students could be expected to resist teaching that emphasised self-discovery and independent learning. Higher-ability students, according to the teachers, would best benefit from inquiry teaching, whereas lower- and middle-ability students could be disadvantaged.

Three concerns that altogether account for 39 % of the references coded were interwoven in the teachers' reflections: lack of class time (15.8 %), assessment conflicts (12.6 %) and heavy content in the curriculum (10.6 %). In Table 6.3, we cluster these in order to highlight their direct connection to the curriculum. Most

Table 6.3 Coding results of teacher's perceived difficulties in inquiry practice

Teaching dimensions	%	Themes	Sub %
Students	16.8	Student's readiness and abilities (middle level)	–
Class time	15.8	Time for conducting and science periods	–
Assessment	12.6	Conflicts with assessment	–
Curriculum content	10.6	Science syllabus (SIOs)	9.5
		No right concept or answer in the end	1.1
Understanding of inquiry	14.8	Not enough experience or training	5.3
		Confusion of meanings or ways of inquiry	5.3
		Lack of knowledge on how to teach or facilitate	4.2
Preparation	12.6	Time for preparation	6.3
		Lack of available resource	6.3
Teachers' science knowledge	7.4	Lack of content knowledge	5.3
		Afraid of having no answer to students' questions	2.1
Community support	5.3	Community support (school and parents)	–
Others	4.3	Class size	2.1
		Noise	1.1
		Giving up power or control	1.1
Total	100.0		

teachers found the currently allotted time for science lessons, typically four 30-min periods per week, to be insufficient. The content scope in the syllabus is already quite lengthy as it is and that the addition of inquiry only worsens the situation. Moreover, the perceived incompatibility of inquiry with current assessment practices adds another layer of challenge for teachers. The teachers see examinations as determining to a large extent the kind of knowledge that must be delivered and/or emphasised in the classroom.

Despite some recent changes in science assessment, such as the inclusion of more process skill questions, teachers continue to emphasise content knowledge to gear students towards the exam. During the discussions, one teacher offered her reasons for this continued emphasis (Excerpt 6.4):

Excerpt 6.4. (Emphasis Added)

Interviewer	So you voice out the concern that at the end of the day there is still need to take exams. So how do you reconcile between test tasking and cross...
Teacher 1	That's why we are rushing on time 'cause at one time, at one hand we are trying to do the inquiry, getting them curious and interested. On the other hand, we are trying to get them ready, to **gear them for the exam**.
Interviewer	So you think inquiry will not gear them to what's in the exam?
Teacher 2	The **alignment is not there**, not apparent
Interviewer	Ok, ok

Teacher 2 Because this is our main concern, at **the end of the day, we need to deliver in terms of results.** Because **we have accountability in terms of results**. I want very much to make lesson fun because that is what I believe in. But at the same time, at the end of the day, like she says, if they don't know how to answer process skills questions or how to go about crafting their answers, my head's on a chopping board

Teacher 1 **Bottom line is still results**. Sad to say

Teachers feel that inquiry may not be the best approach to teaching science under the current curricular framework. The new science curriculum, although prescribing inquiry as the dominant pedagogy, has retained most of the content scope of the old framework. Thus teachers feel overburdened and lacked the time to meet the demands of the syllabus completely. Similarly, the high-stakes national examinations exert great demands on the teachers' time. While some of them believe that an inquiring mind is valuable, it does not guarantee good performance in examinations. In order to meet performance expectations, teachers feel the need to set aside, or do minimal, inquiry teaching and revert to more transmissive, didactic modes of teaching.

Analysis of teachers' reflection surfaced concerns such as teachers' understanding of inquiry (14.8 %) and their lack of content knowledge (7.4 %). Lack of community support (7.4 %), and other concerns, such as class size, noise and giving up power to students (4.3 %) also figured in their writings.

Discussion

We have presented the perceptions in-service teachers hold about science as inquiry two years after the inquiry-based science curriculum was implemented. The findings showed that teachers' perceptions of inquiry were complex and intertwined with their understandings of ideal inquiry lessons and difficulties in its practice. For instance, in their perceptions, they valued evidence and knowledge more than students' explanations or communications. To achieve evidence and right knowledge, they focussed on teachers' facilitation and guidance as an important element of inquiry teaching. Further, they felt the pressure to ensure students' performance in assessment, which was likely to have influenced their perceptions on ideal inquiry teaching. The participating teachers also took into consideration many components of what they perceived to be good teaching in science. Fenstermacher and Richardson (2005) stated that assessment of good teaching must include aspects such as (1) whether the content taught aligns with disciplinary standards and norms and (2) whether the methods used are age-appropriate and intended to enhance the learner's competencies. Clearly, the teachers in this study considered these factors (i.e., curriculum standards, students' attitudes and abilities and conducive environment) and more in their practices. As such, the participating teachers perceived their role and their enactment of inquiry science as a practice that was piqued with dilemma. Based on the findings, we highlight the tensions of inquiry-based teaching that teachers would probably encounter in the current education system.

Positioning the teachers at the centre of the act of teaching, it is clear that issues of assessment, teacher control and content teaching were of concern when the teachers view the reform of science as inquiry. These three aspects highlight some tensions in inquiry teaching between content and process, teacher- and student-centeredness, and curriculum and assessment as teachers tried to enact inquiry under social conditions that remained largely unchanged. In what follows, we attempt to discuss the possibilities and practices of inquiry with regard to the current educational system by focussing on the in-service teachers' perceptions.

Tensions Between Inquiry and Examination

If teaching is viewed from two perspectives – task sense and achievement sense (Fenstermacher & Richardson, 2005), then teachers' actions of planning and facilitating students' learning is teaching in the task sense. Students' performance in examination is teaching in the achievement sense. One cannot be said to be engaged in the activity of teaching in the absence of students' learning. Consequently, our conversations about pedagogies of using science as inquiry cannot be complete without bringing in some measure of students' achievements. Hence, we highlight the tensions and somewhat antagonistic relationship that the participating teachers perceived to exist between the practice of science as inquiry and examination performance. This issue of assessment also links to the issues of teacher control and the heavy demands of learning outcomes. The three issues raised here are intertwined and complex, and despite risking oversimplification of the relationship among them, we attempt to discuss them in detail in a linear and sequential manner.

The current national examination, the Primary School Leaving Examination (PSLE), is structured such that 100 % of the examination is a paper-and-pencil test. Aligning with the proportion of learning outcomes spelt out, about 60 % of the written examination would be focussed on the knowledge, understanding and application. This means that the students will be tested on scientific phenomena, facts, concepts, principles and vocabulary. As such, even though teachers generally agreed on the level of enthusiasm and interest required of students in the classroom learning environment, they still expressed hesitation about the inquiry approach. They were concerned that science as inquiry will not prepare the students adequately to answer questions posed during the examination. After all, there was no doubt that the teachers concurred with the notion that "successful teaching is teaching understood exclusively in its achievement sense" (Fenstermacher & Richardson, 2005, p. 5). Some of the urgent issues that they needed to bear as classroom teachers pertained to the perceived requirements of the current assessment modes, demands made by parents and students about examination preparations, the crowded science syllabus, their preferred mode of inquiry teaching and their concern about students' academic achievement and abilities. They felt that these issues cannot be ignored under the current educational landscape. The teachers' deep concern and practice of assessment have shaped their perceptions on inquiry – inquiry is seen as a means of knowledge acquisition rather than a learning process, teacher guided rather than student centred. Even if there have been efforts on developing more exam questions on process skills, the general sentiment among teachers was that process skill questions still required sound content knowledge to answer. Teachers' ideas and beliefs about the current assessment demands, as well as of inquiry teaching, influenced their perceptions, decision-making and classroom practice consciously and unconsciously. As such, the primary concern of teachers on the final form of science overrides the process of science from which students can learn through experimenting and exploration.

Teacher Control

In order to ensure accuracy and appropriate pacing of lessons to meet assessment demands, teacher control over the topic and content of the inquiry is very strong. In their narratives and discussions, teachers often mentioned the importance of teacher intervention and guidance. They emphasised their roles as facilitators to keep students' learning and the inquiry process on the "right" track. Interestingly, even though teachers mentioned the value of student-centred activities and independent problem-solving, teachers' guidance was always present in their description of classroom scenes. It seems they recognised and put primacy on their responsibility to develop students' knowledge and skills. The teachers' views about their role as facilitator and guide could possibility be derived from the science curriculum document which states the role of teachers as "teacher as the leader of inquiry" and that the teachers are "facilitators and role models of the inquiry process in the classrooms" (MOE, 2007, p. 2). The facilitation that the teachers expressed in this study was one where the teacher maintained tight control over the content to be learnt, how it was learnt and the pace at which it was to be learnt. We argue that this form of high teacher-controlled facilitation on inquiry is the result of (1) the need to ensure that students learn the canonical science as spelt out in the syllabus documents; (2) a continuation of their existing role as a teacher from their current practice of didactic, teacher-fronted pedagogy; and (3) the limited curriculum time that is made available. Similar to the findings of Gyllenpalm, Wickman and Holmgren (2010) on teachers' scientific inquiry discourse, their notion of "facilitating inquiry" seemed to relate more to the *method of teaching* to acquire knowledge of the products of science and their use than to the *method of inquiry* to understand the logic behind producing scientific knowledge. Because of this recognition, their approach to inquiry lesson is teacher guided rather than student centred, which is often assumed to be the norm for inquiry lessons.

As seen in Excerpts 6.1 and 6.2, teachers felt it was vital that they guide students to develop students' thinking, problem solving and achieving knowledge. With this emphasis, it seems difficult to change traditional ways of teaching and give up control of students' learning in science classrooms. Naylor, Keogh and Downing (2007) found that primary school students work as equals and generate productive discussions and argumentation in the absence of their teachers. The presence of the teachers appears to have an inhibitory effect on the students' discussion. This suggests that elementary students are capable of managing themselves in a group. But unless teachers recognise this capacity of elementary students, teacher intervention might limit students' learning in terms of critical and creative ways of thinking, decision making and problem solving. Other researchers such as Cazden (1988) and Mercer, Wegerif and Dawes (1999), however, cautioned that the teachers' presence to set the ground rules and maintain the direction of the group discussions is still important for ensuring productive classroom learning. The participant teachers also highlighted that the preparedness of students to cope with science as inquiry is fundamental to its successful enactment and tight teacher control suggests that the teachers did not think that the students were ready to learn science through inquiry. To discuss the possibility of inquiry teaching, a preliminary step might be to examine the meaning of learning and teachers' responsibilities for students' learning in inquiry-based lessons. Further investigation is needed to understand what would be effective facilitation and teacher actions in inquiry classrooms.

Canonical Science Content

Among the five essential features of inquiry, in-service teachers gave priority to students gathering evidence (38.1 %) and listed the generation of explanation from evidence (7.2 %) as lowest in their practice. This suggests that the teachers spent time ensuring that the students were provided with opportunities to "do science" – working with manipulatives and data gathering. Having students work on manipulatives was perceived as inquiry in practice as this will help to fulfil the 35 learning outcomes that were spelt out under the *Skills and Processes* section of the syllabus documents. The low priority given to getting the students to derive their own explanations from evidence they gathered is suggestive of teachers' eagerness to provide the "correct" answers and interpretations from the students' data. This is evident of a strong emphasis on canonical science knowledge as the learning outcome of inquiry rather than more organic explanations emanating from the evidence at hand as part of the inquiry process. The activities of data gathering and connection to knowledge are likely to be perceived as "tangible" and visible activities related to inquiry, while formulating students' own explanations from evidence is an activity that might not generate the "correct" answers and forms of science for which the teachers planned. As early as 1977, Rachelson (1977) highlighted the need for balance in using science as inquiry in classrooms. Between the two essential processes of hypothesis generation and hypothesis testing, he noticed that there was an over-emphasis on hypothesis testing instead of hypothesis generation. This lopsided practice made evaluation of students' learning simpler but detached the imaginative and creative nature of problem and hypothesis generation from the practice of classroom science. The participant teachers in this study have exhibited a similar preference to the situation highlighted by Rachelson. They gave priority to students gathering evidence as a form of inquiry as this was an activity that was easily observable and measurable. They also preferred students to acquire the more accurate canonical scientific knowledge as reflected in textbooks rather than to be engaged in the creative process of crafting explanations from their evidence, which might ultimately result in learning the "wrong" facts.

This notion of privileging "evidence", "knowledge" and "questions" over "students' explanation, discussion, presentation and communication" also appeared in the narratives on ideals of inquiry lessons of these teachers. The teachers used science as inquiry as a means to modify students' misconceptions and to teach science content knowledge. They valued students' knowledge more than students' process skills in inquiry classrooms. As implied in their conversation (e.g., Excerpt 6.1), it was evident that the in-service teachers were aware of the demands of the science syllabus and of their responsibilities to equip students with the required knowledge to answer questions posed during examinations. Teachers were aware that they have to complete "teaching" the 86 learning outcomes as indicated under the knowledge, understanding and application category of science learning. With a total of 166 learning outcomes spelt out for all three categories in the syllabus documents and an estimated 320 hours over 4 years to achieve these learning outcomes, teachers felt rushed and pressured to complete all the learning outcomes in the limited time available to them. As such, they adopted a more pragmatic approach to teaching science since the syllabus requirements have largely remained unchanged in the new curriculum. These in-service teachers were cognizant of the tension that existed between practising science as inquiry and the learning of canonical science knowledge as required by the curriculum and examination (Tan & Wong, 2011) but were trying to establish a reasonable relationship between the two demands.

Limitations of Study

There are a number of limitations to this study that we like to highlight. Firstly, the use of convenience sampling suggests that the evidence presented could be biased towards teachers or schools that are more receptive to science as inquiry. As such, the ideas presented here have limited generalisability and representation of the whole population. Despite the limited generalisability, we like to add that the issues raised by the participants have been raised by other researches (e.g., Kim & Tan, 2011; Tan & Wong, 2011). The second limitation of this study is that we have no direct measure of teachers' beliefs and hence our analysis and interpretation of teachers' responses to the survey served as the closest measure of their beliefs. The analysis and interpretation of the teachers' responses are also subjected to researcher biases, and hence to reduce the effects of the biases, multiple coders and inter-rater discussions were carried out.

Conclusion

In this study, we surfaced the in-service teachers' perceptions of inquiry as one that focussed on students' abilities, knowledge acquisition, classroom environment and assessment. Their lived experiences with teaching at schools, and their interpretation of the social context of the educational landscape and curricular demands have led them to construct the ideas of inquiry with recognition on assessment performances, students' knowledge and teachers' guidance. We appreciate the fact that teachers perceived their main responsibility as one that is to help students obtain good test results even though they believed the need of inquiry in science education.

Researchers have found that teachers' beliefs strongly influence their actions in the development of inquiry teaching (e.g., Richardson, 1996; Roehrig & Kruse, 2005; Tobin & McRobbie, 1996). The experiences of the participating teachers suggest that their practices are strongly shaped by their ideas about the social and curricular demands more than their beliefs and ideas about science as inquiry. Anecdotal evidence (from teachers' discourses during PD sessions) suggests that the teachers agree with the current change towards inquiry-based teaching. Teachers see it as relevant and necessary for developing students' thinking and problem-solving skills for everyday lives and for the future of society. Yet, they are baffled when it comes to the issue of assessment and completion of syllabus outcomes. The pressure and expectations that they bear from school administrators, students and their parents seem to dictate their thoughts and professional decision making. With reference to the current pen-and-paper examination system, they perceive that inquiry cannot be the best way to increase the scores under the time and curriculum constraints. The examination system will not be removed, nor will the amount of science curriculum time be increased in the near future. And yet, they are expected to practise inquiry in the present classroom contexts as a way of developing scientific minds and future problem solving. This raises questions on how we could narrow the gap between the recent curriculum change and assessment system carried out in schools. For these teachers, is science as inquiry good teaching because it is learner-sensitive or does it help to achieve successful teaching as it is learning-dependent? It is hence fundamental that we question how teachers cope with the struggles in two paradigms that are not clearly aligned in the current systems and how the developers of curriculum and assessment can bring the two closer in school practice so that they could help teachers' decision making and practice in a more integrated realm of inquiry.

There is a multitude of factors that shape teachers' experiences in the classroom. In this study, we presented evidence to show how social context of time, expectations and assessment system affected the fidelity of science inquiry enactment. The complexity of factors affecting this group of in-service teachers' perception of science as inquiry in the curricula reforms should be given serious consideration when planning professional development courses and reviewing the authenticity and fidelity of enactment.

References

Abd-El-Khalick, F., Boujaoude, S., Lederman, N., Mamlok-Naaman, R., Hofstein, A., Niaz, M., et al. (2004). Inquiry in science education: International perspectives. *Science Education, 88*, 397–419.

Appleton, K. (2002). Science activities that work: Perceptions of primary school teachers. *Research in Science Education, 32*, 393–410.

Appleton, K., & Kindt, I. (1999). Why teach primary science? Influence on beginning teacher's practices. *International Journal of Science Education, 21*, 155–168.

Barrow, L. H. (2006). A brief history of inquiry: From Dewey to standards. *Journal of Science Teacher Education, 17*, 265–278.

Blanchard, M. R., Southerland, S. A., & Granger, E. M. (2009). No silver bullet for inquiry: Making sense of teacher change following an inquiry-based research experience for teachers. *Science Education, 93*, 322–360.

Braaten, M., & Windschitl, M. (2011). Working toward a stronger conceptualisation of science explanation for science education. *Science Education, 95*, 639–669.

Britner, S. L., & Finson, K. D. (2005). Preservice teachers' reflections on their growth in an inquiry-oriented science pedagogy course. *Journal of Elementary Science Education, 17*(1), 39–54.

Brown, S. L., & Melear, C. T. (2006). Investigation of secondary science teachers' beliefs and practices after authentic inquiry-based experiences. *Journal of Research in Science Teaching, 43*, 938–962.

Bryan, L. A. (2003). Nestedness of beliefs: Examining a prospective elementary teacher's belief system about science teaching and learning. *Journal of Research in Science Teaching, 40*, 835–868.

Cazden, C. (1988). *Classroom discourse: The language of teaching and learning*. New York: Greenwood Press.

Chin, C., Goh, N. K., Chia, L. S., Lee, K. W. L., & Soh, K. C. (1994). Pre-service teachers' use of problem-solving in primary science teaching. *Research in Science Education, 24*, 41–50.

Cohen, D. K., & Ball, D. L. (1990a). Policy and practice: An overview. *Educational Evaluation and Policy Analysis, 12*, 233–239.

Cohen, D. K., & Ball, D. L. (1990b). Relations between policy and practice: A commentary. *Educational Evaluation and Policy Analysis, 12*(3), 331–338.

Crawford, B. A. (2007). Learning to teach science as inquiry in the rough and tumble of practice. *Journal of Research in Science Teaching, 44*, 613–642.

Dewey, J. (1910). Science as subject matter and as method. *Science, 31*, 121–127.

Dietz, C. M., & Davis, E. A. (2009). Preservice elementary teachers' reflection on narrative images of inquiry. *Journal of Science Teacher Education, 20*, 219–243.

Eick, C. J., & Reed, C. J. (2002). What makes an inquiry-oriented science teacher? The influence of learning histories on student teacher role identity and practice. *Science Education, 86*, 401–416.

Fang, Z. (1996). A review of research on teacher beliefs and practices. *Educational Research, 38*, 47–65.

Fenstermacher, G. D., & Richardson, V. (2005). On making determinations of quality in teaching. *Teachers College Record, 107*, 186–213.

Furtak, E., & Alonzo, A. (2010). The role of content in inquiry-based elementary science lessons: An analysis of teacher beliefs and enactment. *Research in Science Education, 40*, 425–449.
Guskey, T. R. (1985). Staff development and teacher change. *Educational Leadership, 42*, 57–60.
Gustafson, B., Guilbert, S., & MacDonald, D. (2002). Beginning elementary science teachers: Developing professional knowledge during a limited mentoring experience. *Research in Science Education, 32*, 281–302.
Gyllenpalm, J., Wickman, P.-O., & Holmgren, S.-O. (2010). Teachers' language on scientific inquiry: Methods of teaching or methods of inquiry? *International Journal of Science Education, 32*, 1151–1172.
Howitt, C. (2007). Pre-service elementary teachers' perceptions of factors in a holistic methods course influencing their confidence in teaching science. *Research in Science Education, 37*, 41–58.
Hubbard, P., & Abell, S. (2005). Setting sail or missing the boat: Comparing the beliefs of preservice elementary teachers with and without an inquiry-based physics course. *Journal of Science Teacher Education, 16*, 5–25.
Hurd, P. (1998). Scientific literacy: New minds for a changing world. *Science Education, 82*, 407–416.
Johnson, C. C. (2006). Effective professional development and change in practice: Barriers science teachers encounter and implications for reform. *School Science and Mathematics, 106*, 150.
Kagan, D. M. (1992). Implication of research on teacher belief. *Educational Psychologist, 27*, 65–90.
Kang, N.-H., Orgill, M., & Crippen, K. (2008). Understanding teachers' conceptions of classroom inquiry with a teaching scenario survey instrument. *Journal of Science Teacher Education, 19*, 337–354.
Kielborn, T. L., & Gilmer, P. J. (Eds.). (1999). *Meaningful science: Teachers doing inquiry + teaching science*. Tallahassee, FL: SERVE.
Kim, M., & Tan, A.-L. (2011). Rethinking difficulties of teaching inquiry-based practical work: Stories from elementary pre-service teachers. *International Journal of Science Education, 33*, 465–486.
King, K., Shumow, L., & Lietz, S. (2001). Science education in an urban elementary school: Case studies of teacher beliefs and classroom practices. *Science Education, 85*, 89–110.
Kolsto, S. (2000). Consensus projects: Teaching science for citizenship. *International Journal of Science Education, 22*, 654–664.
Leander, K. M., & Osborne, M. D. (2008). Complex positioning: Teachers as agents of curricular and pedagogical reform. *Journal of Curriculum Studies, 40*, 23–46.
Lee, K.-W. L., Tan, L.-L., Goh, N.-K., Chia, L.-S., & Chin, C. (2000). Science teachers and problem solving in elementary schools in Singapore. *Research in Science & Technological Education, 18*, 113–126.
Lee, O., Hart, J. E., Cuevas, P., & Enders, C. (2004). Professional development in inquiry-based science for elementary teachers of diverse student groups. *Journal of Research in Science Teaching, 41*, 1021–1043.
Luft, J. A. (2001). Changing inquiry practices and beliefs: The impact of an inquiry-based professional development programme on beginning and experienced secondary science teachers. *International Journal of Science Education, 23*, 517–534.
Luft, J. A., & Roehrig, G. H. (2007). Capturing science teachers' epistemological beliefs: The development of the teacher beliefs interview. *Electronic Journal of Science Education, 11*(2), 38–63.
Martin-Hauser, L. (2002). Defining inquiry. *The Science Teacher, 69*(2), 34–37.
Mercer, N., Wegerif, R., & Dawes, L. (1999). Children's talk and the development of reasoning in the classroom. *British Educational Research Journal, 25*(1), 95–111.
Ministry of Education. (2007). *Primary science syllabus*. Singapore: Ministry of Education.

Moseley, C., Ramsey, S. J., & Ruff, K. (2004). Science buddies: An authentic context for developing preservice teachers' understandings of learning, teaching, and scientific inquiry. *Journal of Elementary Science Education, 16*(2), 1–17.

Mulholland, J., & Wallace, J. (2003). Crossing borders: Learning and teaching primary science in the pre-service to in-service transition. *International Journal of Science Education, 25*, 879–898.

National Research Council. (1996). *National science education standards*. Washington, DC: National Academy Press.

Naylor, S., Keogh, B., & Downing, B. (2007). Argumentation and primary science. *Research in Science Teaching, 37*, 17–39.

Nespor, J. (1987). The role of beliefs in the practice of teaching. *Journal of Curriculum Studies, 19*, 317–328.

Osborne, J. F., & Patterson, A. (2011). Scientific argument and explanation: A necessary distinction? *Science Education, 95*, 627–638.

Pajares, M. F. (1992). Teachers beliefs and educational research: Cleaning up a messy construct. *Review of Educational Research, 62*, 307–332.

Palmer, D. H. (2006). Sources of self-efficacy in a science methods course for primary teacher education students. *Research in Science Education, 36*, 337–353.

Rachelson, S. (1977). A question of balance: A holistic view of scientific inquiry. *Science Education, 61*, 109–117.

Richardson, V. (1996). The role of attitudes and beliefs in learning to teach. In J. Sikula (Ed.), *The handbook of research in teacher education* (2nd ed., pp. 102–119). New York: Macmillan.

Roehrig, G. H., & Kruse, R. A. (2005). The role of teachers' beliefs and knowledge in the adoption of a reform-based curriculum. *School Science and Mathematics, 105*(8), 412.

Roth, W.-M., McGinn, M. K., & Bowen, G. M. (1998). How prepared are preservice teachers to teach scientific inquiry? Levels of performance in scientific representation practices. *Journal of Science Teacher Education, 9*, 25–48.

Simmons, P. E., Emory, A., Carter, T., Coker, T., Finnegan, B., & Crockett, D. (1999). Beginning teachers: Beliefs and classroom actions. *Journal of Research in Science Teaching, 36*, 930–954.

Smolleck, L., Zembal-Saul, C., & Yoder, E. (2006). The development and validation of an instrument to measure preservice teachers' self-efficacy in regard to the teaching of science as inquiry. *Journal of Science Teacher Education, 17*, 137–163.

Tan, A.-L., & Wong, H.-M. (2011). 'Didn't get expected answer, rectify it.': Teaching science content in an elementary science classroom using hands-on activities. *International Journal of Science Education, 34*, 197–222.

Tobin, K., & McRobbie, C. J. (1996). Cultural myths as constraints to the enacted science curriculum. *Science Education, 80*, 223–241.

Trumbull, D. J., Scarano, G., & Bonney, R. (2006). Relations among two teachers' practices and beliefs, conceptualizations of the nature of science, and their implementation of student independent inquiry projects. *International Journal of Science Education, 28*, 1717–1750.

Volkmann, M. J., & Zgagacz, M. (2004). Learning to teach physics through inquiry: The lived experience of a graduate teaching assistant. *Journal of Research in Science Teaching, 41*, 584–602.

Yerrick, R., Parke, H., & Nugent, J. (1997). Struggling to promote deeply rooted change: The "filtering effect" of teachers' beliefs on understanding transformational views of teaching science. *Science Education, 81*, 137–159.

Yoon, H.-G., & Kim, M. (2010). Collaborative reflection through dilemma cases of science practical work during practicum. *International Journal of Science Education, 32*, 283–301.

Zion, M., Cohen, S., & Amir, R. (2007). The spectrum of dynamic inquiry teaching practice. *Research in Science Education, 37*, 423–447.

Chapter 7
Teaching Inquiry: Global Influences and Local Responses

Benny H.W. Yung

Keywords Global influences • Local responses • Formative assessment • Inquiry-based teaching • Cultural beliefs • Competing belief sets • Professional development • Guided inquiry

In this commentary, I will first sketch an overall impression of my reading of the three preceding chapters, followed by sharing my perspectives on the issues raised in the chapters. In Chap. 6, Tan, Talaue, and Kim successfully put across a portrait of the daily struggles of 41 elementary school teachers as they try to implement inquiry-based teaching in an educational setting that places due emphasis on learning canonical science content. Though it is a portrait based on the teachers' responses to questionnaire items as well as narratives and videos of inquiry lessons, I believe that there is a certain grain of truth based on my experiences with the Hong Kong education system which is also very examination driven (Yung, 2006). In Chap. 4, Tan and Tan examined the factors that 39 elementary school teachers considered as they decided on inquiry tasks for their students. It was disturbing to note that, among the teachers' consideration, there was a lack of systematic and conscious thought about enactment of formative assessment in the classroom. The teachers also showed a lack of understanding about the use of formative assessment as a process to develop students working conceptions of science rather than assessing their content mastery and attainment.

Chapters 4 and 6, together, paint a rather gloomy picture on the future development of inquiry-based teaching in Singapore. Yet, hopes of light are shed in Chap. 5 where Wong and Lau described their successful experiences of helping teachers implement an inquiry curriculum at the junior secondary level. Key to their success was the adaptation of overseas curriculum resources into a guided-inquiry

B.H.W. Yung (✉)
Division of Science, Mathematics and Computing, Faculty of Education, University of Hong Kong, Pokfulam, Hong Kong
e-mail: hwyung@hku.hk

A.-L. Tan et al. (eds.), *Inquiry into the Singapore Science Classroom*, Education Innovation Series, DOI 10.1007/978-981-4585-78-1_7,

curriculum which can cater for the needs and concerns of the educational landscapes in Singapore.

Clearly science education reforms in Singapore are well intentioned and broadly aligned with global trends. The emphases on inquiry-based teaching and assessment for learning reported in the preceding chapters are cases in point. Based on my reading of the chapters, I reckon if these reforms are to take root in Singapore, appropriate local responses and adaptations need to be taken instead of embracing the imported ideas uncritically. This resonates with my own experiences in Hong Kong whereby the educational and sociocultural context is very similar to that in Singapore. What follows is a more detailed account of my perspective on the various issues raised and discussed in the chapters.

A Daily Struggle for Science Teachers

Despite the globally recognised reform vision of enacting inquiry-based teaching in science classrooms, there is abundant evidence pointing to the day-to-day struggles that teachers around the world experience while they are enacting the inquiry-based approaches. Some of their struggles are inherent with factors associated with the teachers themselves such as their lack of the necessary knowledge, beliefs, and attitudes. Other factors are external to the teacher such as the lack of external support, time, materials and laboratory assistance. Perhaps, the most influential external factors that can undermine teachers' motivation to enact inquiry-based teaching are the cultural and institutional expectations of the responsibilities of science teachers. This stands out as a key finding of the work described in Chap. 6, as the authors put it in the abstract:

> Three key inter-related aspects of in-service teachers' perceptions of science as inquiry were revealed: (1) pressure of assessment demands, (2) perceived need for strong teacher intervention in inquiry, and (3) favouring content knowledge rather than process skills. These considerations surface conflicts of teaching science as inquiry that are driven by assessment demands and resulted in a mode of science as inquiry that has tight teacher control with a strong focus on learning the canonical science content. (Tan, Talaue, & Kim, Chap. 6, abstract)

Clearly the 41 Singapore teachers believed that they were held accountable for the performance of their students in the public assessment, which they perceived had a strong focus on canonical science content. As a result, they would privilege content knowledge rather than process skills as they enact inquiry-based teaching. As such, this presents a grim picture of the future development of inquiry-based teaching in Singapore.

However, I would argue that the pressure for teachers to cover the mandated curriculum and to prepare students for public assessment is neither a new phenomenon nor is it unique to Singapore. There has been quite a bit of research on this phenomenon. For example, in Australia, Tobin and McRobbie (1996) characterised these cultural beliefs as 'myths' of science teaching that subvert teachers from enacting reform-oriented pedagogical approaches. In the United States, Wallace and Kang (2004) disagreed with Tobin and McRobbie's characterisation of cultural beliefs as 'myths'. Instead, they believed that these cultural beliefs are 'real' to teachers – i.e., real challenges that teachers have to face in their day-to-day struggles. For this reason, they wanted to explore the possibilities and limitations of inquiry in typical high school classrooms, how teachers would respond to inquiry in the context of school culture with all its ordinary constraints. Thus, they focused on teachers who taught high school classes and did not have special permission to deviate from the mandated curriculum. Their goal was to illuminate the roles of competing belief sets, specifically related to the construct of inquiry, and how teachers come to highlight one belief set over another in particular situations. They also reasoned that teachers who had not already moved to a total adoption of inquiry-based methods might better exhibit these competing belief sets.

Hence they purposefully selected six experienced teachers who attended summer workshops on inquiry and had an interest in implementing inquiry-based teaching for their study. They observed the teachers' classroom teaching and interviewed them for the pedagogical reasoning behind their actions. It was found that this group of motivated teachers, who had been enacting what they believed to be inquiry-based science teaching, held competing belief sets in their mind. The first set of beliefs appeared to have stemmed from school culture and centred on constraining factors that limit inquiry. This made the teachers believe that they had to present canonical science contents in an efficient manner – this echoes the voices of Singapore teachers reported in Chap. 6. However, in addition to the cultural beliefs, each of the six teachers held a second set of beliefs of their own – a core belief set in favour of teaching science as inquiry, such as fostering independent thinking, deep thinking, problem solving, and/or stimulating creativity in science learning.

Based on the enacted inquiry-based teaching activities observed in this group of motivated teachers, Wallace and Kang argued that the belief sets that constrained inquiry-based teaching were more public and culturally based, while the belief sets that promoted inquiry were more private and based on the individual teacher's notion of successful learning. So there were teachers who stayed within the bound of teacher-guided inquiry on the one hand. In contrast, some teachers tried to resolve the tension between the competing belief sets through different means such as conducting most activities in a manner to achieve culturally based learning goals and conducting a few separate activities to foster their private goals for scientific thinking.

Based on their findings, these authors recommended that (1) the sharing of private learning goals with fellow teachers in learning communities may facilitate teachers to act on core beliefs that run in opposition to the cultural beliefs and (2) at the policy level, education authorities or school boards should make explicit the value of rich and meaningful learning goals for students by supporting curriculum standards for scientific thinking (Wallace & Kang, 2004, pp. 958–959). I am not sure how far the second recommendation could be realised in Singapore. Certainly, the first recommendation would be worth considering for colleagues in Singapore given the findings reported in Chap. 6. I wonder how many teachers among the 41 who participated in the study reported in Chap. 6 are motivated to teach science as inquiry. If the answer is 'none', then we may run the risk of perpetuating the cultural beliefs instead of awakening and/or instilling in teachers the core beliefs that can run in opposition to the cultural beliefs. Tan et al.'s recommendation in Chap. 6 – using exemplars for teacher professional development – seems to be more in line with the first recommendation.

Adding to the Struggle

Given the dynamic nature of science inquiry, teachers' understanding and skilful enactment of formative assessment strategies in facilitating students' learning and/or progress along their path of inquiry is of paramount importance. As argued by Tan and Tan in Chap. 4, formative assessment in science must be conceptualised to take into account the present curriculum shift towards inquiry-based teaching and learning. Their work, together with that of Coffey, Hammer, Levin, and Grant (2011), is a timely reminder for science educators who might have overlooked the important role of formative assessment in the context of inquiry-based teaching. This is especially the case when teachers see inquiry only as a means to convey canonical science knowledge to students but overlook the inquiry nature of science as a discipline.

Tan and Tan's work reported in Chap. 4 was based on two premises. First, alignment of the inquiry process to assessment practice is a key to successful inquiry-based teaching. Second, accurate alignment is dependent upon teachers' knowledge bases of science as inquiry and the formative assessment process. They explored the issue by examining if elementary school teachers in their study would give explicit attention to the principles of formative assessment when they are planning and selecting tasks/activities for inquiry-based teaching.

Not surprisingly, their findings pointed to 'a rather disturbing lack of systematic and conscious thought about enactment of formative assessment in the classroom' as well as 'a lack of understanding about the use of formative assessment as a process to develop students' working conceptions of science rather than assessing their content mastery and attainment' among the teachers. This coincides precisely with what was reported in Coffey et al. (2011)'s paper on missing disciplinary substance. Clearly, these represent additional barriers to achieving successful inquiry-based teaching for this group of teachers. They need to develop the concept of integrating formative assessment seamlessly into the teaching and learning process. This is not an easy task. It adds to the daily struggle of science teachers as they enact inquiry-based teaching.

A Grim Picture but with Lights of Hope

Chapters 4 and 6, together, have painted a rather gloomy picture for Singapore in her stride towards inquiry-based teaching. Pervasive in the teachers' accounts are their worries if inquiry-based teaching has to be carried out at the expense of test scores. In Chap. 5, Wong and Lau documented some evidence which may help lay off some of the teachers' concerns. They described their journey in the development and implementation of an inquiry curriculum with three cohorts of secondary one students. A notable feature is that they adapted a US-developed inquiry curriculum originally designed for teacher professional development purposes and turned it into a guided-inquiry curriculum for students. The authors unpacked the challenges they faced in helping the teachers to implement the curriculum. They showed how the teachers made small steps in grappling with the problems they encountered. These included insufficient laboratory provisions, the large class size which makes insufficient time for individual group checkouts during a lesson, etc. Clearly these problems associated with the large class size would not have been considered by designers of the original curriculum. But these are real problems for teachers in the East where a large class size is of a common occurrence. I am delighted to find that Wong and Lau, together with the participant teachers, have worked out a guided-inquiry approach which can cater for the Singapore settings. The main features include (1) guided format of inquiry activity worksheets, (2) integrated use of laboratory-based and classroom-based materials, and (3) use of classroom check-outs and embedded probing questions. In particular, we can see how feature (4) is designed to address many of the local teachers' concern whether students are able to come to grips with the relevant content knowledge underlying the inquiry. In short, this is an example of local responses to science education reforms that are under global influences – hence the subtitle of this commentary – *Teaching inquiry: global influences and local responses.*

The 'Singaporean style' of guided inquiry and the teachers' pervasive concerns about students' test scores that run through the chapters remind me of the book by David Carless (2011) – *From Testing to Productive Student Learning: Implementing Formative Assessment in Confucian-Heritage Settings.* Based on his experiences working with Hong Kong teachers (who are also very concerned about students' test scores) on reform-oriented formative assessment pedagogies, Carless proposed that different variations of formative assessment are needed for different educational settings. He introduced a model comprising a range of formative assessment strategies lying on a continuum between two poles: 'restricted' and 'extended' formative assessment. The former hints at a conception of formative assessment that is largely teacher controlled with teachers providing guidance or feedback to students, and the latter is more learner oriented with student reflecting on their performance and seeking to improve. In short, Carless interpreted the term 'extended' as hinting at an ambitious, but probably not widely attainable, version of formative assessment. This somehow parallels the classification of science inquiry into different levels – e.g., structured, guided and open inquiry (Bell, Smetana, &

Binns, 2005). If I can borrow Carless's argument, this is not to say that open inquiry is not a laudable goal, but merely to acknowledge pragmatically that it is not always readily achievable. In contrast, structured or guided inquiry is likely to be more feasible in educational settings like Singapore. I concur with Carless that educational innovations involve interaction between schools and the wider social sphere, including political and cultural elements. I also maintain that all change is local, and conditions in local contexts prompt teachers to modify reforms in ways that make sense to them as teachers reported in Chap. 5 did.

Nevertheless, Carless was conscious of a possible caveat of advocating too much for local adaptations as he put it: 'local adaptations ... could support the development of contextually grounded models but may also risk undermining its principles' (p. 98). It may lead to what he described as a 'lethal mutation' in which the local reinterpretation of the innovation no longer retains its pedagogical essence. This is especially the case if the learning principles underlying relevant teaching behaviours are not fully understood. Hence, the challenge for Singapore is to develop a guided inquiry that can cater for their specific local and national context while still retaining its pedagogical essence, that is, a contextually grounded version of inquiry rather than a 'lethal mutation'. In labelling their adaptation as guided inquiry, Wong and Lau are acknowledging that guided inquiry is not a full-fledged version of inquiry meeting all its principles but instead a more pragmatic and limited version. This carries a greater potential to stimulate teachers to engage with inquiry-based teaching. As guided inquiry is likely to be within the capacities of a greater number of teachers in a wider variety of contexts, it can increase interest and acceptance of inquiry in settings where it has hitherto been largely neglected. As Carless put it, 'The enactment of classroom practices on the continuum is not viewed statically, but as a seed for further development and innovation'. (p. 205). Hence, I will continue to watch the future development of inquiry teaching in Singapore science classrooms with great interest.

References

Bell, R. L., Smetana, L., & Binns, I. (2005). Simplifying inquiry instruction: Assessing the inquiry level of classroom activities. *The Science Teacher, 72*(7), 30–33.

Carless, D. (2011). *From testing to productive student learning: Implementing formative assessment in Confucian-heritage settings*. London: Routledge.

Coffey, J. E., Hammer, D., Levin, D. M., & Grant, T. (2011). The missing disciplinary substance of formative assessment. *Journal of Research in Science Teaching, 48*, 1109–1136.

Tobin, K., & McRobbie, C. J. (1996). Cultural myths as constraints to the enacted science curriculum. *Science Education, 80*, 223–241.

Wallace, C. S., & Kang, N.-H. (2004). An investigation of experienced secondary science teachers' beliefs about inquiry: An examination of competing belief sets. *Journal of Research in Science Education, 41*, 936–960.

Yung, B. H. W. (2006). *Assessment reform in science: Fairness and fear*. Dordrecht, the Netherlands: Springer.

Chapter 8
Transiting into Inquiry Science Practice: Tales from a Primary School

Chew-Leng Poon and Shirley S.L. Lim

Abstract In this chapter, we describe a case study of the inquiry practices of four science teachers in a primary school in Singapore. We were motivated to carry out this study to address two issues. Firstly, while there has been a significant amount of research into the nature of inquiry science, more recent discussions have pointed to the smaller amount of research on teachers' instructional practices of inquiry in the classrooms. For teachers transiting into inquiry practice, concrete descriptions and analyses of what inquiry looks like in practice, when and how it occurs can be helpful. Secondly, teachers were concerned that inquiry science, as described in literature, was impossible to implement in countries such as Singapore where there are usually large class sizes and a strong focus on preparing students for national examinations. Our study surfaced nuances in the teachers' inquiry practice that sought not only to address local issues but also reflected their tacit experience and beliefs. Importantly, this study also examined the tensions teachers faced and the factors that contributed to their dilemma and decisions as they adapted their teaching practices to their beliefs and interpretations of what constitute inquiry science. This line of inquiry would contribute towards efforts in supporting teachers who are transiting into inquiry science practices.

Keywords Inquiry Science • Practice of inquiry science • Inquiry science in classroom • Pedagogical framework • Factors impacting inquiry practice • Student-directed inquiry • Singapore primary science education

---

C.-L. Poon (✉)
Research and Evaluation, Planning Division, Ministry of Education, 285 Ghim Moh Road, Singapore 279622, Singapore
e-mail: poon_chew_leng@moe.gov.sg

S.S.L. Lim
National Institute of Education, Nanyang Technological University, 1 Nanyang Walk, Singapore 637616, Singapore
e-mail: shirley.lim@nie.edu.sg

A.-L. Tan et al. (eds.), *Inquiry into the Singapore Science Classroom*, Education Innovation Series, DOI 10.1007/978-981-4585-78-1_8,

Introduction

Inquiry science has been an enduring and prominent theme in science education research. Some authors such as Barrow (2006), Bybee, Powell and Trowbridge (2008) and the National Research Council (NRC, 2000) believed that inquiry was introduced into the school science curriculum from as early as the 1900s by Dewey. In his address on 'Science as subject-matter and as method' in 1909, Dewey was already persuading his audience at the American Association for the Advancement of Science (AAAS) that the teaching of science should be approached as a method of inquiry (Bybee et al., 2008). Hodson (1985) observed, however, that inquiry science only gained momentum much later – with the 1958 Rockefeller Report, 'The pursuit of excellence, education and the future of America', following the launch of Sputnik I in 1957. During this period, educationists such as Bruner, Schwab and Rutherford have frequently been cited as instrumental in galvanising a reorientation of science teaching towards inquiry (Bybee et al., 2008; DeBoer, 1991; NRC, 2000; Shulman & Tamir, 1973).

More recently, there have been some criticisms, or perhaps realisations, however that much of the research into inquiry science has stayed at the theoretical level, dwelling on issues such as the construct of inquiry, the benefits and bane of inquiry and the efficacy of inquiry programmes and curricula (see, e.g., Lederman, 2007; Schwab, 1962; Shymansky, Hedges, & Woodworth, 1990; Tobin, 2006). While this is important, discussions have pointed to the gap in research on teachers' instructional practices of inquiry in the classrooms. Flick (2000) has argued that detailed descriptions and analyses of teacher behaviour and teacher-student interactions that occurred in the classroom could help adaptation of inquiry for a wider range of students. In reviewing the progress of research in inquiry science and proposing a direction for future research in this field, Keys and Bryan (2001) suggested that portraits of teachers' practice of inquiry in diverse settings will be 'an extremely valuable addition to the knowledge base' (p. 637). Building a comprehensive knowledge base of classroom instructional practices is therefore critical in advancing inquiry learning across classrooms (McNeill & Krajcik, 2008).

Given the emerging importance of this field in inquiry science, we embarked on a study of the inquiry practices of four teachers in a primary school in Singapore. Our objective was not to capture the range of inquiry practices across classrooms in Singapore. Instead, our research chose a case study approach to examine more deeply the instructional components, sequences and interactions that occurred in the teachers' science classrooms to answer questions about the:

(a) Patterns observed in the instructional components and sequences of teaching by inquiry in the teachers' classrooms
(b) Decisions teachers take on the extent of teacher direction and student direction in the inquiry process
(c) Pursuit of science concepts in the inquiry classroom
(d) Challenges and support in adopting inquiry practice in the classroom

This process of characterising inquiry surfaced core components of inquiry found in theoretically driven models as well as other practice-oriented components that were regarded as integral to the success of day-to-day classroom practice in Singapore. Importantly, this study also examined the tensions teachers faced and the factors that contributed to their dilemma and decisions as they adapted their teaching practices to their beliefs and interpretations of what constitute inquiry science. As inquiry practice is not regarded as widespread in Singapore classrooms (Hogan et al., 2006), we hope this line of inquiry would contribute towards efforts in supporting teachers who are transiting into inquiry science practices.

Literature Review

Importance of Teaching Science as Inquiry

Arguments for teaching science as inquiry can be organised around four areas: inquiry teaching and learning approaches (a) are congruent with the contemporary philosophy that scientific knowledge is tentative, (b) are resonant with cognitive science research on how people learn, (c) have pragmatic values in society and (d) have educational values for the learner. One of the key motivations in teaching science as inquiry is its congruence with the philosophy that scientific knowledge is tentative, as opposed to a fixed body of knowledge (Kuhn, 1970; Lederman, 2004; Popper, 1985). Inquiry science presents scientific knowledge not only as a construct of empirical investigations and inferences but also as a product of human creative efforts in understanding, representing and communicating ideas about physical and natural phenomena. It presents opportunities for students to construct models and explanations derived from their investigations and to evaluate these models and explanations against evidence and alternative ideas presented by others.

A second line of argument for inquiry science is the resonance of its principles with that of cognitive science-based research on how people learn. Research in the field of cognitive science has contributed to theories about the processes of thinking and learning taking place in the brain, with consequential implications for learning and teaching in the classroom (see, e.g., Bransford, Brown, & Cocking, 2000).

Research on learners suggests that students' minds are not empty vessels waiting to be filled – even very young learners come to primary schools with conceptions (and misconceptions) of the natural and physical phenomena around them (Carey, 1979; Wellman & Gelman, 1992). An inquiry approach to science learning provides the environment and tools (e.g., an investigation task) to engage and interrogate these prior concepts.

The prior and often naïve conceptions that learners possess do not limit their abilities to learn science. Rather, when given opportunities to explore deeper understanding of ideas and concepts in a subject matter, learners can develop competence and the ability to organise these concepts in ways that facilitate retrieval and application. 'Competence' here refers to the ability to identify patterns and relationships in new information, to generate reasoned arguments and explanations and to apply what was learned into new situations (Bransford et al., 2000). Inquiry science provides opportunities for students to develop competence in science by emphasising processes that require students to identify patterns and relationships from data collected and to generate explanations based on evidence.

Another emergent area of research that has informed inquiry science approaches is neurocognitive science which based its theories on integrating research on cognitive science, studies of the brain and the neural activities of the central nervous system. Drawing on these neurocognitive studies, Anderson (1997) claimed that inquiry methodologies in science generated two effects. First, the use of hands-on and manipulative experiences in conjunction with group work during inquiry science encouraged the 'learner to construct interpretations of experience that involve multiple pathways and build strengthened and enlarged networks of knowledge representation' (Anderson, p. 79). Secondly, the general principles of scientific inquiry learned in science classroom increased the learners' capacity to transfer these skills to other problem situations. In other words, Anderson claimed that inquiry approaches extend literacy beyond scientific knowledge and skills by providing generalisable schemas for students to gather and communicate evidence to support claims and to explain discrepant events beyond science (Bruner, 1960; von Secker & Lissitz, 1999). Inquiry therefore fosters the development of cognitive processes such as critical thinking and problem-solving (DeBoer, 2004; Shulman & Tamir, 1973; Smith, Maclin, Houghton, & Hennessey, 2000).

Some researchers are concerned, however, that certain forms of inquiry practice may not be well aligned to principles of how students learn. Metz (2004), for example, was concerned with inquiry instruction that pursued scientific process skills outside of subject knowledge goals. These forms of inquiry ran counter to the learning principle that learners develop competence in an area through a deep foundation and understanding of the ideas and knowledge in that domain. Metz contended that a weak grasp of domain knowledge worked against students' ability to reason. There are also criticisms that some prevailing models of inquiry science emphasise discrete concept learning at the expense of scientific knowledge systems that facilitate retrieval, application and a deeper understanding of links among concepts (Grandy & Duschl, 2008; Krajcik, 2008). In examining the four teachers' classrooms, we paid attention to how the teachers viewed the role of knowledge and how they organised knowledge in their classrooms. We also looked out for the presence or absence of evidence and mechanisms where students generated and organised ideas and concepts.

A third line of argument for inquiry science is tied up with the perceptions of the roles science play in society. Beyond providing students with a philosophy for understanding science and its ideas, inquiry science is seen to have pragmatic values to society. Inquiry science could potentially contribute to the development of reasoning and analytical skills and decision and problem-solving abilities (AAAS, 1990; NRC, 1996), skills and abilities that are premium in a knowledge-driven world economy. At the same time, because it seeks to emulate more closely how science is practised, an inquiry-centric science education could therefore play an important role in helping some students develop the aptitudes and attitudes of a good scientist and an inclination towards choosing a career in science-related fields.

An informed political leadership and a knowledgeable public are critical to the processes of democracy. Teaching science as inquiry is one way of equipping citizens with the schema and tools with which to critique, debate and make decisions in areas such as technology, environment, health and medical ethical issues (AAAS, 1990; Bybee, 1997; Driver, Newton, & Osborne, 1998; Smith et al., 2000). There is no lack of examples of how science impact the daily life of citizens: 'Is there a strong scientific basis to ban trans fat in food products?'; 'On what basis should we support or not support genetically modified crops?'; 'Should DNA-mapping technology be used for gene selection in human reproduction?'; 'Is stem-cell research ethical?'; 'What roles can Internet technology play in governance and civic societies?'; and so on. Both political leaders and citizens benefit from understanding how scientific claims can be investigated. Knowing the strengths, limitations and constraints of scientific research could therefore enable greater citizenry participation in debates and decisions that affect individual lives and that of society.

A fourth key argument for teaching science as inquiry is its claims of educational values for the learner. There are numerous studies and meta-analyses on the impact of inquiry science on student learning (Anderson, 2002; Berg, Bergendahl, & Lundberg, 2003; Cianciolo, Flory & Atwell, 2006; Goldston, 2005; NRC, 2000; von Secker, 2002; von Secker & Lissitz, 1999). In terms of student academic achievement, the outcomes have been mixed. While a significant number of studies (e.g., Shymansky et al., 1990; Wise & Okey, 1983) showed gains in student achievement (over traditional teacher-centred instructional strategies such as large-group instruction, recitation and drill), there were others that did not show significant gains (e.g., Pine et al., 2006). Apart from academic achievement, there was also evidence of the development of positive attitudes towards science in an inquiry environment (Ornstein, 2006; Shymansky et al., 1990).

Practice of Inquiry Science in Singapore

While there are clear benefits of a science inquiry curriculum, they can only be realised when enacted in the classroom. The practice of inquiry is therefore an important area of study, which is the focus of this chapter. The research and discussion is set specifically in the context of the Singapore classroom. A 2004 study of primary 5 classrooms by the Centre for Research in Pedagogy and Practice (CRPP) at the National Institute of Education, Singapore (NIE) showed that an inquiry culture has not taken root in the Singapore science classrooms (Hogan et al., 2006). Although this may not be true of every classroom, the general image of teaching and learning that emerged from the study was that of the teacher as the main source of authoritative knowledge and science pedagogy being dominated by whole class lecture and answer checking with 'initiate', 'response' and 'evaluate' patterns of classroom talk (Hogan et al. 2006).

In the 2008 Singapore science curriculum framework, greater emphasis has been given to the use of inquiry as a pedagogical approach to teaching and learning science (Curriculum Planning and Development Division [CPDD], 2007). The curriculum document explicitly advocated inquiry as a pedagogical approach to fuel the natural curiosity of children, providing them with the attitude, skills and tools to explore the natural and physical world. With the stronger emphasis on inquiry teaching and learning, there has been a renewed interest in adopting inquiry practice. An in-depth study of a few Singapore classrooms adopting inquiry could help surface some of the day-to-day issues that teachers face when they stepped up efforts in adopting inquiry practices and suggest how practising teachers resolve some of these dilemmas.

Pedagogical Frameworks for Inquiry Science

In studying the practice of inquiry, pedagogical frameworks are useful to represent and help visualise the instructional moves made by teachers. Researchers have used slightly different conceptions of pedagogical frameworks in different contexts. Kim, Hannafin and Bryan (2007) used a very broad conception that embraced both macro-contexts (e.g., systemic reforms, educational policies and professional development of teachers) and micro-contexts (e.g., tools) to design a pedagogical framework for a technology-supported inquiry class. Others, such as Bybee (2004), Eick, Meadows and Balkcom (2005), and Schwarz and Gwekwerere (2007), took a more focussed perspective and designed their pedagogical framework as an instructional or teaching cycle for the classroom. Schwarz and Gwekwerere (2007) defined a pedagogical framework as an instructional model that provides 'a simplified representation of the process one might engage in and the content one might address while teaching science' (p. 159). Bybee (2004) and Harlen and Osborne (1985) argued that pedagogical frameworks provide a systematic, coherent and focussed model to teaching science by inquiry.

Some documented pedagogical frameworks for inquiry science include the Biological Sciences Curriculum Studies' (BSCS) 5E (Engagement, Exploration, Explanation, Elaboration, Evaluation) instructional model (see Bybee, 2004), the MORE (Model-Observe-Reflect-Explain) Thinking Frame (see Tien, Ricky, & Stacy, 1999) and the heuristic for guided inquiry (see Magnusson, Palincsar, & Templin, 2004) – an overview of each are summarised in Table 8.1.

The five pedagogical frameworks share one common characteristic: all are theoretically derived frameworks. The starting points of these frameworks are theories about contemporary goals of science education, instruction, learning and the learning environment. The strength of these frameworks is therefore their sound theoretical underpinnings. Most of the research on these frameworks has focussed on the theoretical constructs, with few publications on how teachers actually enact

Table 8.1 Summary of some pedagogical frameworks for inquiry science

Instructional model	Purpose of model	Subject and level for which model is designed	Key components in model
Teaching sequence in Children's Learning in Science Project (Driver & Oldham, 1986)	Sequence of teaching	Secondary school science	Orientation Elicitation of ideas Restructuring of ideas Application of ideas
	Design of curriculum materials		Review change in ideas
MORE Thinking Frame (Tien et al., 1999)	Instructional tool for chemistry laboratories	Began as an instructional tool for undergraduate chemistry laboratory courses	Model Observe
		Adapted for secondary school chemistry laboratory	Reflect Explain
BSCS 5E Instructional model (Bybee et al., 2006)	Instructional tool Design of curriculum materials	K-12 science	Engage Explore Explain Extend/elaborate Evaluate
EIMA Instructional Framework (Schwarz & Gwekwerere, 2007)	Tool to guide teachers in designing and implementing inquiry-based lessons	Primary and lower secondary science	Engage students and elicit prior ideas Investigate Model or explanation creation Apply models or explanations to novel situations
Guided inquiry science instruction heuristic (Magnusson et al., 2004)	Instructional tool	K-12 science	Engage Prepare to investigate Investigate Prepare to report Reporting

(continued)

Table 8.1 (continued)

Instructional model	Purpose of model	Subject and level for which model is designed	Key components in model
Pedagogical framework for teaching and learning with inquiry tools (Kim et al., 2007)	Macro perspectives for designing instruction in a technology-supported inquiry environment	K-12 science	Macro-context: national policies such as standards Teacher community and professional development Micro-context: inquiry tools and classroom contexts

the pedagogical frameworks in their classrooms. There is an even bigger gap of knowledge about the pedagogical moves teachers made in inquiry-based classrooms where these theoretically derived pedagogical frameworks are not used. We believe that building a knowledge base of classroom enactment of inquiry, with and without the aid of pedagogical frameworks, can contribute to efforts to bring about a wider adoption of inquiry practices that are not only theoretically sound but are also grounded in the teachers' experiences of the realities of the classroom.

Impact of Teacher Beliefs on Practice of Inquiry Science

Research has established a close link between teacher beliefs and teaching practice (Bryan, 2003; Bryan & Abell, 1999; Pajares, 1992; Richardson, 1996). Beliefs are 'psychologically held understandings, premises or propositions about the world that are felt to be true' (Richardson, 1996, p. 103). As an evaluative construct, beliefs may not be aligned with reality or reason; for example, some teachers hold the belief that boys are better at science than girls. As an affective or emotional construct, beliefs are less open to critical examination and even change. Beliefs that teachers held about science and how it is taught would therefore have an impact on the adoption of teaching strategies that are new to the teacher, in this case, inquiry science.

Specifically, beliefs about the nature and value of science, about science teaching and about children learning science have been shown to influence how science is practised in the classroom. When the beliefs a teacher holds are not in congruent with a teaching approach, a tension may result (Bryan, 2003). For example, a teacher holding the belief that scientific knowledge consists of a body of established truths would have difficulty adopting inquiry approaches that allow for students to generate alternative explanations and to question 'established truths'.

While the idea that beliefs influence the behaviour and decisions of teachers is well established, some studies have reported that practice is not necessarily congruent with expressed beliefs. In Bryan's (2003) study, for example, Barbara's (a pre-service primary school teacher) belief that science teaching practice *should* emphasise a hands-on approach was incongruent with the didactic teaching orientation she adopted in *actual* practice. Bryan suggested that beliefs are a complex system, and it is possible for an individual to hold contradictory sets of beliefs. While Barbara had a vision of practice, her actual practice was mediated by lifelong science learning experiences grounded in didactic approaches.

The study of four pre-service primary and secondary science teachers by Mellado (1998) also found that there were no clear relationships between the teachers' beliefs and their observed practice. In this case, Mellado suggested that the lack of practical knowledge of the classroom and teaching hampered the teachers' ability to deliver the type of lessons they envisioned. Studies of teachers' classroom practices, such as those already mentioned, are therefore important as they illumine the interacting factors (e.g., beliefs interacting with knowledge) in the complexity of the classroom that could explain observed practices.

Methodology

A case study approach was employed, with the collection of detailed empirical data on the classroom practices of four primary science teachers. Data collection involved observations and video recording of lessons as well as interviews with teachers. These were complemented by artefacts such as student activity sheet, experimental kits and lesson plans.

In the first phase of the study, the researcher observed and video recorded 13 inquiry science lessons (26 half-hour school periods) over two academic school terms (6 months). The video-taped lessons were transcribed and subsequently coded by the researcher and a research assistant independently to identify the pedagogical components and sequence observed in the teacher's inquiry practice. We achieved an inter-rater reliability of 0.85. The coding scheme was developed from existing pedagogical frameworks of inquiry (see Appendix A for coding scheme) and calibrated by six researchers. The theoretical basis of the coding methodology is Evers and Wu's (2006) *pattern matching* where observations in the field were matched against components described in existing theoretically driven pedagogical frameworks. Schematics were drawn to represent the instructional sequence of the pedagogical components. A workshop with the collaborating teachers held at the end of this first phase helped validate the broad pedagogical components and sequence that characterised the teachers' inquiry practice.

In the second phase of the study, another 49 inquiry lessons were observed and video recorded over a 6-month period. The dialogue between teachers and students were transcribed verbatim from the videotapes. Interview sessions and workshop discussions were also audiotaped and transcribed. These data were analysed inductively to pick up patterns. Applied conversation analysis (Baker, 1997) was used to tease out and make explicit the nuances embedded in the teacher's inquiry practice.

Through the cycles of classroom observations, reflection and dialogue, the pedagogical framework that characterised the teachers' practice was iteratively defined. A fuller description of the methodology, the coding scheme and processes and limitations of the methodological approach can be found in Poon, Lee, Tan and Lim (2012) or can be obtained from the authors.

Research Site

The four teacher participants were from Tembusu Primary (school and teacher names are all pseudonyms). Tembusu is a co-educational public school serving 2,300 students from a multi-ethnic (Chinese, Malay, Indian and Eurasian) background.

One of the reasons we chose Tembusu was that the teachers had started inquiry-based teaching and learning. Though not pervasive, teachers were implementing inquiry lessons based on their interpretations of inquiry science from various workshops they have attended. We acknowledge that it would have been ideal to include teachers from a few schools to participate in the study. However, we decided that starting with one school was more realistic for our data collection as the case study approach demanded building strong knowledge of the teachers and the school in which they were teaching. Similar classroom-based pedagogical studies have also employed a single school as a field site (e.g., Brown & Campione, 1994). Moreover, the teachers at Tembusu were committed to the research, even though the methodology was intrusive, with both researcher and video cameras deployed in their classrooms.

Research Participants

The four collaborating teachers – Dorothy, Ming, Isa and Jan – have fairly different profiles, in terms of age, years of teaching experience and educational background (summarised in Table 8.2).

Table 8.2 Teacher collaborators and their student participants

	Dorothy	Ming	Isa	Jan
Age	58	45	39	28
Years of teaching experience	40	6	13	4
Years of experience teaching science	40	5	7	4
Educational background	University degree in English language and business administration	BSc in Chemistry	BA in English language and Economics, with no science background beyond secondary school	BA in English with Psychology, with a Diploma in Electronics Engineering
Classes involved in study	One Primary 5 (P5) and one P6 class, each with about 40 students	One P6 class of 40 students	One P3 class with 31 students	One P3 class with 42 students

Findings and Discussions

Pedagogical Framework of Inquiry: As Situated in the Singapore Classroom Context

The pedagogical framework of inquiry science observed from this ground-up study is shown in Fig. 8.1. Several patterns observed in the instructional components and sequences include:

(a) The use of fieldtrips, activities, pertinent questions or narratives of everyday encounters with science to capture the interest and curiosity of the students when introducing students to a major topic or theme
(b) The use of a *PIE* (prepare to investigate, investigate and explain) sequence that constituted the core of the inquiry process in helping students learn and develop science concepts and skills
(c) The iteration within the *PIE* sequence where teachers made decisions to clarify, modify or add investigative activities as they assessed the learning taking place during the investigation and explain phases
(d) The conclusion of a series of lessons on a major topic or theme by consolidating the concepts and ideas learned
(e) Assessment for learning took place not only at the end of an instructional sequence but was also integrated into the *PIE* and consolidation phases

Fig. 8.1 Refined pedagogical framework after second phase of study

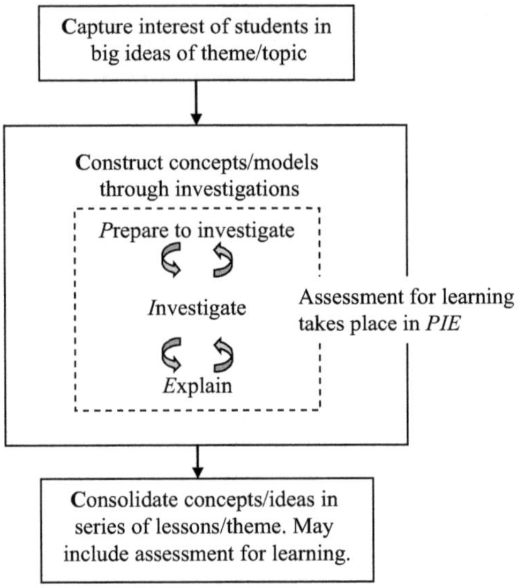

As expected, many of these components observed were also found in the pedagogical frameworks for inquiry learning found in the literature. After all, some of the existing frameworks (e.g., the MORE thinking frame, the BSCS 5E and the EIMA instructional frameworks and the guided inquiry science heuristics that were described in the literature review segment and also reflected in the coding scheme in Appendix A) were used as a lens to examine the lessons. In addition, it was also reminiscent of the fact that the teachers were in the process of adopting inquiry practices and a major departure of practice from the theoretically driven pedagogical frameworks would have raised questions about their practice. Nevertheless, there were nuances of practice characterising the teachers' practice that could be instructive for teaching and learning science by inquiry, particularly for Singapore classrooms and other classrooms with similar contexts.

One of the key motivations driving this study into the actual classroom practice of teachers here was the oft-heard comments by teachers that inquiry science, as described in literature, was impossible to implement in countries such as Singapore where there was a large class size (averaging about 40) and a strong focus on preparing students for national examinations. Teachers have often said that they valued inquiry science but could only do an ad hoc module, often as an extracurricular activity or as a post-examination activity. This study explored the possibility of carrying out inquiry-based lessons on a sustained basis in the Singapore context and found characteristics in the teachers' practice that sought not only to address local issues (such as the need to prepare for national examinations) but also reflected their tacit experience and beliefs. Some of these practices include:

(a) The teachers consistently included a 'consolidation' phase at the end of a series of *PIE* activities. Using strategies such as class tests, debates, concept maps or group projects, the consolidation phase provided opportunities for students to synthesise concepts coherently across the different units of a topic or theme (topics are integrated into major themes such as 'cycles' and 'diversity' in the Singapore national science curriculum). This practice likely reflected the teachers' concern with preparing students for examinations where students must be able to demonstrate their ability to grasp and make connections across science concepts. Nevertheless, it can be argued that instituting such a practice need not detract from inquiry practices as it could serve to help students appreciate scientific knowledge systems rather than learning science as discrete concepts or ideas (Flick, 2000; Grandy & Duschl, 2008; Krajcik, 2008).

(b) The teachers placed importance to preparing students cognitively and procedurally for an inquiry-based investigation. They prepared their students for an investigation cognitively by eliciting their prior concepts and reviewing or introducing concepts relevant to the investigation, thus scaffolding the younger learners to benefit from the investigation. Procedural preparations include teaching students skills for manipulating equipment (e.g., how to use a thermometer), skills for data collection and presentation (e.g., using a table to record and present data) and skills or attitudes for collaboration (e.g., respecting different views). While experienced teachers may regard this component as 'obvious', including 'preparation to investigate' as an explicit integral component of an inquiry, pedagogical framework would raise the awareness of the importance of this component for novice teachers and even teachers transiting into inquiry-based practices. As Dorothy later reflected, it would save many teachers from a 'disastrous' investigative activity.

(c) Another feature of the inquiry practice observed was the cyclic nature of the *PIE* components, which was not evident in the theoretical pedagogical framework found in literature. The iteration of *prepare to investigate-investigate-explain* reflected the dynamic character of teaching and learning in the classroom where helping students understand and construct concepts did not follow a planned linear path but involved continuous assessment for learning that led teachers to plan, modify or introduce activities to address learning problems that arose.

Inquiry practices often involve more student-centred activities where students interact more intensively with materials and with other students during investigations. In addition to monitoring the learning taking place, teachers in an inquiry classroom have to manage more movements of materials and equipment and the social dynamics among students. To manage the challenges of conducting inquiry-based lessons with a larger class size, the collaborating teachers in this study used several strategies, such as (a) creating shared values for learning by inquiry (e.g., students in Dorothy's class were observed to censure group members who threatened to disrupt their activities with off-task behaviours because they did not want Ms Dorothy to stop teaching science through collaborative investigative activities), (b) organising resources for investigative activities (e.g., pre-pack activity kits with all the necessary materials and equipment) and (c) using routines and signals to manage transitions between lesson phases that will gain the attention of a large group of students engaged in their activity. More details and illustrations from the teachers' practice are described in Poon, Tan and Tan (2009).

Issues in Transiting into Inquiry Practices

The next section describes some of the tensions and challenges the collaborating teachers faced and how they resolved these as they transit into inquiry-based practices.

Teacher Direction and Student Direction in Inquiry

The NRC has embraced a wide approach to inquiry, recognising that in practice, each of the five essential features of inquiry could vary from a higher degree of student direction (and a corresponding lower degree of teacher guidance) to a lower extent of student direction (NRC, 1996). In a highly student-directed inquiry environment, students could pose their own questions or problems for investigation, determine what constituted as evidence, design the experimental procedures to collect the evidence and formulate, present and defend their own explanations based on the evidence collected. At the other end of the continuum, a more teacher-guided inquiry environment sees questions being posed by the teacher or the curriculum materials, students being provided with data, and shown ways to formulate and communicate their explanations.

The classrooms of the four teachers reflected a more teacher-directed inquiry environment. Investigation questions and problems were generally posed and planned by the teachers. There were few instances where students designed their experiments or were given opportunities to pursue their own questions of inquiry. The importance the teachers attached to preparation for investigation was also an indication of the strong guidance role they assumed in the inquiry process. During the explanation phase, teachers led the interpretation of experimental data and the active construction of understanding of concepts more often than students.

It would be totally inaccurate, however, to paint the four teachers' classrooms as places where the students were passively receiving knowledge from the teachers or where a repeated pattern of IRE (initiate, respond and evaluate) governed the classroom discourse. For example, while students in Ming's class were solving the same oil spill problem, each of the eight groups designed and used different methods to solve the problem. Dorothy's P6 students designing different parachutes to solve a problem and Jan's P3 students designing experimental procedures to determine the relative strengths of two magnets were also instances where students were given space to direct some of their activities. While Dorothy fielded questions to help students connect experimental evidence to the science concepts, her students were the ones who presented their findings, suggested plausible explanations, evaluated one another's explanations and defended their ideas.

The general image that emerged from the classrooms was that of learners actively interacting with materials and one another and the teacher guiding learning through planned activities, questions, probing as well as explicit instruction. This form of guided inquiry is consistent with findings from recent research on learners and learning that demonstrated the importance and effectiveness of strategies that involved teachers in modelling, scaffolding and explicitly teaching scientific explanations, reasoning and inquiry processes (Holliday, 2004; Kirschner, Sweller & Clark, 2006; Metz, 1995). Holliday has pointed out that allowing students to make mistakes repeatedly, in the absence of explicit teaching and guidance by teachers, may lead to errant learning and entrenchment of misconceptions.

Research on children's performance of investigation tasks showed that while children could use evidence to draw conclusions, they sometimes used unreliable or invalid data and proposed relationships based on two data points (Millar, Lubben, Gott & Duggan, 1994). Metz (1995, 2004) therefore suggested that although children have the capacity to engage in developing hypotheses, designing experiments, manipulating variables, gathering and interpreting data and drawing causal relationships, teachers played an important role in scaffolding the development of these skills and abilities in primary school children. Crawford, Kelly and Brown (2000) and McNeill and Krajcik (2008) also found that teacher modelling of scientific explanations, reasoning processes and the use of evidence to support claims helped students in their own practice of generating explanations and arguments based on evidence.

It was observed that although all the four teachers took on a key role in guiding students in the inquiry process, there were differences in the extent of their guidance. The two teachers teaching upper primary students, Dorothy (P5 and P6) and Ming (P6), provided less guidance than the teachers teaching lower primary students, Jan and Isa (both P3 teachers). One indicator was the relative proportion of time in preparing students for investigation. The *prepare to investigate to investigate ratio* (P:I), obtained by comparing the proportion of time teachers spent in preparing students for an investigation with the actual amount of time spent on investigation, showed a much smaller ratio for Ming and Dorothy (see Table 8.3).

In addition, episodes in the classrooms showed that the students in Ming's and Dorothy's classes had more opportunities to direct some of the processes in their activities and in developing and communicating their explanations than students in Isa's and Jan's classes. The teachers explained that age and ability of the students were considerations that guided their decisions on the extent of teacher guidance, with younger and less-able learners requiring more direction. This is in line with Kirschner et al.'s (2006) arguments that failure to provide strong learning support for less experienced and less-able students could lead to ineffective learning. As Metz (2004) suggested, primary school students cannot be thrown into the deep end of investigation from the start but need to be scaffolded into increasing responsibility for their investigation as they mature in age and experiences with inquiry.

Table 8.3 Relative proportion of time spent on preparing to investigate and investigating

	Ming	Dorothy	Jan	Isa
Level of participating class	P6	P5	P3	P3
Prepare to investigate: investigate	1:4	1:3	1:2	1:1.6
——Increasing proportion of time spent on preparing to investigate ——►				

This is an important implication for the practice of inquiry science in primary school classrooms.

There were indications, however, that age and ability of the students might not be the only consideration in decisions on the extent of teacher guidance. Isa had talked about her lack of science content knowledge as a challenge for her to enact inquiry confidently. Her fear of not knowing how to address students' queries had discouraged her from inviting students to ask questions in the previous school she taught. A more teacher-directed classroom environment gave her greater control over the scope of activities and problems encountered in the classroom.

Jan also started the year with a highly structured learning environment. Towards the end of the academic year, however, as she experienced positive feedback on her students' responses to the inquiry approach, Jan began to move towards a less-guided approach to inquiry. In the topic of magnet taught in the third cycle of this research, Jan designed activities that gave her students the opportunity to design the experimental procedures rather than follow a fixed set of procedures.

There were indications therefore that the teachers' observed practice, in terms of the extent of teacher and student direction, might be persuaded by factors such as their level of confidence in their knowledge about science and inquiry rather than the theoretical soundness of the approach. This resonates with the findings of studies that underscore the importance of subject matter knowledge and pedagogical skills for teaching science as inquiry (Bryan & Abell, 1999; Laplante, 1997; Shulman, 1987; Smith & Neale, 1989). Subject matter knowledge is regarded as important and yet is a major area of concern for primary science teachers (Banilower, Hecks & Weiss, 2007; Schwarz, 2009; Smith & Neale, 1989). The Trends in International Mathematics and Science Study (TIMSS) 2007 International Science Report (Martin, Mullis & Foy, 2008) noted that 58 % of primary 4 students in Singapore were taught science by teachers with at least a university degree (not necessarily science-related degrees). This was lower than the international average of 70 % for countries participating in the study. About half of the students were taught science by teachers without even a postsecondary specialisation in science. Teachers were concerned that their few years of exposure to formal learning of science (learning science in formal institutions like schools and universities) would be inadequate for them to implement inquiry science effectively. We therefore repeat the call for teacher preparation programmes to focus on both content and pedagogical skills in preparing teachers to teach inquiry science effectively in the classroom.

The Pursuit of Science Concepts in the Inquiry Classroom

Across the lessons enacted by the four teachers, inquiry was used primarily as an approach to help students learn scientific concepts and inquiry skills rather than as an approach to build epistemological understanding of the nature of science (see Hodson, 1985; Lederman, 2007) and scientific thinking. The pedagogical framework that emerged from the ground-up study reflected the teachers' curricular goal and focus on concept acquisition. In the *PIE* cycle, activities were designed primarily to clarify, develop and reinforce scientific concepts and explanations. Even in the *preparing to investigate* phase, much emphasis was given to eliciting prior concepts and reviewing concepts that provided the foundation for learning new concepts. The consolidation phase was again a time when teachers guided students to make links across concepts they have learned. The discourse between teacher and students and among students also showed a strong focus on science concepts. Inquiry skills emphasised in the lessons were confined mainly to observation skills, collection of data, inferring from evidence and understanding the need for holding variables constant and repeated measurements.

Interviews with the teachers indicated that their emphasis on concept acquisition was consistent with their beliefs and interpretations of inquiry science. For Ming, inquiry science was a strategy for helping students to restructure their prior concepts. Similarly, Dorothy viewed inquiry science as an exploration by students to discover knowledge. For Isa, teaching and learning science revolved around scientific knowledge, and inadequacy of such knowledge was a predicament for her. For Jan, inquiry science was to be explored as a more effective method to help her students learn and do well in science.

The teachers' interpretation of inquiry science and their emphasis on concept development could possibly be linked to the structure of the national curriculum and examinations. The syllabus specified learning outcomes in terms of content knowledge and a set of process skills. These were also the learning outcomes assessed at the school and national levels. At this stage of adoption of inquiry practice, what was observed in the teachers' classrooms was a focus on teaching science through inquiry, with less attention paid to helping students appreciate and appropriate inquiry as a means for understanding the nature of scientific theories and in problem-solving.

The strong emphasis on concept learning is not to be trivialised as developing a deep understanding of scientific concepts is an important aspect of inquiry-based learning (Bransford et al., 2000; Metz, 2004). The efforts by Dorothy, Ming and Jan to design activities that help students develop problem-solving skills in an inquiry environment are also laudable. However, the little emphasis paid to helping students appropriate inquiry as a means for understanding the nature of scientific theories and problem-solving could have implications on the development of scientific literacy (Abd-el-Khalick et al., 2004; Lederman, 2007). If it is also largely true of other classrooms in Singapore, then there might be a problem in fulfilling the policy intent of a science education that develops flexibility of mind, innovation and creativity in problem-solving in science and beyond that of science. Indications from this study point to the need to further examine how the national curriculum, and particularly the national examinations, could influence the form inquiry takes in the classroom and, if necessary, be modified to bridge the gap identified.

Challenges and Support in Adopting Inquiry Practice

The adoption of inquiry practices is more a dynamic than dichotomised process. Teachers do not switch from one form of practice to another in a day. Teachers' understanding and adoption of inquiry-based approaches are impacted by their beliefs, their grounding in content knowledge and pedagogical knowledge and skills and quite importantly their experiences of challenges and success in enacting inquiry in the classroom.

There have been a number of studies on the challenges teachers faced, both perceived and real, that served as barriers to adopting inquiry practices (e.g., Abd-el-Khalick et al., 2004; Bybee, 1997). These studies generally provided an aggregated (in terms of statistical numbers) and segmented account of issues (as in breaking down into a list of issues) drawn from surveys or interviews of teachers. What would be informative and instructive is to know how the challenges impacted inquiry in the classroom and how teachers coped with these challenges. This case study of four teachers adopting inquiry practice provided the opportunity to look holistically at the challenges from within the classroom. More importantly, it also provided information on the classroom support that motivated the teachers to surmount some of these challenges.

Isa's story is most illustrative of the interplay between the barriers and the support structures in adopting inquiry practice. In Isa's first year of teaching science in her previous school, inquiry-based teaching was furthest from her mind as she struggled for 'survival' in the classroom. There was physical resource support in the form of two brand new science rooms that were supposed to be equipped for laboratory-based activities. Policy-wise, MOE has supported classroom teachers by provisioning for the science rooms to support inquiry-based activities. Inquiry learning, however, did not materialise in Isa's classroom. The materials and equipment for the activities that were 'supposed to have been there' could not be found. There were also no helpful colleagues to turn to for support. Isa took refuge in the prescribed textbook and workbook and shut out discussions in the classroom.

In Tembusu, several push and pull factors worked in concert to motivate Isa to attempt inquiry science in the classroom. For one, resources were in sufficient supply and accessible. According to Isa, 'everything is ready, if you don't make use, it's a waste'. In planning for several of the activities during this study, we observed that the teachers had no problem purchasing the materials required. As Ming remarked: 'This school, I think for primary school, it's very well equipped. The department is very supportive of whatever you want to buy'. The availability of resources lowered but did not remove a barrier to implementation. Besides access to materials and equipment, teachers also needed time to trial and prepare the materials for the lesson. This is a challenge for primary school teachers as they generally do not have many nonteaching periods in between lessons. The provision of resources for facilitating the teaching of inquiry science is without doubt an enabling policy. Nevertheless, at the school level, it is critical that processes be put in place to ensure access to resources, whether it is of equipment and materials or time and manpower.

Even though resource was not a major issue in Tembusu, Isa struggled with her inadequacy of knowledge of science and inquiry. According to Crawford (2007), the lack of subject matter expertise could possibly account for why novice teachers do not adopt inquiry practices. The inadequacy was felt when Isa was selecting, modifying or creating activities to match learning outcomes and inquiry goals. Inadequacy of disciplinary knowledge affected Isa's ability to probe the students' prior concepts and guide them in constructing understanding and explanation of concepts. In contrast, a stronger disciplinary knowledge (in comparison with Isa) helped sharpen Jan's ability to use more open questions, while Isa was more comfortable with convergent questions. The issue of subject matter knowledge has important implications for policymakers and teacher educators. While it may not be possible to devote a substantial amount of time to the learning of subject matter knowledge for prospective primary school teachers, ways must be found to strengthen the subject matter knowledge of teachers. In Singapore where the majority of serving primary science teachers do not have a science-related degree, teacher educators, policymakers and teachers must work together to find ways to meet the science knowledge needs of teachers.

Even though Isa was hampered in her ability to enact inquiry science because of her lack of subject matter knowledge, a suite of support prevented her from retreating away from inquiry. As head of the science department, Dorothy provided opportunities for Isa to attend professional development courses related to science and inquiry, both outside and within the school. Isa also accepted Dorothy's invitation to observe her inquiry lessons which provided useful pointers for classroom management skills, questioning skills and group-working skills integral to the day-to-day practice of inquiry. The constant encouragement by the principal and Dorothy served as impetus for Isa to attempt inquiry-based lessons. There was tolerance of noise and failures by management. Isa also had a group of colleagues who were starting their journey in adopting inquiry practices, and there were frequent sharing of incidences of success and failures that bolstered confidence to try.

Concerns over the ability to complete the syllabus, given that more time was required for inquiry-based lessons, were addressed at the departmental level. Dorothy worked with her teachers to plan the scheme of work so that the curriculum was spread out across the 4 years (P3 to P6) and the teachers were assured they have a feasible schedule to complete the syllabus. The number of pen and paper worksheets was also reduced to make way for inquiry activities. Dorothy shouldered the responsibility of this decision, assuring parents and school management that this would not negatively affect the students' performance at the national examinations.

In helping teachers embrace inquiry as an everyday practice, a large part of the 'battleground' is in the everyday issues found in the classroom. For Isa and also for Jan and Ming, the process of adopting inquiry practice did not start with a big change in belief of teaching and learning. In part, they saw it as a 'top-down' initiative by Dorothy and supported by management, albeit not as an enforced policy. The foray into inquiry began with a little push from their Head of Department, but was sustained by concrete support that addressed the real issues encountered in the classroom. As illustrated by the case stories of the teachers, it is possible that change in belief could follow after experiences of success in enacting inquiry. While there are studies that suggest that teachers' beliefs about teaching and learning are an overriding factor influencing choice and success of inquiry teaching (e.g., Crawford, 2007; Kagan, 1992), this study provides some evidence that support for teachers in addressing real issues of implementation in the classroom can serve as mitigating factors in influencing adoption of inquiry practice *in spite of* beliefs.

Implications and Conclusion

Given the small number of teachers observed in this study, there is a limitation to the statistical generalisability of the findings. However, Donmoyer (1990), Lincoln and Guba (1985) and Stake (2000) have argued that the rich, detailed, in-depth and holistic investigations in case studies enable readers to learn vicariously. This concept of knowledge transferability by vicarious learning, as an alternative lens to statistical generalisability, offers value for a wide range of stakeholders to reflect on and apply in their own practices (Stevenson, 2004). Locally, as well as internationally, there is increasing attention paid to research into the organic classroom, where policies, milieu, matter and the participants (such as teachers and students) interact in a complex way to impact practice. This case study of the four classroom teachers is therefore significant in its contribution to the intimate knowledge of inquiry science practice. In spite of the small number of cases, we hope that the descriptions and findings hold rich lessons for both policymakers, teacher educators and teacher practitioners.

Preparation for investigation, consolidation of concepts and iteration within the *PIE* processes were components and sequences that were generally not articulated in existing theoretically driven pedagogical frameworks of inquiry but emerged from observation of classroom practices in this study. The importance the teachers attached to these components in practice has implications for teaching and learning and for teacher education. Curriculum designers and teacher educators should pay attention to these practice-oriented components in designing materials and teacher education programmes for inquiry science. Characterisation of such practices can provide both pre-service and in-service teachers with concrete images of how these components are enacted in practice.

The tension between teacher direction and student direction is a typical scenario played out in classrooms transiting from traditional models of teaching to the constructivist inquiry model of teaching and learning (Bryan & Abell, 1999). Richardson (1996) suggested that teacher beliefs can be modulated by allowing teachers to trial these beliefs in the classroom and most importantly to be able to reflect on how the translation of beliefs into practice impacted student learning. Rather than prescribing a fixed position on the extent of teacher guidance, teacher education programmes could incorporate opportunities for teachers to identify their beliefs, test the beliefs in their classrooms and think about how this impacted their student learning. This can be an iterative process, with teachers modifying their beliefs with each cycle of implementation.

This study has also indicated that in striving to achieve a more widespread adoption of inquiry practices, besides working on the macro issues such as teacher training, beliefs and mindsets, curriculum and assessment, perhaps it is just as expedient to work with schools to lower the barriers faced by teachers within the realities of the classroom. We hope that this study has contributed to a better understanding of some of the classroom-based issues in transiting into inquiry practice.

Acknowledgements This study was funded by a research grant awarded to Shirley Lim (RI 6/06 SL) from the National Institute of Education, Nanyang Technological University.

Appendix A: Coding Scheme Developed to Code the Components of Inquiry

Rater ID: _____	Unit number: _____		Time unit start: _______	Time unit end: _______	
CLIS	**MORE**	**BSCS**	**EIMA**	**Heuristic for inquiry**	**Others**
Orientation	*Model*	*Engage*	*Engage*	*Engage*	
Initiate learning through a question, a problem or discrepant event	Construct model of phenomenon based on existing ideas	Elicit prior concepts and arouse curiosity through short activity/ questions	Engage students prior knowledge and interest in a topic	Engage students around a question or problem	
Elicitation	*Observe*	*Explore*	*Investigate*	*Prepare to investigate*	
Articulate or make explicit existing/prior ideas/ concept	Design and carry out investigations of phenomenon, make and record observations	Help students use prior concepts, generate new ideas, explore possibilities, design and conduct investigation through activities	Collect and analyse data to generate patterns or rules	Students plan investigations Teachers guide thinking through key issues of investigation, may introduce skills, attitudes and beliefs	
Restructuring	*Reflect*	*Explain*	*Model*	*Investigate*	
Conceptual change through clarifications, exchange of ideas or conceptual conflicts	Reflect on goals, methodology of experiment and consistency between observations and proposed model	Learners explain their understanding of concept	Use data to create models that provide coherent explanations for the observations	Students carry out activities to investigate the problem or to derive knowledge claims	
Alt ideas constructed through expt or thinking through implications		Teachers may introduce concept or skill to guide students to deeper understanding			
Application of ideas	*Explain*	*Elaborate*	*Apply*	*Prepare to report*	
Test new ideas in similar and/or novel situations	Explain to class how data served as evidence to support or refine model	Extend understanding and apply concept through additional activities	Test and apply models in novel situations	Small group prepare to report Teacher may guide students' articulation of ideas and impart skills	
Review change of ideas		*Evaluate*		*Report*	
Compare new knowledge with prior knowledge		Students and teachers assess understanding and progress towards achieving educational objectives		Students report their claims and findings to class, defend ideas Whole class (teacher and student) evaluation	

References

Abd-el-Khalick, F., Boujaoude, S., Duschl, R., Lederman, N. G., Hofstein, A., Mamlock-Naaman, R., et al. (2004). Inquiry in science education: International perspectives. *Science Education, 88*, 397–419.

American Association for the Advancement of Science. (1990). *Science for all Americans*. New York: Oxford University Press.

Anderson, O. R. (1997). A neurocognitive perspective on current learning theory and science instructional strategies. *Science Education, 81*, 67–89.

Anderson, R. D. (2002). Reforming science teaching: What research says about inquiry. *Journal of Science Teacher Education, 13*(1), 1–12.

Baker, C. (1997). Ethnomethodological studies of talk in educational settings. In B. Davies & D. Corson (Eds.), *Oral discourse and education* (pp. 43–52). Dordrecht, The Netherlands: Kluwer.

Banilower, E. R., Hecks, D. J., & Weiss, I. R. (2007). Can professional development make the vision of the standards a reality? The impact of the National Science Foundation's local systemic change through teacher enhancement initiative. *Journal of Research in Science Teaching, 44*, 375–395.

Barrow, L. H. (2006). A brief history of inquiry: From Dewey to standards. *Journal of Science Teacher Education, 17*, 265–278.

Berg, C. A. R., Bergendahl, V. C. B., & Lundberg, B. K. S. (2003). Benefiting from an open-ended experiment? A comparison of attitudes to, and outcomes of, an expository versus an open-inquiry version of the same experiment. *International Journal of Science Education, 24*, 351–372.

Bransford, J. D., Brown, A. L., & Cocking, R. R. (Eds.). (2000). *How people learn: Brain, mind, experience, and school* (Expanded ed.). Washington, DC: National Academy Press.

Brown, A. L., & Campione, J. C. (1994). Guided discovery in a community of learners. In K. McGilly (Ed.), *Classroom lessons: Integrating cognitive theory and classroom practice* (pp. 229–270). Cambridge, MA: MIT Press.

Bruner, J. S. (1960). *The process of education*. Cambridge, MA: Harvard University Press.

Bryan, L. A. (2003). Nestedness of beliefs: Examining a prospective elementary teacher's belief system about science teaching and learning. *Journal of Research in Science Teaching, 40*, 835–868.

Bryan, L. A., & Abell, S. (1999). Development of professional knowledge in learning to teach elementary science. *Journal of Research in Science Teaching, 36*, 121–139.

Bybee, R. W. (1997). *Achieving scientific literacy: From purposes to practices*. Portsmouth, NH: Heinemann.

Bybee, R. W. (2004). Scientific inquiry and science teaching. In L. B. Flick & N. G. Lederman (Eds.), *Scientific inquiry and nature of science: Implications for teaching, learning and teacher education* (pp. 1–14). Dordrecht, The Netherlands: Kluwer.

Bybee, R. W., Powell, J. C., & Trowbridge, L. W. (2008). *Teaching secondary school science* (9th ed.). Upper Saddle River, NJ: Pearson Prentice Hall.

Bybee, R. W., Taylor, J. A., Gardner, A., Van Scotter, P., Powell, J. C., Westbrook, A., et al. (2006). *The BSCS 5E instructional model: Origins and effectiveness*. Colorado Springs, CO: BSCS.

Carey, S. (1979). Cognitive competence. In A. Floyd (Ed.), *Cognitive development in the school years* (pp. 45–66). London: The Open University Press.

Cianciolo, J., Flory, L., & Atwell, J. (2006). Evaluating the use of inquiry-based activities: Do student and teacher behaviors really change? *Journal of College Science Teaching, 36*(3), 50–55.

Crawford, B. A. (2007). Learning to teach science as inquiry in the rough and tumble of practice. *Journal of Research in Science Teaching, 44*, 613–642.

Crawford, T., Kelly, G. J., & Brown, C. (2000). Ways of knowing beyond facts and laws of science: An ethnographic investigation of student engagement in scientific practices. *Journal of Research in Science Teaching, 37*, 237–258.

Curriculum Planning and Development Division. (2007). *Science syllabus: Primary, 2008*. Singapore, Singapore: Curriculum Planning and Development Division, Ministry of Education.

DeBoer, G. E. (1991). *A history of ideas in science education*. New York: Teachers College Press.

DeBoer, G. E. (2004). Historical perspectives on inquiry teaching in schools. In L. B. Flick & N. G. Lederman (Eds.), *Scientific inquiry and nature of science: Implications for teaching, learning and teacher education* (pp. 17–35). Dordrecht, The Netherlands: Kluwer.

Donmoyer, R. (1990). Generalizability and the single-case study. In E. W. Eisner & A. Peshkin (Eds.), *Qualitative inquiry in education: A continuing debate* (pp. 175–200). New York: Teachers College Press.

Driver, R., Newton, P., & Osborne, J. (1998). Establishing the norms of scientific argumentation in classrooms. In B. J. Fraser & K. G. Tobin (Eds.), *International handbook of science education* (pp. 287–312). Dordrecht, The Netherlands: Kluwer.

Driver, R., & Oldham, V. (1986). A constructivist approach to curriculum development in science. *Studies in Science Education, 13*, 105–122.

Eick, C., Meadows, L., & Balkcom, R. (2005). Breaking into inquiry. *The Science Teacher, 72*(7), 49–53.

Evers, C. W., & Wu, E. H. (2006). On generalising from single case studies: Epistemological reflections. *Journal of Philosophy of Education, 40*(4), 511–526.

Flick, L. B. (2000). Cognitive scaffolding that fosters scientific inquiry in middle level science. *Journal of Science Teacher Education, 11*, 109–129.

Goldston, D. (2005). Elementary science: Left behind? *Journal of Science Teacher Education, 16*, 185–187.

Grandy, R. E., & Duschl, R. A. (2008). Consensus: Expanding the scientific method and school science. In R. A. Duschl & R. E. Grandy (Eds.), *Teaching scientific inquiry: Recommendations for research and implementation* (pp. 304–325). Rotterdam, The Netherlands: Sense Publishers.

Harlen, W., & Osborne, R. (1985). A model for learning and teaching applied to primary science. *Journal of Curriculum Studies, 17*, 133–146.

Hodson, D. (1985). Philosophy of science, science and science education. *Studies in Science Education, 12*, 25–57.

Hogan, D. J., Luke, A., Kramer-Dahl, A., Lau, S., Liau, A., Kang, T., et al. (2006). *Core research program: Year two progress report* (Unpublished CRPP Technical Report). Singapore, Singapore: National Institute of Education, Nanyang Technological University.

Holliday, W. G. (2004). A balanced approach to science inquiry teaching. In L. B. Flick & N. G. Lederman (Eds.), *Scientific inquiry and nature of science: Implications for teaching, learning and teacher education* (pp. 201–217). Dordrecht, The Netherlands/Boston, MA: Kluwer.

Kagan, D. M. (1992). Implications of research on teacher belief. *Educational Psychologist, 27*, 65–90.

Keys, C. W., & Bryan, L. A. (2001). Co-constructing inquiry-based science with teachers: Essential research for lasting reform. *Journal of Research in Science Teaching, 38*, 631–645.

Kim, M. C., Hannafin, M. J., & Bryan, L. A. (2007). Technology-enhanced inquiry tools in science education: An emerging pedagogical framework for classroom practice. *Science Education, 91*, 1010–1030.

Kirschner, P. A., Sweller, J., & Clark, R. E. (2006). Why minimal guidance during instruction does not work: An analysis of the failure of constructivist, discovery, problem-based, experiential, and inquiry-based teaching. *Educational Psychologist, 41*, 75–86.

Krajcik, J. (2008). Commentary on Chinn's and Samarapungavan's paper. In R. A. Duschl & R. E. Grandy (Eds.), *Teaching scientific inquiry: Recommendations for research and implementation* (pp. 226–232). Rotterdam, The Netherlands: Sense Publishers.

Kuhn, T. S. (1970). *The structure of scientific revolutions* (2nd ed.). Chicago: The University of Chicago Press.
Laplante, B. (1997). Teachers' beliefs and instructional strategies in science: Pushing analysis further. *Science Education, 81*, 277–294.
Lederman, N. G. (2004). Syntax of nature of science within inquiry and science instruction. In L. B. Flick & N. G. Lederman (Eds.), *Scientific inquiry and nature of science: Implications for teaching, learning and teacher education* (pp. 301–317). Dordrecht, The Netherlands: Kluwer.
Lederman, N. G. (2007). Nature of science: Past, present, and future. In S. K. Abell & N. G. Lederman (Eds.), *Handbook of research on science education* (pp. 831–879). Mahwah, NJ: Lawrence Erlbaum Associates.
Lincoln, Y. S., & Guba, E. G. (1985). *Naturalistic inquiry*. Newbury Park, CA: Sage Publications.
Magnusson, S. J., Palincsar, A. M., & Templin, M. (2004). Community, culture, and conversation in inquiry-based science instruction. In L. B. Flick & N. G. Lederman (Eds.), *Scientific inquiry and nature of science: Implications for teaching, learning and teacher education* (pp. 131–155). Dordrecht, The Netherlands: Kluwer.
Martin, M. O., Mullis, I. V. S., & Foy, P. (2008). *TIMSS 2007 international science report*. Boston, MA: TIMSS & PIRLS International Study Center.
McNeill, K. L., & Krajcik, J. (2008). Scientific explanations: Characterizing and evaluating the effects of teachers' instructional practices on student learning. *Journal of Research in Science Teaching, 45*, 53–78.
Mellado, V. (1998). The classroom practice of preservice teachers and their conceptions of teaching and learning science. *Science Education, 82*, 197–214.
Metz, K. E. (1995). Reassessment of developmental constraints on children's science instruction. *Review of Educational Research, 65*, 93–127.
Metz, K. E. (2004). Children's understanding of scientific inquiry: Their conceptualization of uncertainty in investigations of their own design. *Cognition and Instruction, 22*, 219–290.
Millar, R., Lubben, F., Gott, R., & Duggan, S. (1994). Investigating in the school science laboratory: Conceptual and procedural knowledge and their influence on performance. *Research Papers in Education, 9*, 207–248.
National Research Council. (1996). *National science education standards*. Washington, DC: National Academy Press.
National Research Council. (2000). *Inquiry and the national science education standards*. Washington, DC: National Academy Press.
Ornstein, A. (2006). The frequency of hands-on experimentation and student attitudes toward science: A statistically significant relation. *Journal of Science Education and Technology, 15*, 285–297.
Pajares, M. F. (1992). Teachers' beliefs and educational research: Cleaning up a messy construct. *Review of Educational Research, 62*, 307–332.
Pine, J., Aschbacher, P., Roth, E., Jones, M., McPhee, C., Martin, C., et al. (2006). Fifth graders' science inquiry abilities: A comparative study of students in hands-on and textbook curricula. *Journal of Research in Science Teaching, 43*, 467–484.
Poon, C.-L., Lee, Y.-J., Tan, A.-L., & Lim, S. S. L. (2012). Knowing inquiry as practice and theory: Developing a pedagogical framework with elementary school teachers. *Research in Science Education, 42*, 303–327.
Poon, C.-L., Tan, D., & Tan, A.-L. (2009). Classroom management and inquiry-based learning: Finding the balance. *Science Scope, 32*, 18–21.
Popper, K. R. (1985). The beginnings of rationalism. In D. Miller (Ed.), *Popper selections* (pp. 25–32). Princeton, NJ: Princeton University Press.
Richardson, V. (1996). The role of attitudes and beliefs in learning to teach. In J. Sikula (Ed.), *Handbook of research in teacher education* (2nd ed., pp. 102–119). New York: Macmillan Library Reference.
Schwab, J. J. (1962). The teaching of science as enquiry. In J. J. Schwab & P. F. Brandwein (Eds.), *The teaching of science* (pp. 1–103). Cambridge, MA: Harvard University Press.

Schwarz, C. (2009). Developing preservice elementary teachers' knowledge and practices through modeling-centered scientific inquiry. *Science Education, 93*, 720–744.
Schwarz, C. V., & Gwekwerere, Y. N. (2007). Using a guided inquiry and modelling instructional framework (EIMA) to support preservice K-8 science teaching. *Science Education, 91*, 158–186.
Shulman, L. S. (1987). Knowledge and teaching: Foundations of the new reform. *Harvard Educational Review, 57*, 1–22.
Shulman, L. S., & Tamir, P. (1973). Research on teaching in the natural sciences. In R. M. W. Travers (Ed.), *Second handbook of research on teaching* (pp. 1098–1148). Chicago: Rand McNally & Company.
Shymansky, J. A., Hedges, L. V., & Woodworth, G. (1990). A reassessment of the effects of inquiry-based science curricula of the 60's on student performance. *Journal of Research in Science Teaching, 27*, 127–144.
Smith, C. S., Maclin, D., Houghton, C., & Hennessey, M. G. (2000). Sixth-grade students' epistemologies of science: The impact of school science experiences on epistemological development. *Cognition and Instruction, 18*, 349–422.
Smith, D. C., & Neale, D. C. (1989). The construction of subject matter knowledge in primary science teaching. *Teaching and Teacher Education, 5*, 1–20.
Stake, R. E. (2000). Case studies. In N. K. Denzin & Y. S. Lincoln (Eds.), *Handbook of qualitative research* (2nd ed., pp. 435–454). Thousand Oaks, CA: Sage Publications.
Stevenson, R. B. (2004). Constructing knowledge of educational practices from case studies. *Environmental Education Research, 10*, 39–51.
Tien, L. T., Ricky, D., & Stacy, A. M. (1999). The MORE thinking frame: Guiding students' thinking in the laboratory. *Journal of College Science Teaching, 28*, 318–324.
Tobin, K. (2006). Analyses of current trends and practices in science education. In K. Tobin (Ed.), *Teaching and learning science: A handbook* (Vol. 1, pp. 3–16). Westport, CT: Praeger Publishers.
von Secker, C. E. (2002). Effects of inquiry-based teacher practices on science excellence and equity. *The Journal of Educational Research, 95*, 151–160.
von Secker, C. E., & Lissitz, R. W. (1999). Estimating the impact of instructional practices on student achievement in science. *Journal of Research in Science Teaching, 36*, 1110–1126.
Wellman, H. M., & Gelman, S. A. (1992). Cognitive development: Foundational theories of core domains. *Annual Review of Psychology, 43*, 337–375.
Wise, K. C., & Okey, J. R. (1983). A meta-analysis of the effects of various science teaching strategies on achievement. *Journal of Research in Science Education, 20*, 419–435.

Chapter 9
Science Education in a Straightjacket: The Interplay of People, Policies, and Place in an East Asian Developmental State

Yew-Jin Lee

Abstract It is widely agreed that approaches based on inquiry or process skills are desirable in science instruction, but past efforts at their implementation in classrooms have often been problematic. This is because implementation processes confront many challenges and usually map imperfectly from developers' intentions to actual practice. Using data from a recent school-based science curriculum development in Singapore, I show how this programme for Grade 7 and 8 students was ultimately compromised in its overall aims to teach content and higher-order process skills. Despite the explicit provision of much hands-on experiences here, a high degree of epistemic agency could not be ascertained although students did indicate that they found aspects of it enjoyable and relevant to daily life. I argue that this situation was partly due to internal shortcomings and practical difficulties within the programme as well as tensions with other educational policies and sociocultural norms in the country. In addition, there were interacting levels that influenced—both positively and negatively—the implementation process and thus eventual programme success. This study has important implications for science educators who prioritise the planned or intrinsic aspects of science curricula (e.g., ensuring an appropriate amount of inquiry activities) but neglect programme implementation and understanding educational change as inevitable complex ecologies of people, policies, and place. Finally, I confirm the presence of educational change problems in an internationally recognised high-achieving system that are not inherently different from those in other locations or subject areas.

Keywords Curriculum reform • Curriculum implementation • Curriculum history • Science inquiry • Process skills • Educational policy • Educational change • Cultural-historical activity theory • East-Asian development state • Singapore

Y.-J. Lee (✉)
National Institute of Education, Nanyang Technological University, Singapore, Singapore
e-mail: yewjin.lee@nie.edu.sg

A.-L. Tan et al. (eds.), *Inquiry into the Singapore Science Classroom*, Education Innovation Series, DOI 10.1007/978-981-4585-78-1_9,

Introduction

The inquiry approach in science education boasts of a long history and a pervasive influence in school curricula worldwide. In the UK, its beginnings can be traced to efforts in the mid-nineteenth century to incorporate laboratory work into classrooms that later culminated in large-scale science reforms initiated by the Schools Council and Nuffield Foundation a century later. Across the Atlantic, the 1960s saw the vigorous promotion of discovery learning and associated inquiry-oriented experiences from distinguished educators such as Jerome Bruner and Joseph Schwab (Abd-El-Khalick et al., 2004). Despite the brief lifespans of most of these curricula promoting inquiry and/or process skills (c.f. the "alphabet soup" programmes), it is not too off the mark to state that nearly all present-day school science teaching owes something to these ambitious reform movements (Rudolph, 2002). Practising inquiry and process skills (which I deem to be equivalent here although see Atkin & Black, 2003) has become a notable intellectual hallmark, separating and indeed elevating the subject of science from nearly all other ways of knowing within school settings. Without exaggeration, inquiry is *the* battle cry that has dominated modern science education and is virtually synonymous with what is considered as good science instruction (Anderson, 2002).

While policymakers, academics, and science teachers regard inquiry or process skills approaches as highly desirable, their sustained implementation has been problematic, an outcome that has surfaced repeatedly across time and place (e.g., Welch, 1981). Any person remotely familiar with schooling as an institution can list many possibilities that can mitigate against the use of such cognitively demanding pedagogies. These factors include low comfort levels with inquiry teaching, textbooks, and laboratory materials that emphasise content coverage or surface understanding, various assessment challenges, and the absence of quality student collaboration among others. The failure of large-scale inquiry curricula in schools can also persist despite ongoing training at the pre-service and professional development levels (Harlen, 2004). From the perspective of some science educators, these problems in the enacted curriculum revolve around four cultural myths regarding how and what kinds of knowledge are to be transmitted, being efficient as teachers, maintaining rigour in the curriculum, and preparing students for examinations (Tobin & McRobbie, 1996). Any curriculum that attempts to impart advanced disciplinary thinking will certainly experience a lag implementation period whereby teacher resistance has to be overcome, student attitudes changed, resource provisioning and physical infrastructures built up, as well as various sociotechnical obstacles removed (Rogan, 2007). Yet one wonders why given the length of time that inquiry has been within school curricula and the sheer amount of knowledge gained from adopting this approach that educators still face enormous challenges when teaching inquiry.

One likely explanation is due to the complexity and difficulty of teaching inquiry. As an activity involving critical reasoning in science (not merely hands-on activities), it is a contingent phenomenon that does not casually follow from the mere presence of a good lesson plan (Chue & Lee, 2013). The intentions of curriculum developers or teachers, as is increasingly conceded, often does not map well onto actual classroom practices. Planning for rigour is a necessary first step, but it is never a sufficient condition by itself. Attention has to be paid to another aspect of curriculum development that has been less emphasised—programme/curriculum implementation (Lee & Chue, 2013). Especially critical for new interventions (including inquiry teaching that has been more an intent rather than a concrete reality in classrooms), testing a programme under ideal conditions has been called efficacy studies, and its adoption under real-world contexts where issues of scaling-up are at stake is known as effectiveness studies. In either case, monitoring implementation issues with diligence is the precursor for delivering evidence-based decisions about the intrinsic worth of a programme, its operational levers and specifications, and whether it should be disseminated widely. This study that critically interrogates the implementation of inquiry in one school therefore lends weight to answering these important questions regarding the value of new curricular innovations.

Specifically, I foreground the interplay of people, policies, and place in the East Asian developmental state of Singapore that both facilitated and constrained an inquiry science programme from achieving its stated aims. Such states, according to political scientists, regard education/training as the vehicle *par excellence* to serve national economic needs through its provision of skilled workers carefully slotted into predetermined labour sectors (e.g., Pereira, 2008). Planned with the best of intentions by teachers and enjoying unbridled support from the government, I discuss how internal programme shortcomings, practical difficulties of implementation, and tensions with other existing educational policies and sociocultural norms in the country had inadvertently compromised the learning of content and higher-order process skills and inquiry science. Interacting levels in the system also influenced—both positively and negatively—implementation processes and thus eventual programme success. The implications are nonetheless unambiguous and speak to all educators: Studying implementation is not secondary to studying curriculum development. Instead, it is an obligatory feature for illuminating which aspects of and why certain curricula can succeed and why others are destined to mediocrity or failure in schools. In this chapter, it is an express purpose to better understand what happened during inquiry science implementation, specifically in one recent school-based curriculum that was nationally lauded as being highly innovative in Singapore. Known as *SciencePlus*, it focused on raising the overall relevance and knowledge of science through the hands-on application of process skills for Grade 7–8 pupils in Chester Secondary (pseudonyms used throughout). What would deep knowledge about one local curricular innovation offer to (science) educators elsewhere?

It might come as a surprise to learn that not only are the vagaries of classroom inquiry teaching not well documented (Crawford, 2000; McNeill & Krajcik, 2008) but that research on the enacted science curriculum in Singapore is equally scarce (Lee, 2008; Poon, Lee, Tan, & Lim, 2012; Tan & Wong, 2012; Yeo, Tan, & Lee, 2012). This black box exists despite student achievement levels in the subject ranking among the best in the world or that state science curricula here have been long praised for their coherency and conceptual focus (Ministry of Education [MOE], 2012). This study therefore yields fresh insights into the underbelly of a seemingly successful system and is essential if one regards learning from the past as integral for school transformation. Findings here will moreover inform programme/ policy implementation and educational change writ large because there is little reason to suspect that knowing what happened in *SciencePlus* cannot be of benefit to those in other subject areas or educational systems. Educational problems, when viewed beyond its outer or superficial veneer, often bear striking resemblances to each other regardless of their contexts although I do not claim that their solutions should be identical. In short, *SciencePlus* is offered as a condensation symbol—a dense, richly layered object that is instructive (e.g., on reform, school improvement, implementation dilemmas) as well as evoking multiple meanings for stakeholders. The next sections detail the theoretical frameworks that are informed by cultural-historical activity theory and the reasons for adopting a narrative mode of representation. After descriptions of Chester, the *SciencePlus* programme, and about educational reforms in Singapore, this story of curriculum implementation interweaves germane science education concepts with empirical data. I conclude with implications for researchers who prioritise the planned over the enacted curriculum and those that fail to appreciate the gravity of the so-called external or sociocultural factors that penetrate into every programme in school.

Theoretical Framework

It is my belief that popular technical-rational models describing process-product relationships in education are patently weak if not wrong altogether. No longer can we assume that we have the ability to completely identify and rectify where the problem(s) of educational change are, no longer can we claim full knowledge from some lofty vantage point (Hargreaves, 2005). In place of analytic certainty, I acknowledge what some are already calling a mangle of practice, an intertwining of various human and non-human actors in social phenomenon that can only be untangled with great difficulty (Pickering, 1995). These sociomaterial concepts that resonate with complexity, open-endedness, and non-linearity are the price that we have to pay for a truer reflection of reality and a hope for more phronetic ways of acting in the world. For my theoretical lenses, I therefore rely on cultural-historical activity theory, which is an underutilised but powerful methodology to examine educational change (see Lee, 2011; Roth & Lee, 2007; Roth, Lee, & Boyer, 2008).

Activity systems such as schooling and the schools that constitute schooling have to be properly analysed in a holistic and dynamic manner; subjects at all levels through their everyday interactions are contributing towards making policy or curricula. There is no "time-out" whereby subjects can escape how they interpret, transform, and adapt—actively and passively—teaching and learning. This is why the terms policy-in-action and curriculum-in-action more appropriately capture how these are verbs rather than mere nouns. People do not simply take orders from the top; instead, they constitute policy and plans through their mundane interactions and pursuit of meaning-making (i.e., pursuit of the object of activity). Sometimes, their shaping actions favour or enhance original policy intent, whereas on other occasions this is frustrated (Pressman & Wildavsky, 1973). That is, there might be overlapping or different objects of activity that motivate people in certain projects. The direction of change can move in the opposite direction too; policy can be altered by implementers. Within education, one key subject in the activity system is the teacher, whom Lipsky (1982) characterised as a street-level bureaucrat—these are the people on the ground that *make* policy and whose collective actions add up to agency behaviour.

According to activity theory, it is difficult to pinpoint which factor or moment in an activity system was the most consequential because they all act in concert, as an ecology. However, actors will have differential access to power and thus differing exercising of agency. And in common with studies of a historical nature, it is almost impossible to identify with precision which of the episodes in the data were the most salient or, for that matter, being able to uncover the full extent of the local contingencies involved. Given these formidable difficulties regarding causation, we therefore have to lean towards "parsimony in consequences, but not causes" (Gaddis, 2002, p. 105) together with a conscious awareness of the interdependency of such events in our theorising. Apart from surfacing the contradictions that present obstacles for change, activity theory can help us make visible normally invisible structures, processes, relations, and configurations in educational change. It is for these advantages that activity theory as a dialectical theory to describe, explain, and perhaps effect change was the framework of choice here.

Analyses and Representation

A narrative approach informed by historiographic principles appeared to be the best way of representation, organising reality and giving meaning to events in this account of curriculum implementation (Munslow, 2007). The story space has been informed by the author's synoptic interpretation of past temporal events alongside my declared (and hidden) biases and the inevitable viewpoints that readers bring to the text.[1] Narrative modes are thoroughly reconstructionist in nature (unlike a constructionist stance which aims to be science-like with covering law explanations). One can get closer to or further from the phenomena, but it must be acknowledged that in the end all accounts or meanings including the very data

[1] As a former high school science teacher in Singapore now turned teacher educator, I have come to cherish gaining scientific literacy over amassing content knowledge for an active citizenry. Over the years, I too have come to be wary of cosmopolitanism and the spread of neoliberalism that have influenced state policies.

sources themselves are constructions of language and simplifications of reality (Milne, 1998; Tosh, 2010).

Writing about the past in this way even though the events barely happened a decade ago raises a number of epistemological questions (e.g., the status of historical "facts" and their meaning, selection of sources, tension between objectivity and subjectivity) that educational historians and historians of science education have taken for granted as part of their craft. For example, in an otherwise thoroughly researched and convincing account of why scientists became involved in curriculum development in post-war America by John Rudolph (2002), there was no mention of the reasons or trade-offs of adopting a narrative approach. However, I have taken a step back in order to make explicit how and why narratives are sociocultural constructions that undergird *all* human meaning-making, including that within science education, which is understood by many to be a highly objective field of study. Lest narrative modes are accused of dubious "truthfulness" or even irresponsible fictive plots that entertain without increasing analytic knowledge (Megill, 1989), we do well to remember that:

> narrative itself is a kind of explanation, and. . .narrative is the only form of explanation suitable for history. Narrative makes no assumptions about laws of human behavior or essential human nature. It does not rely on generalization. Rather, it highlights human individuality and allows for free will. Historical events are explained simply in virtue of fitting into a narrative. They are explained when they are situated in their context. Description in narrative context is explanation, and in this way, there is no difference between description and explanation. . ..We see what it means when we see that it fits. (Kosso, 2011, p. 24)

Site and Data

Chester Secondary in the western part of Singapore was what was known as a "neighbourhood school" that was distinguished by her students with average academic scores. Located in a working-class district, this state-run co-educational school attracted students from grades 7 to 10 from the vicinity over the past 30 years. By 2007 however, students here were reportedly more motivated and hardworking compared to other nearby schools. From the middle of 2007 to the end of 2009, we wrote field notes, video-recorded over 60 h of classroom teaching, conducted interviews with teachers and students, and analysed various documents such as school curricular material,[2] student artefacts, government policies, and official speeches.

---

[2] Disclosure of the exact references from public documents would reveal the identity of the school and hence this has been avoided. I recognise that this concealment goes contrary to historical writing norms but protecting anonymity has taken precedence here. This ethical compromise is akin to readers placing faith in the veracity of ethnographic accounts where independent verification of data by readers is impossible because field notes are never open for public inspection (Lee

The Dynamics of Curricular Reforms: *SciencePlus* in Singapore

In this section, I show how the *SciencePlus* programme was compromised in its aims of teaching content and higher-order process skills in the discipline. There were a variety of factors both within and outside of the control of its teacher-developers that led to this outcome, which I claim placed science inquiry within a straightjacket. Rather than offering students a radically new way of experiencing science in Singapore, it largely failed in achieving these curricular aims although less so with respect to increasing knowledge about the relevance of science to everyday life. Before this story can be continued, one has to situate this innovation among the slew of major educational reforms on the landscape over the last decade.

Teach Less, Learn More (TLLM) was a major reform movement[3] that was launched in 2004 by the Prime Minister, an unusual move for the top politician in the land, but it signalled its singular importance in the minds of policymakers. *Teach Less, Learn More* represented a sea change in how education was to be conducted: Holistic and engaged learning was to replace a crowded curriculum that emphasised rote learning and memorisation. In place of stressful school assessment techniques and rigid teacher-centred pedagogies, more collaborative and frequent student projects, open-book tests, and higher-order thinking skills were to be infused into every subject. For example, the (Interdisciplinary) Project Work programme valued teaching the transfer and application of learning within authentic problem spaces.[4] Together with the parallel *Innovation and Entrepreneurship* initiatives that sought to cultivate risk-taking and creativity, young people were completely immersed in a discourse that sought to prepare citizens and workers for life in the global village.[5]

Against this frenzy of reforms and new curricula, science teachers in Chester in 2005 felt that the mandatory Project Work programme had failed, in practical terms, to develop sufficient depth of content knowledge and investigative skills among learners. On one hand, Project Work demanded some form of

& Roth, 2006). I would nonetheless refer to these school documents in the most general descriptive terms where possible.

[3] Earlier in 1997, the *Thinking Schools, Learning Nation* policy had already laid the foundation for the shift away from exam scores towards cultivating a genuine love for learning (see MOE, 2007).

[4] Interdisciplinary Project Work that was later named Project Work was part of Ability-Driven Education initiatives introduced at the turn of the millennium to shift attention away from merely doing well on paper-and-pencil examinations towards more holistic forms of learning and a wider definition of talent (Yeong, 2005). Besides requiring groups of students to collaboratively synthesise findings from different subjects/disciplines (see MOE, 2013), assessment criteria here focused on the processes and products of inquiry that made this a source of confusion, anxiety, and debate not only among teachers but also with parents especially with regard to the high-stakes preuniversity examinations. As will be explained later, *SciencePlus* as the offshoot of Project Work inherited many of the implementation issues and tensions that plagued its parent.

[5] Chan (2011) is helpful for understanding these multiple and overlapping policies.

interdisciplinary investigation done with rigour and communicated with persuasive evidence, a tall order for school children in Singapore who were unfamiliar with critiquing authoritative sources or dealing with academic controversies.[6] In contrast, due to the nature of weekly lessons that interrupted the flow of sustained inquiry, the curricular objectives in Project Work were difficult to fulfil. As such, the origin of *SciencePlus* in 2006 was none other than a pragmatic decision to merge allocated curriculum time for Project Work with the pre-existing emphasis on inquiry from the lower secondary science curriculum.[7] With this new rearrangement of the school timetable that was described by the school as "curriculum compacting", it meant that all students in the cohort were now able to meaningfully pursue a particular science inquiry project over 10 weeks besides having opportunities for interdisciplinary learning in an authentic discovery context.

With the logistical constraints that jointly faced Project Work and the lower secondary science syllabus overcome in one stroke,[8] the teachers were then eager to "teach science as it was meant to be taught" as they informed us. *SciencePlus* was accordingly designed to make learning more interesting, enjoyable, and relevant than what the pre-existing national science syllabus afforded. Framed within what is popularly known as guided inquiry in science,[9] *SciencePlus* consisted of four student-chosen modules[10] (each helmed by a single teacher with relevant expertise)

[6] Science teachers around the island were often asked to supervise these projects because their discipline above all other school subjects best fitted the exacting aims of Project Work so it was claimed. For example, knowledge about making clear arguments, reasoning with data, and dealing with variables were skill sets that found parallels in science, and thus, it became the disciplinary anchor for conducting Project Work activities at the secondary level.

[7] A national curriculum that was spread over Grades 7–8, schools had the freedom to tweak it to suit their needs. It was meant to be a continuation of the primary science syllabus and was usually distinguished by a small group/individual activity at the end of Grade 8 that consisted of either an investigative, review or synthesis, or model-building project. These choices had narrowed down to carrying out student-chosen mini-investigations a decade later at the time of *SciencePlus*. Again, due to the lack of sufficient curriculum time or lack of knowledge and resources by teachers, this was often undertaken in a shallow manner (in the author's experience) by students with mundane research questions (e.g., "Which paper towel is most absorbent/best?" "How to make soap/paper?" "Which pH is best to grow plants?"). In addition, all schools were encouraged to conduct practical tests to develop a variety of science process skills (MOE, 2004).

[8] Curriculum time in *SciencePlus* during the elective modules amounted to 270 min per week that was 2 h longer than the prescribed allocation for lower secondary science.

[9] In guided inquiry, learners pursue given research questions or at least make choices within a bounded space of inquiry without knowing the preferred investigative methods and solutions. During open inquiry, however, the questions, methods of investigation, and the answers are all unknown "at the beginning."

[10] These included the chemistry of scents, ecology/conservation, biotechnology/microbiology, and rocketry (i.e., projectile science). There was a major and nationwide push for learning the life sciences beginning from 2001 that lasted a few years. The lower secondary science curriculum in Chester was in fact aligned (i.e., had increased emphases) with the former from 2003 to 2005.

that raised students' appreciation of science process skills[11] through student-centred teaching and hands-on approaches including field trips to factories and nature parks. Learning scientific concepts no longer depended upon frontal teaching but through meaningful and practical engagement with contemporary concerns in science. This was believed to be motivating for students thereby encouraging fuller forms of participation and room for creativity and discovery. By balancing content learning and process skills, this was said to be able to generate high levels of intellectual satisfaction among learners too. Moreover, it was hoped that *SciencePlus* would facilitate the higher-order inquiry skills such as the interpretation and evaluation of data, critical reasoning, problem-solving, and so forth that are captured under the shorthand notation of scientific literacy.[12] It was clear that right from the start, there were high expectations arising from the *SciencePlus* programme from its stakeholders, not least because it inherited all the intellectual demands made on students from both Project Work and inquiry science. There was a final lynchpin decision made that aligned *SciencePlus* very closely with the spirit of the new TLLM reforms: No longer was there any need to assess students based on written exams. In its place, oral presentations, reflection journals/reports, self- and peer evaluations, and making a product or prototype[13] were deemed sufficient for science teachers in Chester to evaluate student progress at the end of Grade 8. For an East Asian state, these modes of assessment were major changes that went against what society had traditionally used as its chief mechanism for determining the objective worth of an individual.

A high point soon came when *SciencePlus* was recognised as a promising TLLM curriculum at an event that was attended by more than a thousand local school leaders and educators. During this high-visibility and high-profile seminar in February 2007 organised by the MOE, the school was given an award (only one of three) for the quality of its research design and reporting about *SciencePlus*.[14]

[11] There was also a conscious attempt to align the process skills with the newly introduced Science Practical Assessment (SPA) requirements (e.g., revising experimental design, testing variables) that was being tested at Grades 9–10. In the words of the then science Head of Department that vividly demonstrated their obsession with maximising resources, "being in the school we know that we have limited time. I mean, teachers are all like octopus. . .one spoon feed many many birds, that's the best for us. . .That's the philosophy" (Interview with Carol, 18 July 2008). Another senior teacher involved in *SciencePlus* concurred that they always sought the highest rates of returns for their teaching efforts.

[12] It was envisaged that some character values that the school emphasised could also be developed such as adaptability, self-discipline, and perseverance. Again, these found no conflict with what is known as the scientific habits of mind. See also footnote 18.

[13] Product development was discontinued in 2007 (except for those in the rocketry module) as it was felt too demanding for students. Note that paper-and-pencil tests were not abolished completely as the former comprised 60 % of the overall science score for students, that is, *SciencePlus* accounted for the remaining 40 %. These tests were also the chief mode of assessment in Grade 7 science.

[14] Benefits of being a "TLLM prototype school" such as Chester included monetary (up to about S $15,000), administrative, and various pedagogical supports (e.g., workshops). Local capacity building also came in the form of a member of school staff specially trained for action

Table 9.1 Students self-reported changes in their perceived levels of skills competencies before and after participation in the 10 week SciencePlus programme in 2006

	Mean value (scale 1–4)	
Process skill	Pre-course	Post-course
1. Elaborating (Research)	3.1	3.2
2. Conducting scientific investigations (Planning investigations)	2.4	2.6
3. Conducting scientific investigations (Using scientific apparatus)	2.5	3.0
4. Conducting scientific investigations (Analysing data)	2.6	3.0
5. Communicating (Writing scientific reports)	2.7	2.7
6. Reflecting	3.1	3.1
7. Questioning (Learning by asking questions)	2.8	3.2

Note: This was adapted from the 2006 action research report from Chester Secondary. Skill 1 was related to searching and evaluating information; skill 2 was about knowledge of planning investigations; skill 3 was related to knowing how to use scientific equipment; skill 4 was about knowing how to analyse data and make conclusions; skill 5 was related to learning how to write reports; skill 6 tested the ability to reflect on the lessons and the last skill 7 involved asking more frequent and relevant questions related to the project at hand. Interesting here is that from this list we see what the school regarded as a process skill and its definition

Given how the first cohort of 147 students passing through the programme reported seeing greater relevancy of science—socially, professionally, and personally[15]—there was even talk of gradually introducing a similar programme in literature for Grade 7 students in the school. In terms of the learning of process skills, the school's action research report gave similarly encouraging survey results about the gains in perceived levels of skills competency (Table 9.1).

Confident that *SciencePlus* was the right way forward together with the higher ability scores of the Grade 7 student intake, elements of process skills learning were formally introduced into Grade 7 and first half-year of the Grade 8 science syllabus for all students in the Express and Normal (academic) streams.[16] Thus, by 2007, *SciencePlus* had come to maturity with participants from both Grades 7 (learning content and process skills[17]) and 8 (content and process skills learning for 20 weeks

research—the "Research Activist". This person spent two days a week for about 40 weeks learning from and partnering with curriculum specialists and education experts employed by the MOE.

[15] When asked if the programme had made them more aware of the relevance of science in their lives, the "Agree" and "Strongly Agree" responses ranged from 84 % for rocketry to 97 % for ecology/conservation modules although these results did not indicate the nuances of interpretation among the students which seemed content specific. That is, relevance was often articulated within a narrow context of what students saw, investigated, and experienced in person but not beyond (i.e., they had narrow domain interest as opposed to broader subject interest). Professional relevance also received the lowest response rate—science as a career appeared a distant prospect for these 14-year-olds.

[16] This chapter confines itself to describing events that happened in the Express stream whose students complete their secondary education in 4 years as opposed to those in the Normal (Academic) stream who finish in 5 years.

[17] Based on the lower secondary syllabus, these included observing, measuring, inferring, predicting, interpreting data, and so forth.

followed by a capstone inquiry project for 10 weeks). Information about *SciencePlus* was disseminated to other schools through sharing sessions at the cluster[18] level as well as on-site presentations to visitors eager to learn about the secrets of the programme. Local educational conference presentations were another avenue for dissemination. Summaries of what the 29 TLLM prototype schools had accomplished were compiled into a booklet that was distributed to all schools.[19] Although teachers from Chester were modest and quick to concede that there were a number of outstanding problems such as the lack of time in the programme, the standardisation of assessment across elective modules, and lesson delivery issues, the MOE as well as the general public had little reason to doubt that *SciencePlus* was truly a winning TLLM curriculum.

Following this landmark TLLM seminar, the research team became involved with Chester upon the invitation of Carol (the then science Head of Department) to provide an unbiased evaluation of the programme. To her credit, Carol as the programme leader genuinely wanted an objective judgement as she believed that there was always room for improvement and that the researcher's point of view would be valuable in this respect. From July 2007, we observed and video-recorded two modules out of the four available; this was partly due to the lack of manpower as well as the unwillingness (of both parties) to disrupt everyday teaching and learning through our presence. Despite being lauded by the MOE for being an excellent programme in the spirit of TLLM, I have arrived at the conclusion that it was fundamentally constrained in the amount of epistemic agency it could afford.

Obstacles to Epistemic Agency from Within and Without

Recall that *SciencePlus* by 2007 consisted of two distinct parts: a longer period spanning Grade 7 and half of Grade 8 where various process skills were infused into science lessons and an intensive 10-week period at the end of Grade 8 where the students were supposed to bring what they had learnt to bear on a capstone research investigation in one of four possible modules. These dwelt on different science content areas, but what united them all was instruction in four core process skills that the school felt was vital at this stage:

- Information research (collecting reliable data)
- Questioning (applying Bloom's revised taxonomy to ask researchable questions)
- Investigating (organising a rigorous project)
- Report writing (producing a scientific report)

[18] Schools were grouped by the MOE into "clusters" with a mix of primary, secondary, and colleges to facilitate sharing of knowledge and resources.

[19] See MOE (2008).

I will focus attention on this final phase of *SciencePlus* as it most favoured the emergence of mature scientific thinking. Students were told to collaborate in groups of three for their projects and then present their findings in a written report to the teacher as well as give an oral presentation to the class followed by a short question-and-answer session. What happened during actual implementation was, however, problematic at various junctures, which did not enable students to achieve a high level of epistemic agency, at least in the two elective modules that we observed. By this I refer to the learning by a student of complex skills and reasoning associated with scientific literacy from a curricular perspective such as evaluating theories, models, and data with reasons/evidence. Such forms of knowing eschew passive content accumulation, which school science has traditionally rewarded but instead focuses on the active, constructionist interpretation of fallible/tentative results and the multiple disciplinary (i.e., epistemic) criteria for establishing "facts" and explanations. Someone with epistemic agency in science will thus have opportunities to generate, understand, and apply the reasons why and how knowledge becomes accepted in a manner appropriate to the discipline.[20] Neither does the provision of hands-on activities qualify as especially rewarding in terms of enhancing student's epistemic agency unless these are accompanied by the aforementioned behaviours of understanding that some have called "minds-on" learning.

Even though there were 10 weeks devoted to inquiry activities during the elective phase, the pressures of supervising nearly 13 groups of students by a single teacher had effectively precluded close support or sustained discussions with students. This was a common lament garnered from both teachers and students in the programme that we observed. For Ken who was in charge of the rocketry module, he also said that in 1 year, "two groups [of students] were always not on-task and when I come to them during their group discussion, they don't have much to explore", which in his opinion was "such a waste of learning opportunity". Compared to the rocketry module, those in the biotechnology module under Kelly had slightly more freedom in what they wanted to investigate. For instance, the choices of additives or varying physical conditions such as temperature that inhibited the growth of bacteria were focal areas that aligned with the overarching guided inquiry framework in *SciencePlus*. Yet, student investigations into fresh areas or reconfirming their data with certainty (and reflection on their [ambiguous] meanings) was nigh impossible given the tight schedule and the specialised equipment (i.e., the incubators as well as the rocket launchers) that was confined to the laboratory. In the rubrics for the group written report (for all modules), students were aware that they merely had to pose "new relevant questions with suggestions on further questions" rather than carrying these out, which again artificially limited the practical understanding of science. Finally, the number of lessons devoted to learning the fundamentals of biochemistry (although basic) had prevented the exploration of emergent research questions, thinking through and evaluating the

[20] See Chinn and Malhotra (2002), Duschl, Schweingruber, and Shouse (2007), and Ford and Forman (2006).

results, or creation of tentative models/theories during the course of investigations. All these omissions in exercising epistemic agency—by assessment design and indirect effect of time shortages—were done at the expense of learning concepts and doing other tasks. Arguably, these were all consequential in a negative sense as illustrated in the following three interview excerpts from the biotechnology participants.

Interviewer: The Powerpoint slides to prepare, you find that the time wasn't enough, what about carrying out the experiment? The time for proper investigation, do you think that was enough?
Student A: Yes, was enough, because the experiment only took three days to carry out and we were given about one week to carry out, so, we did two experiments, so was enough.
Interviewer: You mean two, you mean you did it twice?
Student A: Ya, we did twice. (23 September 2008)

Interviewer: Did you need to modify your plan?
Student A: Ya.
Interviewer: How many times?
Students: Once, once. (2 October 2008)

Student A: We found that bacteria grows best in room temperature.
Student B: No! It's body temperature.
Student A: Body temperature.
Interviewer: What temperature did you use? Below room temperature and above room temperature right? To check on the effect on bacteria growth.
Student B: Yes, incubator.
Interviewer: But the incubator's temperature is higher than room temperature! Bacteria grew more at room temperature?
Student A: No, bacteria grew more in the incubator.
Interviewer: Incubator?
Student A: Body temperature, incubator was at body temperature.
Student B: Thirty degrees.
Interviewer: So that that is lower than our body temperature. If you increase the temperature more than thirty-seven degrees, did you test what can happen?
Student A: No, we didn't really try it out. (2 October 2008)

Conversely, students in the rocketry module encountered a predominantly hands-on mode of inquiry where carrying out the activities oftentimes amounted to trial-and-error learning or routine troubleshooting that took precedence over studying the physics of flight (Chue & Lee, 2013).

> We make the rocket, but after two or three tries, the rocket is spoilt, destroyed and daily we have to make three to four rockets and there is water leakage from the launcher. It is very difficult. We had to change the length of the rocket many times. (Interview with Group NA, Student 3)

Shallow scientific knowledge arising from the lopsided emphasis on problem-solving or design work was apparent when one group of students shared their findings during the oral presentation with their module teacher Ken; they explained away their poor results as due to "unforeseen circumstances" instead of applying adequate scientific reasoning.

> The blue line graph is the rocket without the nose, it was directly proportion and cut through all the points. The yellow line is the tennis ball one and the pink line is the sharp nose rocket. Both the sharp nose and tennis nose did not cut through the points proportionately. Meaning that due to unforeseen circumstances, like the wind, therefore it's not accurate. (Group 7)

When probed by their peers, this same group was unable to reasonably defend their findings, which were truly missed learning opportunities as seen in the following exchange.

Student:	I thought the aim of your experiment was to find the stability of the rocket instead of the distance.
Presenter:	We were told to find the distance.
Ken:	Ok, next question.
Student:	Why three rockets?
Presenter:	To find the most stable one and we use that one to fly.
Ken:	Because you say average but you did not tell the class what is stable. Can you explain?
Presenter:	Stable means fly over more consistent distance.
Ken:	Ok, let's thank the group for their presentation. Let's have the next group.

In part a function of the project—successfully launching a plastic rocket to hit a predetermined target in the school field—it privileged design activities over wrestling with the underlying scientific concepts that was perhaps due to Ken's own disciplinary training as an engineer and his underestimation of the complexity of inquiry teaching in general. Criticism of learning by trial and error is not the primary motive here[21]; I wish to make the point that opportunities to meaningfully reflect upon or engage in deep learning were downplayed as students (and teacher) sought to build a visible and working product.

> I think my strength lies in the practical aspects, but to help students to understand why things go, doesn't work, I think the theory part, my own background, I don't think it's enough. So especially the mechanics, because I am an electronics communications, my specialization was communication. To help students understand the mechanics is slightly

[21] In the real world, technological design might follow its own autonomous trajectory just as it can even lead the development of certain aspects of scientific knowledge. However, one should recall that this episode was located within school contexts where the learning of scientific abstractions and representations is the normative end point. As well, science and technology are now seen as mutually symbiotic. See Layton (1993).

difficult so let them play with it and then and show them. But in terms of theory behind, I think that's the difficulty. (Interview with Ken, 18 July 2008)

Faced with multiple burdens of large class sizes, student unfamiliarity with inquiry, and the lack of curriculum time, the teachers employed mind maps, Bloom's revised taxonomy (to learn how to categorise and ask questions), and worksheets on the scientific method to expedite and strengthen the planning of student investigations. As one student from an unidentified module said, "I am not sure how to design an experiment on my own". However, the projects now seemed prescribed to the point of being more "cookbook-like" than what was originally envisaged as shown in the next interview excerpt. This was surely a difficult trade-off: Liberalising epistemic agency was undoubtedly ideal from the perspective of the teachers but weighed against the possibility of mass student confusion and classroom management issues during open discovery modes; the learners eventually received substantial forms of teacher direction.[22] Encountering all these various constraints, the teachers lowered the frequency and quality of learning critical skills in science and the epistemic agency of learners suffered in these two modules, and in all likelihood are features cutting across the entire programme.

Student A: Like initially it was hard because we weren't exposed to this kind of things at first. So after we trained in that then it was easier, we ask guidance from teachers.

Interviewer: Oh ok, do you think you need more guidance for this project of yours? What do you think?

Student A: No, because now all this kind of like, like we are so scared of experiments right, we were all given this helpline...I mean this kind of question they guide us through along our whole our experiments. So its easier because they already state out and then we just have to fill in. (Interview with biotechnology students on 7 July 2007)

It should be already obvious that these four elective modules gave students differing experiences and learning of science despite the presence of four core process skills (see footnote 9). This diversity created a double-edged sword: It catered to varying interests and motivations among students, but grading projects reliably in an equitable manner now became increasingly challenging for teachers, more so when dealing with scientific investigations and higher-order process skills. Knowledge of content—a familiar and key learning objective of most curricula—could not be tested directly in *SciencePlus* through a conventional test as all

[22] According to the school's action research report from 2006, the biotechnology students were taught how to design experiments step by step with given examples. As a teacher remarked, students still needed a lot of hand-holding and practice in order to be competent. As the report noted, teachers needed to find a balance between "providing students support and allowing them to be independent learners". See Bencze and Alsop (2009) for a distinctive view about science inquiry and advice for teacher supports.

modules covered different topical areas.[23] Insofar as content and process skills were supposed to be accorded equal emphasis[24] in *SciencePlus*, its developers could never had foreseen how incredibly challenging it was now to assess all *four modules together with fairness*.[25] This dilemma that haunted the programme right from the beginning took on extra significance in Grade 8 as students were to be tracked/streamed into subject specialisations the following year. Once near the end of a school year, the average grades in the ecology module were found to be about 10 marks higher than that of other modules thereby prompting the coordinator to contemplate moderating those grades downwards or pushing the others upwards so as to narrow the gap.[26] Such considerations prevailed over intellectual discussions about the quality of student work in the curriculum precisely because this occurred in Singapore, which is an unashamedly developmental state where limited resources and rewards lead to competition among students for academic success. These sociocultural norms and beliefs about meritocracy are long entrenched and have penetrated into every school on the island, including Chester.

Their effects on everyday teaching are subtle but nonetheless powerful. Teachers in the upper grades had previously complained that students they taught had not learnt sufficient content from earlier years.[27] Thus, there was some indirect pressure to prepare students for a demanding curriculum in the coming year. One should also not forget that there were other forms of washback from the high-stakes examination in Grade 10 when some of the skills of planning and dealing with experimental variables were features that were consciously inserted into *SciencePlus*. Some students too were fully cognisant of what was at stake. During an interview with a 14-year-old student who was asked if she preferred the *SciencePlus* way of learning over the typical format of written examinations, she replied that the latter was superior because it presumably rewarded the assiduous individual more precisely than work from collaborating with a heterogeneous group of students drawn

[23] By 2009, the biotechnology module and presumably the other modules as well allocated 9 % of the marks in *SciencePlus* (Grade 8) to a content-based quiz. This might have been a token concession to increase objectivity into the assessments here.

[24] Because the speed of content coverage was uneven across all modules, this also created some anxiety for teachers who wanted to devote equal time when teaching the common process skills in their modules.

[25] A suggestion by the research team to minimise the subjectivity in the assessment was to introduce a content-independent test of science process skills. This proposal was welcomed by the teachers. We thus created a test with an open-ended comprehension question involving the control of variables together with multiple-choice items adapted from the high popular quizzes administered by the University of New South Wales.

[26] Teachers had in fact spent nearly an hour every week (as part of their action research project) over the better part of 2008 to devise equitable marking schemes and rubrics (for the report and presentation) to deflect any conceivable complaints from students or parents.

[27] By 2010, the head of science was pondering whether to implement a modular system for lower secondary science where upper secondary science specialists were brought down to teach content in which they had expertise. This was an increasingly common practice in schools to ensure that students learnt from specialists and that fewer misconceptions would be spread.

from different classes. This then was the hidden curriculum that students oriented towards as well as one that shaped their habitus of getting ahead in Singapore.

To summarise the account thus far, having different topical areas in an important tracking year coupled with a societal fixation with obtaining good grades created yet more tensions that limited both the amount and quality of interest-driven content and inquiry that students could freely pursue. While the choice of investigations (i.e., research questions) was bounded, this was not altogether an entirely negative scenario as learners probably had to be scaffolded from low levels of knowledge. However, at the same time, it bolted on an epistemic ceiling on the trajectory of student learning. Oddly, student anxiety about performing well was, in the end, misplaced because it was unlikely for a student to obtain poor grades given the sheer variety of staggered assessments, helpful teachers that monitored progress, and the nature of group work in *SciencePlus*, itself a mandatory programme for everyone in Chester. It comes as no surprise to learn, as far as is known, that passing *SciencePlus* was generally assured for most students.

Be that as it may, handling inquiry science even at the best of times is an uphill struggle that requires, among other things, the prerequisites of mastery of content and relevant pedagogical knowledge.[28] Thus, asking whether the teachers were adequately trained and prepared for *SciencePlus* is a very relevant question; Ken and Kelly were 5 years into the profession with Ken in Chester as his first school and Kelly newly transferred here. From their reflection journals in 2006, teachers mentioned that there were only three sessions of TLLM training completed in April of that year and that it was "very pressuring to develop the whole module in such a short duration" and "mostly single-handed". Another teacher mentioned her own lack of knowledge and experience in teaching chemistry, having not taught it for the last 12 years. During our research, new staff members that taught *SciencePlus* similarly mentioned training for the programme that lasted a day or less prior to teaching the modules all on their own.[29] The Research Activist teacher who was part of the original team that conceptualised *SciencePlus* had taken family leave prior to our involvement with the school, which denied her colleagues the benefit of her newly gained knowledge for she never returned to Chester while we were there.

A major criticism of the *SciencePlus* programme perhaps lies in its lack of a clear programme theory—a theoretically informed plan of action whereby deliberate activities lead towards projected short- and long-term outcomes. Not just the responsibility of researchers, school practitioners can no longer ignore awkward questions over *exactly* how each programme component will result in the desired objectives. One acknowledges, however, that practitioners often do not possess expertise in programme evaluation and thus rely on a blend of folk- and evidence-based theories of action such as what was observed in *SciencePlus*. As far as can be ascertained, programme theory in *SciencePlus* consisted of process skills and

---

[28] These very concerns were also articulated by the teachers in *SciencePlus*.

[29] In the biotechnology module, a different teacher took over the running of the module in each year that we were present in the school.

interesting hands-on investigations that would *eventually* help develop content knowledge, seeing the relevance of science, and a number of important reasoning skills in the discipline.[30] While these components frequently appear alone or in combination among policy documents and recommendations for best practices, whether these planned outcomes would occur under these specific operating conditions (i.e., the sequence, quality, type, duration and variety of activities, sociocultural contexts, etc.) in *SciencePlus* remains an empirical question. In other words, the design of the curriculum had to be proven to work with specific causal mechanism(s) rather than assuming that the cumulative merits of all these worthy activities would guarantee programme success. This has been a distinct shortcoming of the programme despite my emphasis in this chapter on programme implementation that involves interactions with other levels in the system and with prevailing sociocultural norms and educational policies. As I claim later, this absence of a theory of action nonetheless seems commonplace among school innovations in Singapore.

Readers are probably now impatient to know the outcomes from the programme and what it had achieved. Apart from perceived increased competencies in process skills, students in 2006 had reported feeling an increased relevance of science after their engagement in *SciencePlus*.[31] The latter was, in my opinion, a known by-product of simply doing practical work in schools: Students worldwide enjoy going to the laboratory or doing fieldwork. They can usually accomplish what is expected during these hands-on lessons, but learning the science behind these phenomena is never as tightly correlated with the "doing" of science as we witness in the two following excerpts and what has been described in earlier paragraphs.[32] If enjoyment of hands-on experiences were the chief yardstick by which one determined success, then *SciencePlus* as a curriculum innovation indubitably qualified as an accomplishment. Asking for something beyond those criteria, however, one cannot reply without qualifications. Using the language of activity theory, the students in the second excerpt were apparently pursuing a different object (fun and

[30] Relevance was believed to be a direct consequence of choosing one's elective module and research questions although student choice was not always respected due to practical limits on module enrolments. Further inferences about the causal mechanisms in the programme theory cannot be determined.

[31] Seeing relevance or taking an interest in science is notoriously multifarious, highly contextual in nature, and hard to operationalise psychometrically and measure with certainty. Furthermore, it has been claimed that the "research evidence shows clearly that it is the teacher variables that are the most significant factor determining attitude [towards science], not curriculum variables" (Osborne, Simon, & Collins, 2003, p. 1070).

[32] See Abrahams and Reiss (2012) and Windschitl, Thompson, and Braaten (2008, p. 942) who claimed "[a]ctivity without understanding seems to be a regular feature of classroom life for science students in American schools". The only known reliable programme linking process skills and higher-order thinking has been Cognitive Acceleration through Science Education (CASE) from the UK, which has a robust activity structure based on Piagetian and Vygotskian learning theory.

enjoyment) as compared to the formal learning aims in the object of the science teachers in Chester.

Interviewer: So far...your impression of *SciencePlus*...you think it's achieving its stated objectives?
Coordinator: *SciencePlus* whether it, if is not for academic, I say yes.
Interviewer: Umm.
Coordinator: If the whole idea of *SciencePlus* is to equip the students with more skills, that means to let them know exactly how a scientist operate...to get them interested in science, then I think yes.
Interviewer: Umm.
Coordinator: But in academic-wise, I can't tell as for now...I wouldn't want to say, I can't tell whether *SciencePlus* did make our results better or not.
Interviewer: We had some surveys of people's attitudes right? So generally, they were quite positive.
Coordinator: Generally they are positive but as all Asians will write in their survey forms, I...
Interviewer: Generally they won't put negative...either is neutral or plus.
Coordinator: Yah neutral means no good already. So far having got a lot of neutral.
Interviewer: Okay.
Coordinator: They are not like super, it's not that extreme, it's not the super good.
Interviewer: Umm umm.
Coordinator: So that's what we say. So don't know.
Interviewer: Ummm still a question mark.
Coordinator: Question mark lah. Whether the students do benefit or not, perhaps they just see it just as another project work.
(Excerpt from interview with *SciencePlus* coordinator on 9 July 2008)

Interviewer: Okay so that one helped a bit is it? Okay have you heard of this thing called process skills or not?
Students: Ah a bit.
Interviewer: Quite a bit, but I'm not too sure about it. Okay, ahm. ... I'm sure in primary school your teacher would have mentioned process skill examples, like measuring, classifying, [inaudible] observing. Okay so did you remember any of these when you are doing rocketry?
Student 1: No.
Student 2: Don't have.
Interviewer: Okay that's fine, if you want to tell [a] sec one boy or sec one girl[33] about *SciencePlus*, what will you tell him or her?
Student 1: They will find it interesting.

[33] Secondary one or Grade 7.

Interviewer: Pardon?
Student 1: Fun and interesting.
Interviewer: It is fun and interesting.
Student 2: Depending on what module it is.
Interviewer: Okay which one is the most fun module?
Student 4: Actually everything is, actually everything is fun depending on yourself.
(Excerpt from student focus-group interview on 30 September 2008)

Development of higher-order epistemic skills (other than practice in manipulating one experimental variable at a time) could not be ascertained with any certainty, nor was it explicitly tested within this curriculum that in all likelihood set a boundary on what students could learn. Controlling variables was the chief focus and probably the most sophisticated process skill to be tested, at least in the elective modules that were observed.[34] Other than the ecology module, the skill of analysing or inferring (see #4 in Table 9.1) enjoyed the largest increases in perceived competencies as reported by students in 2006 although through their observations, teachers had reported that students' analyses lacked depth. Some of the process skills such as report writing (#5) and reflecting (#6) seem unique to this programme. Some of the other skills did not appear to be well operationalised, which might have led to teaching/assessing them in a vague or arbitrary manner. What did "using apparatus" actually mean in concrete terms—mere usage or proficiency? How was teaching Bloom's taxonomy useful in the skill of questioning—to know which category a question belonged or to generate more questions across a spectrum of complexity? Is questioning equivalent to predicting or hypothesising? And without knowledge of the pathways by which the essential components—the process skills, hands-on activities, content teaching, and field trips—in *SciencePlus* are believed to function, any potential or intrinsic benefits of the programme cannot be comprehensively endorsed. As promising as it sounds, this curriculum has to remain as an experimental model until enough evidence has been gathered that it truly "works" in the way that its developers had envisaged.

As a final irony, one would imagine that other schools would have eagerly adapted or borrowed this programme for their own purposes, but this has not happened. In fact, *SciencePlus* has remained confined to Chester precisely because it was widely celebrated as being out of the ordinary. In Singapore, competition among schools to excel not only in the academic arena but also in other niche areas such as sports and the arts is unusually intense. Seeking to distinguish their institution from all others and to foreground their own educational innovations,

[34] Writing the group report (25 % of final science marks in Grade 8) followed a prescribed template reminiscent of the so-called scientific method. Teachers were also on the lookout for some key inquiry skills such as "shows control of all involved variables", "data are collected from several repetitions. . .and are consistent within a reasonable range", "identifies patterns", and "explanation of the results is clear and consistent with the data obtained and supported by research". These were very similar to the skill sets from the pre-existing lower secondary science syllabus.

school leaders have been anecdotally known to eschew replicating what others have already implemented or created as these programmes, regardless of whatever intrinsic merit, are not at the vanguard nor regarded as completely "innovative" by MOE. In this landscape of policy contradictions where centralisation and decentralisation, independent student thought and obedience to authority, and interschool collaboration and competition are simultaneously championed, it was regrettable and yet understandable that the *SciencePlus* programme has not travelled beyond the gates of Chester since its inception in 2007. One can speculate whether other schools might have revised or transformed it according to local contexts, which would arguably lend greater credibility and perhaps improvements. Unless it can be tested in other ways or contexts, *SciencePlus* remains a programme in a relatively youthful stage of development.

Discussion and Conclusion

SciencePlus was a member of an elite group of school-based curriculum innovations that was honoured by the MOE in 2007. With its emphasis on inquiry science through authentic, hands-on investigations and increasing the relevance of the subject for young people, this programme quickly caught the attention of policymakers. Because it was so well supported from the top, the programme flourished as it also became the poster boy of TLLM and a host of allied policy reforms at a time when other schools were still struggling to make sense of what these new and relatively unfamiliar ways of teaching, learning, and assessment meant on an everyday basis.

What happened when our research team spent time in the school, however, did not inspire us with equal confidence as the MOE's evaluation. While we found the provision of investigative projects and an emphasis on learning various process skills as identified by the school, these efforts were not believed to have generated rich experiences in developing epistemic agency, which are the higher-order and critical reasoning skills in science education. Students, nonetheless, did appear to have enjoyed the programme, claimed competence in some process skills, and reported seeing an increased relevance of science in their lives that indicated that the programme has had some success.

Many reasons are believed to have led to this uneven situation. Although the 10-week elective module phase had increased hours from the consolidation of Project Work and the existing science curriculum, this was in reality insufficient to allow learners room to exercise more sustained or deeper forms of inquiry. Instruction thus remained at a guided inquiry level that often amounted to directed forms of teaching given the large class sizes and need to cover basic concepts in the biotechnology module. For the rocketry module, the reverse was true whereby there was a dominance of practical, hands-on design activities that overshadowed the conceptual learning of the science of flight. The differing nature of the elective topics although possibly motivating for students also presented enormous dilemmas with respect to equitable marking across the four modules. Given societal pressures from the upcoming high-stakes examinations to be experienced two years later and streaming beginning in Grade 9, grading across the modules was subjected to intense moderation exercises that kept final scores within narrow bands, which might not have reflected students' actual abilities (for better or for worse) during this mandatory science programme. Therein we see how various interacting levels of society and the TLLM policy regime subjected *SciencePlus* to conflicting pressures, which both teachers and students in Chester had to negotiate during the learning of science.

Other teacher-created resources such as inquiry templates and mind maps served ambiguous roles for while they did indeed help students unfamiliar with inquiry gain a foothold into their projects, some students became overreliant on these scaffolds. Again, negative societal pressures to do well in school, articulated from teachers and even the students themselves, served as barriers on the amount and quality of epistemic agency during the implementation of *SciencePlus*. As the offspring of Project Work and the lower secondary science syllabus, *SciencePlus* inherited the emphasis on independent student inquiry during collaborative group work, but one must not forget that the programme was implemented within an East Asian developmental state, which prized efficiency and meritocracy including within the educational arena. Activity theory sheds more light on this; individuals always enact and maintain the prevailing social structures/norms just as these very structures also shape and constrain what people will do. In this case, the wriggle room for *SciencePlus* was constrained by the ecologies of various people, policies, and places although it was a legitimate TLLM innovation in itself. This aspect, above all else, shows how the enacted curriculum can thwart the best intentions of curriculum developers.

The teachers in the school were professionals who were sincere and hardworking that made partnering them a joy. Yet, their training in TLLM was hurried as was the experience of some teachers who had to teach the modules with minimal training about the conduct of inquiry science. At no time are they to be blamed for devising some process skills that were peculiar to this programme, nor were they reasonably expected to have knowledge of programme theory. Yet, this latter aspect of curriculum development was a crucial step that prevented us from fully validating *SciencePlus* as we are unsure how the components in this programme actually relate to the stated objectives. Even though the programme has been implemented with satisfactory fidelity, it does not follow that the planned programme was inherently sound in the first place (Lee & Chue, 2013). Examination of other TLLM curriculum innovations and MOE documents neglected mentioning anything about programme theory, which is a real cause for concern. As well, strong competition with other schools to create their own niches prevented this programme from being tested in other contexts; it would have been most valuable to ascertain if the learning of content and process skills could have been as favourable as what Chester Secondary had reported.

What implications do I offer for science educators? This somewhat negative judgement of *SciencePlus* brings to attention that there is an underbelly within educational change on the island that has not often been reported, which if it had remained hidden would discourage its amelioration. All educators therefore need to know what happened during programme implementation before one can perform an outcome evaluation and scaling-up if it is found beneficial. Second, the problems with *SciencePlus* share much in common with other systems and subject areas; thus, solutions or strategies that have been found viable elsewhere can be applied or tested here and vice versa. Rather than hubris from those education systems that apparently lack these kinds of implementation troubles, it is likely that such scenarios indicate an absence of hard scrutiny. Third, this chapter highlights the extreme complexity of programme implementation in schools; curricula can be planned with the best of intentions but fail or are thwarted by a mangle of external and internal forces. Indeed, the so-called external social forces are always present in every classroom. This does not trivialise the theoretical aspects of science curriculum development (e.g., ensuring a right or adequate amount of inquiry) but suggests that we really neglect the practical aspects of programme implementation and understanding educational change as inevitable complex ecologies at our peril.

References

Abd-El-Khalick, F., BouJaoude, S., Duschl, R., Lederman, N., Momlok-Naaman, R., Hofstein, A., et al. (2004). Inquiry in science education: International perspectives. *Science Education, 88*, 397–419.

Abrahams, I., & Reiss, M. J. (2012). Practical work: Its effectiveness in primary and secondary schools in England. *Journal of Research in Science Teaching, 49*, 1035–1055.

Anderson, R. D. (2002). Reforming science teaching: What research says about inquiry. *Journal of Science Teacher Education, 13*, 1–12.

Atkin, M. J., & Black, P. (2003). *Inside science education reform: A history of curricular and policy change*. New York: Teachers College Press.

Bencze, J. L., & Alsop, S. (2009). A critical and creative inquiry into school science inquiry. In W.-M. Roth & K. Tobin (Eds.), *The world of science education: North America* (pp. 27–47). Rotterdam, the Netherlands: Sense Publishers.

Chan, C. M. L. (2011). *An analysis of the 2004/05 teach less, learn more policy in Singapore*. Unpublished doctoral dissertation, The University of Western Australia, Perth, Australia.

Chinn, C. A., & Malhotra, B. A. (2002). Epistemologically authentic inquiry in schools: A theoretical framework for evaluating inquiry tasks. *Science Education, 86*, 175–218.

Chue, S., & Lee, Y.-J. (2013). The proof of the pudding? A case study of an "at-risk" design-based inquiry science curriculum. *Research in Science Education*. doi:10.1007/s11165-013-9366-x.

Crawford, B. (2000). Embracing the essence of inquiry: New roles for science teachers. *Journal of Research in Science Teaching, 37*, 916–937.

Duschl, R. A., Schweingruber, H. A., & Shouse, A. W. (2007). *Taking science to school: Learning and teaching science in grades K-8*. Washington, DC: The National Academies Press.

Ford, M. J., & Forman, E. A. (2006). Redefining disciplinary learning in classroom contexts. *Review of Research in Education, 30*, 1–32.

Gaddis, J. L. (2002). *The landscape of history: How historians map the past*. New York: Oxford University Press.

Hargreaves, A. (2005). Pushing the boundaries of educational change. In A. Hargreaves (Ed.), *Extending educational change* (pp. 1–14). Dordrecht, the Netherlands: Springer.

Harlen, W. (2004). *Evaluating inquiry-based science developments*. Retrieved June 12, 2007, from http://www7.nationalacademies.org/bose/WHarlen_Inquiry_Mtg_Paper.pdf
Kosso, P. (2011). Philosophy of historiography. In A. Tucker (Ed.), *A companion to the philosophy of history and historiography* (pp. 9–25). Malden, MA: Wiley-Blackwell.
Layton, D. (1993). *Technology's challenge to science education: Cathedral, Quarry or company store?* Buckingham, UK: Open University Press.
Lee, Y.-J. (2008). Thriving in-between the cracks: Deleuze and Guerrilla science teaching in Singapore. *Cultural Studies of Science Education, 3*, 917–935.
Lee, Y. J. (2011). More than just story-telling: Cultural-historical activity theory as underutilised methodology for educational change research. *Journal of Curriculum Studies, 43*, 403–424.
Lee, Y. J., & Chue, S. (2013). The value of fidelity of implementation criteria to evaluate school-based science curriculum innovations. *International Journal of Science Education, 35*, 2508–2537.
Lee, Y.-J., & Roth, W.-M. (2006). Learning about workplace learning and expertise from Jack: A discourse analytic study. *Journal of Workplace Learning, 18*, 205–219.
Lipsky, M. (1982). *Street level bureaucracy*. Boston: Allyn and Bacon.
McNeill, K. L., & Krajcik, J. (2008). Scientific explanations: Characterizing and evaluating the effects of teachers' instructional practices on student learning. *Journal of Research in Science Teaching, 45*, 53–78.
Megill, A. (1989). Recounting the past: Description, explanation, and narrative in historiography. *The American Historical Review, 94*, 627–653.
Milne, C. (1998). Philosophically correct science stories? Examining the implications of heroic science stories for school science. *Journal of Research in Science Teaching, 35*, 175–187.
Ministry of Education [MOE]. (2004). *Science syllabus: Lower secondary [Special/Express/Normal (Academic)]*. Singapore: Curriculum Planning & Development Division.
MOE. (2007). *Teach Less, Learn More*. Retrieved December 20, 2012, from http://www3.moe.edu.sg/bluesky/tllm.htm
MOE. (2008). *More support for school's "Teach Less, Learn More" initiatives*. Retrieved January 6, 2013, from http://www.moe.gov.sg/media/press/2008/01/more-support-for-schools-teach.php
MOE. (2012). *International studies affirm Singapore students' strengths in reading, mathematics & science*. Retrieved December 21, 2012, from http://www.moe.gov.sg/media/press/2012/12/international-studies-affirm-s.php
MOE. (2013). *Project work*. Retrieved January 4, 2013, from http://www.moe.gov.sg/education/programmes/project-work/
Munslow, A. (2007). *Narrative and history*. Basingstoke, UK: Palgrave Macmillan.
Osborne, J., Simon, S., & Collins, S. (2003). Attitudes towards science: A review of the literature and its implications. *International Journal of Science Education, 25*, 1049–1079.
Pereira, A. A. (2008). Whither the developmental state? Explaining Singapore's continued developmentalism. *Third World Quarterly, 29*, 1189–1203.
Pickering, A. (1995). *The mangle of practice: Time, agency, & science*. Chicago: University of Chicago Press.
Poon, C. L., Lee, Y.-J., Tan, A.-L., & Lim, S. S. L. (2012). Knowing inquiry as practice and theory: Developing a pedagogical framework with elementary school teachers. *Research in Science Education, 42*, 303–327.
Pressman, J. L., & Wildavsky, A. (1973). *Implementation*. Berkeley, CA: University of California Press.
Rogan, J. M. (2007). How much curriculum change is appropriate? Defining a zone of feasible innovation. *Science Education, 91*, 439–460.
Roth, W.-M., & Lee, Y.-J. (2007). "Vygotsky's neglected legacy": Cultural-historical activity theory. *Review of Educational Research, 77*, 186–232.

Roth, W.-M., Lee, Y.-J., & Boyer, L. (2008). *The eternal return: Reproduction and change in complex activity systems—The case of salmon enhancement*. Berlin, Germany: Lehmanns Media.

Rudolph, J. S. (2002). *Scientists in the classroom: The cold war reconstruction of American science education*. New York: Palgrave.

Tan, A.-L., & Wong, H.-M. (2012). "Didn't get expected answer, rectify it": Teaching science content in an elementary science classroom using hands-on activities. *International Journal of Science Education, 34*, 197–222.

Tobin, K., & McRobbie, C. J. (1996). Cultural myths as constraints to the enacted science curriculum. *Science Education, 80*, 223–241.

Tosh, J. (2010). *The pursuit of history: Aims, methods and new directions in the study of modern history*. New York: Longman.

Welch, W. W. (1981). Inquiry and school science. *Science Education, 65*, 33–50.

Windschitl, M., Thompson, J., & Braaten, M. (2008). Beyond the scientific method: Model-based inquiry as a new paradigm of preference for school science investigations. *Science Education, 92*, 941–967.

Yeo, J., Tan, S. C., & Lee, Y.-J. (2012). A learning journey in problem-based learning in a physics classroom. *Asia Pacific Education Review, 20*, 39–50.

Yeong, A. (2005). *The promise of the project work initiative in Singaporean education: A critical reflection*. Retrieved January 4, 2013, from http://conference.nie.edu.sg/paper/new%20converted/ab00286.pdf

[illegible]

Chapter 10
Implementing Inquiry Science with Knowledge Creation Approaches

Seng-Chee Tan and Jennifer Yeo

Abstract This chapter focuses on the theme of fostering science meaning-making among students using approaches that are anchored on the knowledge creation paradigm of learning. This paradigm views learning as a collaborative process of knowledge creation that focuses on improvement of common knowledge artefacts. Drawing from two research projects that studied two science classrooms in two Singaporean schools, this chapter explicates how teachers could help students achieve deeper understanding of science using two approaches: (1) a problem-based learning approach of a knowledge creation paradigm (KC-PBL) in a high school classroom and (2) a knowledge building (KB) approach in a primary school classroom. Despite the differences, the two approaches shared similar sociocultural theoretical foundation. They also share three pedagogical principles: (1) a focus on eliciting and improving students' epistemic artefacts, (2) interacting with physical phenomenon and interacting with peers in epistemic discourse and (3) drawing from distributed knowledge and expertise among students.

Keywords Inquiry science • Knowledge creation • Knowledge building • Idea improvement • Epistemic artefacts • Epistemic discourse • Distributed cognition • Technology

Learning Science Through Inquiry

This chapter describes two approaches of learning science through inquiry, both of which incorporate the knowledge creation metaphor of learning. Teaching science through inquiry has been gaining momentum since the early 1960s. Endorsed by authoritative organisations such as the National Research Council [NRC] (2000) in

S.-C. Tan (✉) • J. Yeo
National Institute of Education, Nanyang Technological University, Singapore, Singapore
e-mail: sengchee.tan@nie.edu.sg

A.-L. Tan et al. (eds.), *Inquiry into the Singapore Science Classroom*, Education Innovation Series, DOI 10.1007/978-981-4585-78-1_10,

the United States, inquiry science is commonly known as a process approach for learning science through investigation of a natural phenomenon, and it engages learners in processes such as questioning, conducting scientific investigation, collecting evidence, interpreting results, constructing explanations for the findings, reporting results, asserting claims and presenting findings. The goal of inquiry science is directed at finding coherent, mechanistic accounts of the natural phenomenon (Hammer, Russ, Mikeska, & Scherr, 2008). Advocates for inquiry science regard the processes as important process skills that contemporary scientists practise and, therefore, are central aspects of science learning among school children. Various instructional approaches have been proposed for inquiry-based lessons, inter alia, the Biological Sciences Curriculum Study (BSCS) 5E (Engage, Explore, Explain, Elaborate, Evaluate) instructional model by Bybee et al. (2006), the instruction sequence suggested by Driver and Oldham (1986) in their Children's Learning in Science project (orientation, elicitation, restructuring, application of ideas, review, change of ideas) and the Predict-Observe-Explain sequence suggested by Tien, Ricky and Stacy (1999).

Inquiry science, however, is not spared from criticisms. The gulf between professional practice of scientists and school science is one major concern. Grandy and Duschl (2008) questioned the feasibility for school children to engage in authentic socio-epistemic practices of scientists, while Kirschner, Sweller and Clark (2006) held that the failure of many constructivist, ill-guided instructional approaches are consequences of the confusion between practising a discipline (epistemology) and learning a discipline (pedagogy). This has motivated Poon, Lee, Tan and Lim (2010) to examine the enacted inquiry science practices in classrooms in Singapore. They analysed various inquiry science activities by examining sequencing of key pedagogical components (e.g., engage, explore, investigate and evaluate). Their study uncovered "nuances of everyday practice that can influence the implementation of theoretically-driven forms of inquiry science in the classroom" (p. 305). They found, for example, teachers devoted a lot of effort in preparing students for investigation by reviewing relevant conceptual knowledge and procedural knowledge of skills and processes necessary for the subsequent investigative tasks. The Prepare-Investigate-Explain cycle was also found to be largely iterative, rather than linear. Their study shed light on the enacted professional practices of educators in classrooms.

Moving forward, the practice of inquiry science could gain from the field of learning sciences that focuses on deeper understanding of the conditions and processes for effective learning and the emphasis on redesigning learning environments to bring about deep learning. In other words, we could form better theory-practice nexus of inquiry science by subjecting lessons designed with inquiry science through iterative cycles of investigations in authentic classroom environments. In this chapter, we examine inquiry science from the knowledge creation perspective, which is one of the research foci in the field of learning sciences.

Inquiry Science in the Knowledge Creation Paradigm of Learning

Knowledge creation perspective of learning refers to approaches that privilege social cultural perspective of learning and give primacy to the mediating roles of knowledge artefacts or epistemic artefacts (Sterelny, 2004). Paavola and Hakkarainen (2005) called this a new metaphor of learning since it builds on the two predominant metaphors of learning – acquisition metaphor and participation metaphor – both of which are suggested by Sfard (1998). Acquisition metaphor of learning treats knowledge as transferable commodities and focuses on effective transmission of declarative knowledge. It is exemplified by traditional instructional design that focuses on instructors presenting information, eliciting performance and providing feedback to learners to help them gain knowledge. "Cookbook" approach to inquiry learning, which follows blindly a standard set of instructional procedures, would fall into this metaphor of learning. Participation metaphor of learning, in contrast, views learning as an enculturation process. Learning is not the acquisition of knowledge in individual minds, but rather it is about becoming a member of a community by participation. Becoming a member of a community entails "social experience of living in the world in terms of membership in social communities and active involvement in social enterprises" (Wenger, 1998, p. 55). An example of inquiry science in this paradigm would include science apprenticeship camp described by Barab and Hay (2001) where middle school students participated in science-based research with scientists.

Paavola, Lipponen and Hakkarainen (2004) argued that the two metaphors could not adequately account for the intentional process of innovative knowledge creation that we see in a knowledge society. They examined three models of innovative knowledge creation – the knowledge building pedagogy (Scardamalia & Bereiter, 2006), the organisational knowledge creation model (Nonaka & Takeuchi, 1995) and the expansive learning approach (Engeström, 1999) – and found that these models not only stress the importance of social cultural experience but also involve the creation of materials and conceptual artefacts. Paavola et al. suggest "social interaction provides essential cognitive resources for human cognitive accomplishment" and that "new ideas and innovations emerge between rather than within people" (p. 564). Yet, they recognise the roles of individuals in the knowledge creation process such that "individuals' hunches and tacit knowledge should be explicated for communal and organizational use" (p. 565). The knowledge creation metaphor suggests knowledge as objects, but it is different from the acquisition metaphor that treats knowledge as commodities to be acquired or transferred. Rather, knowledge objects are epistemic artefacts (Sterelny, 2004), or Popper's World 3 objects (Popper, 1978), which refers to human creative products that are sharable and accessible. An example of an epistemic artefact is a text describing a theory. Interactions among people occur around and through these epistemic artefacts. In other words, epistemic artefacts are representations of knowledge, when shared among learners, provide the focal point for discussion and are therefore subject to query, critique and improvement. Perhaps the most distinctive characteristic of the knowledge creation metaphor of learning is the "pursuit of newness" (Paavola et al., p. 562). Knowledge creation encourages learners to always build on one another's ideas for better ideas, which leads to a deeper understanding of a phenomenon under investigation. The collaborative improvement of the knowledge artefacts by learners would eventually benefit individual learners.

Knowledge building, pioneered by Scardamalia and Bereiter (2006), is a specific pedagogical approach within the knowledge creation perspective of learning. It has been studied extensively in K-12 classrooms. Knowledge building involves students working on a problem (a problem of understanding a phenomenon), presenting their ideas in a collective space (e.g., an online forum) and collaboratively work on improving these ideas. To Scardamalia (2002), ideas are "systemically interconnected – one idea subsumes, contradicts, constrains, or otherwise relates to a number of others" (p. 72). Knowledge building develops students' disposition and capacity in knowledge work, by engaging learners in collaborative improvement of ideas related to a specific topic. Although students are not discovering knowledge new to the world like scientists do, they are engaged in the processes of improving joint knowledge artefacts by using epistemic criteria to assess the quality of the artefacts.

This chapter reports two interventional studies conducted in Singaporean classrooms. These studies rest at the nexus of learning sciences research and science education since concepts in learning sciences are applied to help students learn science in schools. Aligning to the theme of this book, the emphasis is given to the designs of the inquiry science learning environment and how they were implemented. While there are differences between these two cases, this chapter highlights their similarities, particularly the components that encourage knowledge creation among students. The first intervention study that we report here follows the pedagogical principles of knowledge building proposed by Scardamalia and Bereiter (2006). The students created knowledge artefacts as notes in an online forum. These notes reified students' ideas so that they could be accessed, critiqued and improved. Through this process, the students gained deeper understanding of the digestive systems of animals. More importantly, they engaged in epistemic discourse of linking evidence to their claims, even though they did not participate in the collection of evidence. The second case is a design experiment that started with a problem-based learning method, which, through iterative cycles of research, was integrated with knowledge creation perspective of learning. It included a knowledge building phase that entails the essential element of creating and improving knowledge artefacts. Likewise, the students attempted to make sense of the relevant theories through improving their knowledge artefacts as notes in an online forum. In contrast, it has differences from the knowledge building approach in that it is triggered by a problem scenario (rather than an inquiry task) and is ultimately oriented towards developing a solution to a problem. We call this the problem-based learning method of the knowledge creation paradigm or in short, KC-PBL.

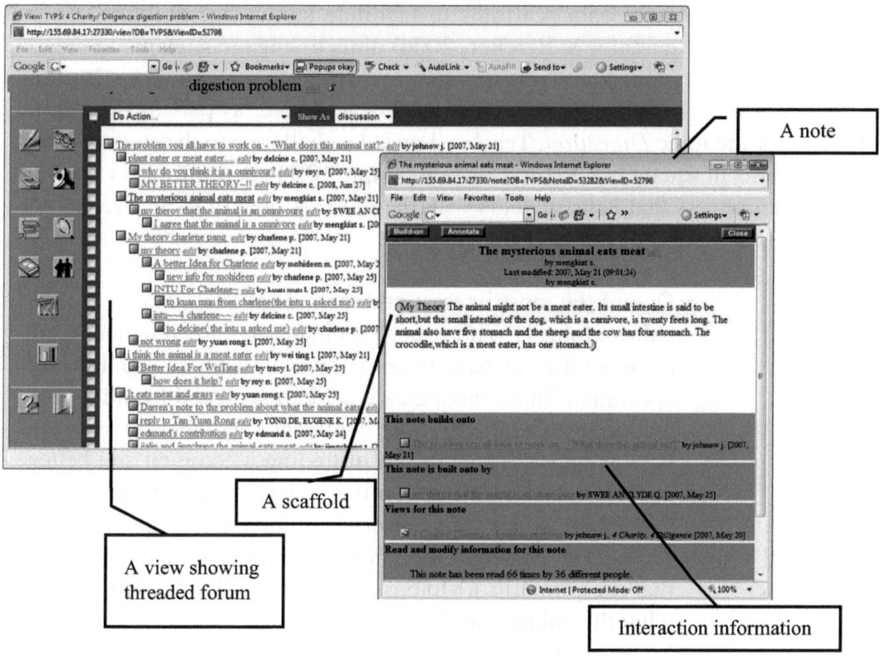

Fig. 10.1 Knowledge forum that supports knowledge building talks

Knowledge Building for Inquiry Science

This section describes a knowledge building approach to inquiry science, as the main intervention in a research study in a mainstream Singaporean primary school. Teacher J, who was also the Science Head of Department (HoD) of the school, taught two Primary Four classes using the knowledge building approach supported by Knowledge Forum (KF) (an online forum). KF (Fig. 10.1) is a web-based forum that supports asynchronous online discussion. It is developed to support knowledge building pedagogy. For example, to support idea-centric work, students can compose notes with scaffolds in the form of sentence starters. To support idea organisation, "Views" can be created for a topic of discussion and notes can be moved or copied across Views. A graphic interface can be selected to show related notes linked by lines, forming an interaction web. The students could also subsume notes under a "Rise-Above" note as they reflect on their discussion. To support social interactions, besides "Build on" function to reply to notes, there is also a "Reference" feature for students to point to other relevant notes. There is a "Group note" function that allows joint authorship for the same note. To support embedded assessment, analytics are accessible to the students and teachers. These analytics provide data on contributions by each member, social network analysis among a group and semantic analysis for vocabulary growth and comparisons of words with a reference base.

KF provides the necessary scaffolds to facilitate knowledge building. Knowledge building (Scardamalia & Bereiter, 2006) aims to engage students in creating and improving knowledge artefacts of value to the learning community. It usually starts with a question about a phenomenon. For example, as part of the learning activities for the topic *Digestion*, Teacher J posed a challenge to the students on the KF. The terms in square brackets (refer to Example 10.1) are scaffolds in the forms of sentence openers that help students with the development of their ideas.

Example 10.1. Scaffolds as Sentence Openers

[New information]	One of the most interesting branches of Science is palaeontology. With every specimen dug up and uncovered, palaeontologists have to come up with theories about the specimen. These theories are seldom accepted from the beginning. The theories are challenged by other theories and new evidence. This process of finding a common and better understanding continues until a new improved theory is built.
[I need to understand]	The remains of a well-preserved animal have recently been found. The palaeontologist who found the animal has a theory that the animal is a vicious meat eater. What is your theory of what the animal eats?
[New information]	Some of the findings on which he based his theory are given as follows. It had a set of teeth with many shapes. Teeth with flat cutting edges in the front, teeth with jagged edges at the sides and flat teeth at the back. It had a short small intestines. It had a long large intestines. It had five stomachs. It had stones in one stomach.

Teacher J also provided students with links to Internet sites on the digestive systems of herbivores, carnivores and omnivores. This was a hypothetical problem with no correct answers. The main objective was to get the students to make meaning of the resources, compare with the evidence provided and make a claim based on the evidence.

The inquiry task helps to elicit students' ideas, which are knowledge-in-the-making. Ideas, when systematically related to other ideas, form a network of coherent ideas that represent knowledge of a specific topic (Scardamalia, 2002). Knowledge building is a social collaborative process that requires students to deliberately attempt to improve one another's ideas face-to-face and in a shared database (the online forum), culminating in collective idea improvement

(Scardamalia & Bereiter, 2006). The key mechanism to achieving idea improvement lies in knowledge building discourse (Bereiter, 1994), which is talk directed at proposing and assessing knowledge claims. Scaffolds for knowledge building discourse can be provided in face-to-face discussion and online forum. When supported by a technology, for example, a discussion forum, the students compose notes to represent their ideas or conceptual understanding of a topic, these written notes are epistemic artefacts (Sterelny, 2004) that capture students' ideas and make them available in a public space; these reified ideas become objects of inquiry to be discussed, interconnected, revised and superseded (Scardamalia, 2002; Stahl, 2004).

Teacher J used the following default starting phrases in KF for the students to compose notes: [My theory], [I need to understand], [My theory doesn't explain] or [A better theory is]. These are cognitive supports that model and encourage learners to take part in interactions that lead to productive inquiry and deepening of understanding rather than interactions of a social nature (e.g., chatting). Table 10.1 shows an example of idea improvement. It started with a simple claim by S1 on Day 1. On Day 2, S2 introduced the scientific term "omnivore". On Day 2, another two students began to question S1, with an attempt to use evidence. On Day 3, several students began to question S1 on the evidence and the source of information. In the meantime, after making sense of the information, S1 posed a more elaborate answer, using evidence and new information to support the claims.

Although knowledge building emphasises epistemic discourse, interactions with the phenomenon (e.g., working with manipulative) could be incorporated so that empirical evidence could become part of resources for knowledge building discourse. For example, as part of the learning activities for the digestive system, students examined the internal organs of a fish when discussing about systems in our body. Teacher J then asked the students to pull their ideas together for a rise above, which refers to attempts to put pieces of information together to form a coherent scientific explanation. Teacher J guided the students to form the concept of system, linking what they found in the fish (empirical evidence) and the concept of a system.

In short, the inquiry activities designed by Teacher J can be divided into these phases: (1) generating puzzling questions related to a topic, (2) putting students into groups to discuss face-to-face and to compose notes in KF to answer the puzzling questions, (3) scaffolding the use of authoritative sources of knowledge (e.g., textbook, Internet resources) or empirical evidence or both in inquiry activities and (4) helping students to rise above by summarising their understanding of the topic and linking different ideas to a coherent scientific explanation.

Table 10.1 Examples of idea improvement through knowledge building discourse

Notes	Interpretation
[My Theory] I think the mysterious animal eats plants. It also eats meat (S1)	S1 proposed an answer (Day 1)
[New information] animal that eat both meat and plants is an omnivore (S2)	S2 provided a scientific name "omnivore" (Day 2)
[I need to understand] how does the animal eat meat and plants together? [My Theory] animal eating meat have short small intestine and long large intestines (S3, S4)	S3 and S4 asked "how" questions (Day 2)
[I need to understand] why do you think so? (S5)	S5, S6 and S7 asked for evidence and for sources of knowledge (Day 3)
[I need to understand] why do you think the animal eats both plant and meat (S6)	
[I need to understand] how do you know that the mysterious animal eats plants and meat? (S7)	
[A better theory] is that the crocodile may be the mysterious animal because I found out that crocodiles eat stones. The mysterious animal has stones in one of the stomach. I think it is a carnivore as if the animal eats the meat the stones could digest the meat faster. But I think it is also a plant eater because it has 5 stomach and meat (eater) only needs 1 stomach. So I think it may not be a crocodile, and it may be another animal which I am trying to find out (S1)	S1 suggested a more concrete answer (Day 3)

Knowledge Creation Problem-Based Learning (KC-PBL) for Inquiry Science

This section describes another design of inquiry science learning environment that forms the key intervention in the second research study. It involved a design experiment where researchers worked with teachers in a Singapore high school with Integrated Programme (for 9th to 12th grades) to iteratively improve the problem-based learning (PBL) approach for inquiry lessons in physics. This study reports the third phase of the design experiment, which culminates in an improved design framework that incorporates the knowledge creation approach into PBL.

KC-PBL Design

The initial design framework was a five-stage instructional model for problem-solving, known as the THINK cycle, which stands for trigger, harness, investigate, network and know. The trigger consists of simulated problems of the real world. Students, working in group size of 4–6, will identify questions that they need to investigate (harness) before embarking on a series of investigation. Investigations may include searching for information from authoritative sources or working with manipulative to conduct experimental investigation. The social collaborative process entailed networking with team members and experts. The consequential task required the students to present their solution to teachers, who acted as a panel of judges, and thus showcase their understanding of physics (know) through the problem-solving process. In each cycle, each group was supported by a teacher facilitator who functioned as a metacognitive coach.

In the first design experiment cycle, it was found that the problem scenario was used as a mediating tool for the students to learn content knowledge. That is, the focus was not on problem-solving skills, but prevalent practices of lecture and tutorial were used to help students acquire physics knowledge. This seems contradictory to the goals of practice-based PBL strategy rooted in medical education (Savery, 2006). In the second design experiment cycle, the teachers emphasised problem-solving processes, but that generated an unforeseen outcome: the students were eager to propose solutions but neglected the learning of physics principles and theories necessary for problem-solving. Some students relied on trial-and-error methods that provided a practical solution, but they failed to demonstrate their physics knowledge.

To overcome this content-process divide, elements of knowledge creation were introduced as a bridging activity in the THINK cycle, as presented in Fig. 10.3.

One key difference between the THINK framework (Fig. 10.2) and the KC-PBL framework (Fig. 10.3) is the inclusion of a knowledge building phase. Unlike the harness phase that focuses on identifying questions and gathering information, students are engaged in making meaning of the relevant theories and principles, which would eventually help the students in the problem-solving process.

To support the process of knowledge building, an online forum known as Knowledge Constructor was developed by a local company. Knowledge Constructor represents discussion interactions in a graphical form (Fig. 10.4).

Unlike the Knowledge Forum, the scaffolds are represented with different icons in the Knowledge Constructor. Otherwise, they serve similar functions of supporting interactions among students and providing a record of the epistemic artefacts so that the ideas can be revised, critiqued and improved. This KC-PBL activity was conducted over a period of three lessons of two hours each. At the end of the activity, each group submitted a joint report and an individual reflection journal to the teacher.

The following activity illustrates how this KC-PBL design works for the topic Law of Conservation of Energy. The trigger problem revolved around a fictitious roller coaster accident in an offshore island in Singapore. One day, a malfunction cart plunged down a slope at the end of the ride and crashed into a barrier, instead of stopping just in time. The students were asked to investigate the cause of the accident and recommend safety measures. Information, such as newspaper reports, police reports and maintenance reports, was provided. Hence, the task required students to assume the role of an investigator and work in groups of five to investigate the cause of the accident. They were also tasked to suggest recommendations to improve the safety aspects of the ride. Resources provided to the students

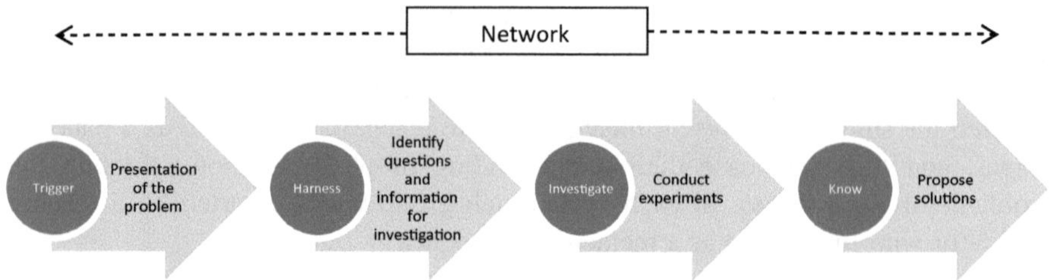

Fig. 10.2 The THINK framework

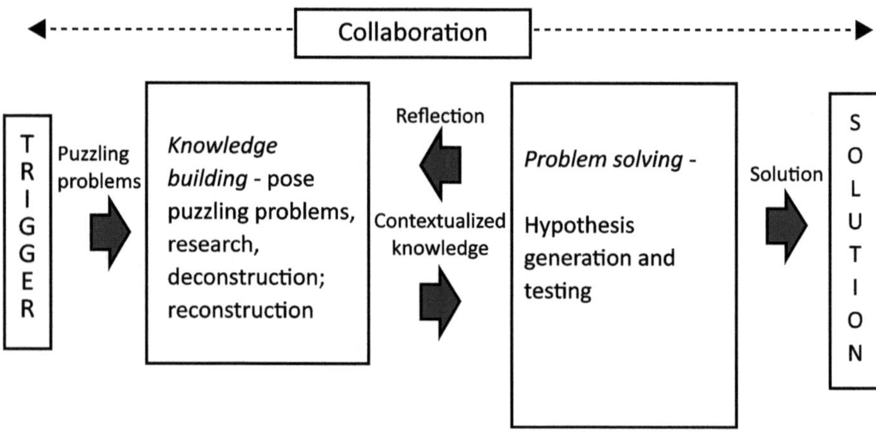

Fig. 10.3 The KC-PBL design framework

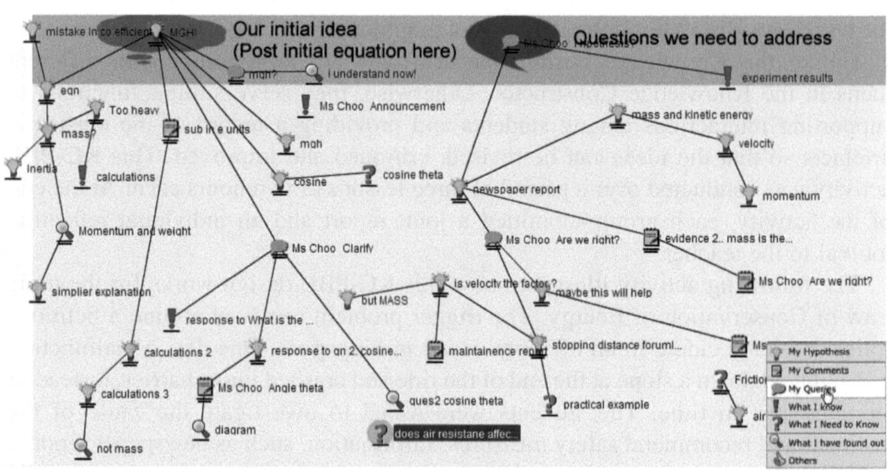

Fig. 10.4 Screenshot of knowledge constructor environment

included (1) a physical model of the last section of the ride where the accident happened, (2) "evidence" gathered from the scene of the accident and (3) the online discussion platform, Knowledge Constructor, to support the collaborative effort of sharing, building and reflection of ideas constructed as a group. The harness stage, that required the students to define the problem space, incorporated an additional element of knowledge creation. The students had to discuss ideas focusing on the law of conservation of energy, interpret the laws in the context of the problem and construct a mathematical expression for the roller coaster plunging motion. The students then hypothesised the cause of the problem and performed investigations with a model of the roller coaster provided by the teacher. Unlike the linear THINK cycle, throughout the process, the students were guided to build on one another's ideas and make revisions to their ideas which include revisions to ideas identified at the harness stage.

The following three excerpts demonstrate the iterative process of knowledge creation of a group of five 14-year-olds as they worked collaboratively to solve the problem described earlier. In Excerpt 10.1, students M and J were of the opinion that only friction (when comparing with air resistance and friction) affected the roller coaster ride because the "air resistance is smaller than friction" (turn 37).

Excerpt 10.1. Students' Interpretation of Problem Context Using Their Prior Knowledge

Note	Author	Date/time	Content
21	M	2006-07-26 10:18:45	If u eliminate other factors like air resistance, friction, etc., the roller coaster will never stop in the first place since energy is neither gained nor lost
30	J	2006-07-26 10:30:13	Friction would affect the roller coaster more than the air resistance and cannot be negligent. Friction would cause energy to be converted to heat energy and thus lesser energy would be available for kinetic energy, and distanced object moves is reduced
34	Teacher	2006-07-26 10:35:57	Why do you say that the effect of air resistance is less than friction?
37	J	2006-07-26 10:39:15	The air resistance is smaller than friction because the roller coaster has more contact with the track than the air. Furthermore, there is gravity acting on the train, pulling it to the track which makes friction greater as gravity has to be overcome to create horizontal motion

Through an Internet search, they found that "a roller coaster ride uses the work-energy theorem" as posted on Knowledge Constructor shown in Excerpt 10.2.

Excerpt 10.2. Work-Energy Theorem

$$\mathrm{KE}_{\mathrm{initial}} + \mathrm{PE}_{\mathrm{initial}} + W_{\mathrm{external}} = \mathrm{KE}_{\mathrm{final}} + \mathrm{PE}_{\mathrm{final}}$$

The kinetic energy the cart has at first is 0 as it is not moving, added to the potential energy it has due to its height above the ground and added to the external forces before it starts moving (which is 0); the sum would equal to the final kinetic energy (which would increase as the cart goes down due to motion) and the final potential energy, which would go decrease as the cart goes down due to decrease in height.

With these two pieces of information – that only friction would affect the motion of the roller coaster and the work-energy theorem – they derived an expression to explain the workings of the roller coaster ride in the problem, as shown in Excerpt 10.3.

Excerpt 10.3. Derivation of Expression Describing the Roller Coaster Ride

23	D	2006-08-04	As the $KE_{initial} + PE_{initial}$ is always known. In the case of the roller coaster, $PE = mgh$ $KE = (1/2)\ mv^2$
		16:38:13	PE can be calculated from the measurements and KE would be 0. By means of working backwards, we can find the force which occurs in the whole process $0 + mgh + [-\text{force} \times \text{displacement} \times \text{cosine (theta)}] = 0 + 0$ $mgh = \text{force} \times \text{displacement} \times \text{cosine (theta)}$
27	M	2006-08-05	Actually W_{ext} = work done by external forces How do we find that? work done by ext factors = force × displacement × cosine (theta) Expand out = (frictional force) × displacement × cosine (theta) $(FN = -mg)$ × displacement × cosine (theta) Where: FN is the normal force in Newton (N),
		12:50:57	m is the mass in kg and g is the gravitational force in m/s^2 Force of friction can be calculated by the formula $Ff = \mu\ FN$ Ff is the force of friction in N, μ is the coefficient of friction and FN is the normal force in N

Based on the information given about the roller coaster accident, the students hypothesised that the accident was caused by overloading of passengers into the roller coaster cart. Testing their hypothesis with the model set-up of the roller coaster given as well as the derived expression, they obtained conflicting results. This led one student to wonder aloud on the Knowledge Constructor "mass DOES not affect the stopping distance. What have we neglected in the process of deriving the final equation?"

The discrepancy between the empirical and theoretical evidences led to another round of knowledge building in their interpretation of work-energy theorem in the context of the problem. Finally, in their written report, they wrote, "Without adding in air resistance into external forces, the stopping distance is the same, so air resistance does affect the stopping distance, ..." and further refined the derived expression to take into account the effect of air resistance. This marked another significant advancement in their scientific conceptions, which resulted from a reflection of their own understanding of the work-energy theorem.

Discussion

We have described two approaches of inquiry science learning environments, which are premised on the knowledge creation metaphor of learning. This section further explicates their potential contributions to the field of inquiry science by analysing the enacted inquiry science pedagogical components and discussing its implications to the design of inquiry science learning environment.

Enacted Inquiry Science Pedagogical Components

Table 10.2 describes the key pedagogical components (Poon et al., 2010) of the two cases to highlight the inquiry science learning activities. The intention is not to describe the detailed sequences of these components, which vary from session to session, but to show how the two cases afford science learning through inquiry. We do this by using the language and pedagogical components of common inquiry science framework (Poon et al.). There are, however, other salient features of these two cases that suggest alternative ways of how these pedagogical components could be achieved.

Interacting with Physical Phenomena and Epistemic Discourse for Inquiry Science

Kelly (2008) suggests that inquiry science should not only engage students in interacting with physical phenomena through empirical investigations but also engage students in interacting with others in epistemic discourse and reasoning around the phenomena. We will describe how the two cases described engage students in interacting with others.

For the first case, Table 10.1 shows examples of students' epistemic discourse. We further examined the students' discourse move in terms of their questioning behaviours (Tan & Seah, 2011). The ability to ask questions is also an important component of scientific literacy (Hofstein, Navon, Kipnis, & Mamlok-Naaman, 2005) and knowledge building discourse, as good questions form the critical initial step for construction of scientific knowledge (Shodell, 1995). Students' questions could help to resolve their puzzlement and narrow their knowledge gaps (Chin & Osborne, 2008), which could eventually lead to deeper conceptual understandings (Harper, Etkina, & Lin, 2003). From the notes in the online forums populated by the students (including forums created for other topics not described in the first case), three main categories of questions emerged: scientific, epistemological and meta-discoursal questions (Tan & Seah). Scientific questions seek ideas related to science contents that could address different aspects of the same topic, for example, "which

Table 10.2 Key pedagogical components of the inquiry science activities in the two cases

Pedagogical components	Knowledge building	KC-PBL	Key features
Engage	Puzzling questions of a phenomenon	Problem scenario	Trigger for inquiry
Prepare to investigate	Explanation of the inquiry task and suggested resources	Presentation of the problem scenario and the main tasks	Scaffolds provided by teachers
Explore	Students making meaning of authoritative sources of knowledge to answer the puzzling questions	Students making meaning of authoritative sources of knowledge that could eventually help in the problem-solving process	Interacting with peers, engage in epistemic discourse in a social collaborative setting, creation of knowledge artefacts (notes in a forum) and use of technology to support epistemic discourse
Investigate	Students design and conduct experiments that have consequential values to explanation of the phenomenon	Students conduct experiment and test their hypothesis with a physical model	Interacting with the phenomenon, collecting empirical data
Explain	Students offer explanations for the puzzling phenomenon	Students develop solutions to the problem and provide justification for their solutions	Iterative process of improving ideas and explanations through epistemic discourse, difficult to differentiate from explore phase, creation of knowledge artefacts (notes in a forum) and use of technology to support epistemic discourse
Evaluate	Students engage in self-evaluation in their epistemic discourse. Teachers provide evaluation at relevant junctures	Students engage in self-evaluation in their epistemic discourse. Teachers provide evaluation at relevant junctures and evaluate the final solutions	Epistemic discourse suggests a continual evaluation of the claims made in the discussion, creation of knowledge artefacts (notes in a forum) and use of technology to support epistemic discourse

parts of the digestive system in our mouth help grinds the food?" Scientific questions also indicate the types of scientific ideas that students deem as relevant to the topic or task as well as the conceptual challenges that students are grappling with. Epistemological-type questions seek for the resources, evidences and criteria of the claims made, for example, "why is the circulatory system the most important system of all?" These questions suggest that the students adopt a critical stance towards the claims posted by their peers. The quantity and range of questions of this nature thus provide an indicator of the level of critical inquiry that students engage in as part of the knowledge building process. Meta-discoursal questions seek clarification about what has been discussed, for example, the meaning of a post or possible link to alternatives. For example, "what do you mean by soft food?" or "do you mean muscles don't need energy to relax?" The quantity of meta-discoursal questions indicates the level of interactivity between peers and engagement with peers' ideas. This kind of questions, which promote interaction and engagement among participants, is critical for social collaborative inquiry to work. The nature of the teacher's questions also plays an important role in the students' questioning. For example, among the forums for various topics, the forum used in the first case generated the most number of epistemological questions. The teacher posed a problem scenario with no single correct answer. It seemed that the students were engaged in a hypothesis-generating discussion, and they challenged one another to provide evidence and support for their claims. In another forum for the topic of digestive system, the teacher asked "do you see similarities between household items and parts of the [digestive] system [?]" This forum generated the highest frequency of meta-discoursal questions. One possible explanation is that the teacher's question requires students to use analogies relating to their daily life rather than to use authoritative sources of information. This reference to daily life rather than more formal canonical knowledge could have encouraged students to challenge their peer's ideas.

The second case illustrates how epistemic discourse and interacting with physical phenomenon are intertwined. Adopting the lens of cultural-historical activity theory, or CHAT (Engeström, 1999), the two key activities – model construction and problem-solving – were observed in this enacted KC-PBL. From the model construction activity, students developed a deeper understanding of the law of conservation of energy, while from the problem-solving activity, they developed investigative and argumentative skills through manipulating with a physical set-up. Table 10.3 summarises the two activities.

While the two activity systems have different objects and were driven by different motives, they are interconnected. In the first model-building episode, the students derived a mathematical expression to describe the roller coaster ride based on their naive interpretation of work-energy theorem in the context of the problem. The problem-solving episode, whereby they tested their hypothesis of the cause of the accident by generating both theoretical and empirical evidences, introduced conflict into their initial theory building, which triggered a second model-building episode. It was the second model-building exercise that they found resolution to their problem and refined their theory of the roller coaster ride. This represents another significant advancement of their scientific conception about the generalised work-energy theorem and the problem context.

In short, the derived mathematical expression constructed during the knowledge building activity served as the mediating tool for problem-solving. The process of problem-solving, in turn, generated puzzling discrepancy which triggers another

Table 10.3 The model construction and problem-solving activities

	Model construction through knowledge building	Problem-solving
Object	Theory building about roller coaster	Problem of roller coaster
Outcomes	A generalised expression to describe the motion of a roller coaster	More puzzling questions (1st iteration), problem solution (2nd iteration)
Tools	Internet, knowledge constructor, problem context and scaled model	The derived expression constructed during the modelling activity, evidence created by the teachers and researchers and the model set-up
Rules	Sharing of information as students posted the information they had found on the Internet, negotiating information found on the Internet and co-interpreting work-energy theorem in the context of the problem	Sharing of information as students posted the information they had found on the Internet, negotiating information found on the Internet and co-interpreting work-energy theorem in the context of the problem
Division of labour	Teacher as facilitators and designers and students as collaborative knowledge builders	Teacher as facilitators and designers and students as collaborative knowledge builders

round of knowledge building. The two processes are interdependent and the iterations between these two processes help to resolve the content-skills divide that has plagued the inquiry process during the first two research cycles. This interdependence of the knowledge construction and problem-solving processes provides a possible solution to resolving the tension between practising and learning a discipline mentioned by Kirschner et al. (2006).

Social Collaborative Process of Inquiry

Many inquiry science approaches are premised on cognitive theories of learning. For example, the BSCS 5E inquiry model (http://www.bscs.org/our-values) is based on constructivist theories of Ausubel (Ausubel, Novak, & Hanesian, 1978) and Vygotsky (1968). It recognises the importance of integrating new knowledge into a learner's prior knowledge and teachers as important agent in scaffolding students in developing scientific understanding within the students' zones of proximal development. Historical development of scientific theories, however, shows individual scientists reasoned with empirical "decision vectors" (Solomon, 2008, p. 88) that privilege certain points of view. Consequently, at the collective level, competing theories might arise. This is illustrated by Solomon, who provided an example that in the formative years of the continental drift theory, supporters of this theory congregated in the southern hemisphere of the earth. Over time, the theory that enjoys empirical successes across distributed settings might become the consensual theory. This example illustrates the social epistemologies of science (Solomon).

In our two research projects, it is important to provide a platform (the online forum) where diversity of ideas is possible and visible to all students. The differing ideas to the same problem or question provide the necessary tension for students to probe deeply into their understanding and engage in epistemic discourse. It creates intersubjectivity among the participants where differing points of view are subjected to questioning and scrutiny. In knowledge building pedagogy, idea diversity (Scardamalia, 2002) is one of the key pedagogical principles. To resolve differences, other pedagogical principles are invoked. For example, it is also necessary for the students to assume collective cognitive responsibility to improve one another's ideas, to demonstrate epistemic agency in taking ownership in learning and to engage in knowledge building discourse in idea improvement. In other words, idea diversity initiates the dissentious tension, while epistemic discourse provides the rational bases towards building consensus.

Supportive Technologies for Knowledge Creation

In the two research projects, we used online discussion forums as supportive technologies. Consistent with the social collaborative process of inquiry, these technologies provide additional communication channel for students to exchange ideas, thus extending the discussion beyond the temporal constraints of face-to-face meeting.

Suthers (2006) summarised the key affordances of technologies that support online collaborative learning. In essence, these technologies afford possibilities for knowledge actions (e.g., discussion of claims and supporting evidence), provide referential capabilities (e.g., linking of related notes), allow mobility of inscriptions (e.g., export discussion to note reader) and promote interaction patterns (e.g., customisable scaffolds).

Specifically, these technologies help to capture students' ideas. As discussed, knowledge artefact is a key mediating tool in the knowledge creation process. Reification of students' ideas allows the knowledge artefacts to be worked on and improved, much like design of physical objects like cars. These technologies also include features that support epistemic work on the students' ideas. For example, customisable scaffolds – [My theory is], [A better theory is] – encourage epistemic discourse. Rise-above functions engage the students in metacognitive actions of reflecting on initial ideas and moving towards higher level of explanation. In addition, these technologies track the development of ideas that result from interactions among the students, thus making the students' thinking visible, and keeping a record of the historical development of ideas.

Conclusion

The two cases of inquiry science learning environments sit at the nexus of learning sciences research and science education. One of their potential contributions to the field of inquiry science is the integration of knowledge creation approaches to learning. The knowledge creation approaches – particularly the knowledge building pedagogy pioneered by Scardamalia and Bereiter (2006) – hold promise to enrich inquiry science approaches for the following reasons: first, scientists are among the most recognised knowledge workers; their key mission is to generate new knowledge. Thus, engaging students in knowledge creation puts them on a journey of taking up socio-epistemic practices of scientists. Second, cognizant of the differences between epistemology and pedagogy, knowledge building pedagogy (Scardamalia & Bereiter) has been implemented and studied in educational settings, in particular, in K-12 schools. It provides an instructional model that could foster knowledge creation capacity of learners. Beyond the socio-cognitive outcomes, knowledge building aims to develop in students the epistemic agency, which refers to the capacity and volition to participate in the knowledge creation process. This existential transformation of the students' identities towards a knowledge builder is regarded as an ontological change (Packer & Goicoechea, 2000). Third, knowledge building provides instructional principles that engage learners in epistemic discourse and reasoning around the phenomena (Kelly, 2008), which complements the interactions with phenomenon through empirical investigation that are suggested by other inquiry science approaches.

For educators who are keen to implement and study inquiry science environments with knowledge creation approaches, these design principles could be useful to (a) provide opportunities for students to create epistemic artefacts that represent their ideas related to the science inquiry tasks, (b) provide a platform that captures and tracks the development of these ideas, (c) provide scaffolds that help students engage in productive epistemic discourse to improve their ideas, (d) provide support that helps introduce the appropriate semiotic resources into students' meaning-making processes and (e) integrate interaction with physical phenomenon and epistemic discourse in the inquiry science activities.

References

Ausubel, D., Novak, J., & Hanesian, H. (1978). *Educational psychology: A cognitive view* (2nd ed.). New York: Holt, Rinehart & Winston.

Barab, S. A., & Hay, K. E. (2001). Doing science at the elbows of experts: Issues related to the science apprenticeship camp. *Journal of Research in Science Teaching, 38*, 70–102.

Bereiter, C. (1994). Implications of postmodernism for science, or, science as a progressive discourse. *Educational Psychologist, 29*(1), 3–12.

Bybee, R. W., Taylor, J. A., Gardner, A., Van Scotter, P., Powell, J. C., Westbrook, A., et al. (2006). *The BSCS 5E instructional model: Origins and effectiveness*. Colorado Springs, CO: BSCS.

Chin, C., & Osborne, J. (2008). Students' questions: A potential resource for teaching and learning science. *Studies in Science Education, 44*, 1–39.

Driver, R., & Oldham, V. (1986). A constructivist approach to curriculum development in science. *Studies in Science Education, 13*, 105–122.

Engeström, Y. (1999). Innovative learning in work teams: Analyzing cycles of knowledge creation in practice. In Y. Engestrom, R. Miettinen, & R. L. Punamaki (Eds.), *Perspectives on activity theory* (pp. 377–404). Cambridge, UK: Cambridge University Press.

Grandy, R. E., & Duschl, R. A. (2008). Consensus: Expanding the scientific method and school science. In R. A. Duschl & R. E. Grandy (Eds.), *Teaching scientific inquiry: Recommendations for research and implementation* (pp. 304–325). Rotterdam, The Netherlands: Sense Publishers.

Hammer, D., Russ, R., Mikeska, J., & Scherr, R. (2008). Identifying inquiry and conceptualizing students' abilities. In R. A. Duschl & R. E. Grandy (Eds.), *Teaching scientific inquiry: Recommendations for research and implementations* (pp. 138–156). Rotterdam, The Netherlands: Sense Publishers.

Harper, K. A., Etkina, E., & Lin, Y. (2003). Encouraging and analyzing student questions in a large physics course: Meaningful patterns for instructors. *Journal of Research in Science Teaching, 40*, 776–791.

Hofstein, A., Navon, O., Kipnis, M., & Mamlok-Naaman, R. (2005). Developing students' ability to ask more and better questions resulting from inquiry-type chemistry laboratories. *Journal of Research in Science Teaching, 42*, 791–806.

Kelly, G. J. (2008). Inquiry, activity, and epistemic practice. In R. A. Dusch & R. E. Grandy (Eds.), *Teaching scientific inquiry: Recommendations for research and implementation* (pp. 99–117). Rotterdam, The Netherlands: Sense Publishers.

Kirschner, P. A., Sweller, J., & Clark, R. E. (2006). Why minimal guidance during instruction does not work: An analysis of the failure of constructivist, discovery, problem-based, experiential, and inquiry-based teaching. *Educational Psychologist, 41*, 75–86.

National Research Council. (2000). *Inquiry and the national science education standards*. Washington, DC: National Academy Press.

Nonaka, I., & Takeuchi, H. (1995). *The knowledge creating company: How Japanese companies create the dynamics of innovation*. New York: Oxford University Press.

Paavola, S., & Hakkarainen, K. (2005). The knowledge creation metaphor – An emergent epistemological approach to learning. *Science & Education, 14*, 535–557.

Paavola, S., Lipponen, L., & Hakkarainen, K. (2004). Models of innovative knowledge communities and three metaphors of learning. *Review of Educational Research, 74*(4), 557–576.

Packer, M. J., & Goicoechea, J. (2000). Sociocultural and constructivist theories of learning: Ontology, not just epistemology. *Educational Psychologist, 35*, 227–241.

Poon, C. L., Lee, Y. J., Tan, A. L., & Lim, S. S. L. (2010). Knowing inquiry as practice and theory: Developing a pedagogical framework with elementary school teachers. *Research in Science Education, 42*, 303–327.

Popper, K. (1978, April 7). *Three worlds*. Retrieved September 9, 2013, from http://www.tannerlectures.utah.edu/lectures/documents/popper80.pdf

Savery, J. R. (2006). Overview of problem-based learning: Definitions and distinctions. *The Interdisciplinary Journal of Problem-based Learning, 1*(1), 9–20. Retrieved September 9, 2013, from http://docs.lib.purdue.edu/cgi/viewcontent.cgi?article=1002&context=ijpbl

Scardamalia, M. (2002). Collective cognitive responsibility for the advancement of knowledge. In B. Smith (Ed.), *Liberal education in a knowledge society* (pp. 67–98). Chicago: Open Court.

Scardamalia, M., & Bereiter, C. (2006). Knowledge building: Theory, pedagogy, and technology. In R. K. Sawyer (Ed.), *The Cambridge handbook of the learning sciences* (pp. 97–118). New York: Cambridge University Press.

Sfard, A. (1998). On two metaphors for learning and the dangers of choosing just one. *Educational Researcher, 27*(2), 4–13.

Shodell, M. (1995). The question-driven classroom. *American Biology Teacher, 57*, 278–281.

Solomon, M. (2008). Knowledge as social processes. In R. A. Duschl & R. E. Grandy (Eds.), *Teaching scientific inquiry: Recommendations for research and implementation* (pp. 86–94). Rotterdam, The Netherlands: Sense Publishers.

Stahl, G. (2004). Building collaborative knowing: Elements of a social theory of CSCL. In J. W. Strijbos, P. A. Kirschner, & R. L. Martens (Eds.), *What we know about CSCL: And implementing it in higher education* (pp. 53–86). Dordrecht, The Netherlands: Kluwer Academic Publishers.

Sterelny, K. (2004). Externalism, epistemic artefacts and the extended mind. In R. Schantz (Ed.), *The externalist challenge: new studies on cognition and intentionality* (pp. 239–254). Berlin, Germany/New York: de Gruyter.

Suthers, D. (2006). Technology affordances for intersubjective meaning-making: A research agenda for CSCL. *International Journal of Computer Supported Collaborative Learning, 1* (3), 315–337.

Tan, S. C., & Seah, L. H. (2011). Exploring relationship between students' questioning behaviours and inquiry task in an online forum through analysis of ideational function of questions. *Computers & Education, 57*, 1675–1685.

Tien, L. T., Ricky, D., & Stacy, A. M. (1999). The MORE thinking frame: Guiding students' thinking in the laboratory. *Journal of College Science Teaching, 28*, 318–324.

Vygotsky, L. (1968). *Mind in society*. Cambridge, MA: Harvard University Press.

Wenger, E. (1998). *Communities of practice: Learning, meaning, and identity*. New York: Cambridge University Press.

Chapter 11
Using Inquiry to Facilitate Meaningful Learning in Inorganic Chemistry Qualitative Analysis

Kim Chwee Daniel Tan

Abstract Many teachers find that their students do not seem to be able to link what they do in the laboratory with what they learn in the classroom. As such, students are unaware of what to think about and take note of during experiments; the same situation exists in Singapore. For example, research using a two-tier multiple choice diagnostic instrument has shown that Singapore secondary and junior college students, as well as graduate pre-service teachers, have difficulty understanding the procedures and reactions involved in basic inorganic chemistry qualitative analysis practical work. To address these difficulties and make learning during practical work more meaningful, an inquiry-based approach to learning basic qualitative analysis was developed to engage students in learning the key procedures, reactions and concepts involved through empirical investigations. The inquiry-based instructional package was implemented in one school and the same diagnostic instrument was administered to the students involved 4 months later. The results indicated that the secondary 4 students who had undergone the instructional package obtained better scores than the secondary 4 students involved in the first study and had lower levels of alternative conceptions. This indicated the feasibility of using inquiry to facilitate students' understanding and application of science concepts during practical work.

Keywords Inquiry • Inorganic chemistry • Qualitative analysis • Acid-base reactions • Practical work • Alternative conceptions • Secondary chemistry • Science laboratory

K.C.D. Tan (✉)
National Institute of Education, Nanyang Technological University, 1 Nanyang Walk, Singapore 637616, Singapore
e-mail: daniel.tan@nie.edu.sg

A.-L. Tan et al. (eds.), *Inquiry into the Singapore Science Classroom*, Education Innovation Series, DOI 10.1007/978-981-4585-78-1_11,

Introduction

Practical work 'can be effective in promoting cognitive skills, metacognitive skills, practical skills, and attitude and interest towards chemistry' (Hofstein, 2004, p. 259). However, studies have shown that students find practical work difficult and that they have little understanding of what they do during practical work (Hart, Mulhall, Berry, Loughran, & Gunstone, 2000; Hodson, 1992). Sere (2002) argues that students do not understand the purpose of apparatus and how to use them, as well as what to observe during experiments, because they do not draw upon the theory that they have learned in class to guide what they do in the laboratory. Students in Singapore have similar difficulties with practical work, for example, when conducting qualitative analysis experiments.

In secondary inorganic chemistry qualitative analysis (QA), Singapore students are generally required to identify cations, anions and gases using reagents and procedures as specified in the General Certificate of Education Ordinary Level (O-level) chemistry syllabus (University of Cambridge Local Examinations Syndicate [UCLES], 2011) or to deduce the properties of unknown substances based on the results obtained from experiments using the specified reagents (see Fig. 11.1).

Qualitative analysis experiments require students to be able to carry out the procedures stated in Fig. 11.1 proficiently, know what to observe and record, and apply the concepts of '*acids, bases and salts*', '*oxidation and reduction*', '*reactivity of metals*' and '*periodicity*' to make inferences on the properties and/or identities of unknown compounds given in the experiments. Thus QA experiments involve manipulative, observational, analytical and inferential skills, as well as the understanding, integration and application of concepts, making QA an important topic in secondary school chemistry. Teachers normally introduce QA by demonstrating and explaining the procedures that the students need to carry out and teach them how to write the observation as required in the chemistry examinations. The students then do a series of tests for the various cations, anions and gases before being given past years' examination papers where they follow the procedures given to determine the identities or properties of unknown substances. The focus is on getting the 'correct' observations and using the observations to get the 'correct' identities and properties as marks are awarded for these, and explanations of what happened when the procedures are carried out are seldom required.

Thus, it is not surprising that students admitted that they merely followed instructions during QA practical work with little idea about what they were doing and that they were more concerned with getting the correct results than understanding the procedures and reactions involved (Tan, Goh, Chia, & Treagust, 2001). Teachers also have commented that their students adopted a recipe-approach during QA practical work, could not carry out the experimental procedures properly and did not understand the purpose of the experimental procedures (Tan, 2000). The students' lack of understanding in QA has been noted by the O-level examiners. For example, in the 1996 examination report (UCLES, 1997), the examiners highlighted that many students invented 'observations' for several tests in which

a) describe the use of aqueous sodium hydroxide and aqueous ammonia to identify the following aqueous cations: aluminium, ammonium, calcium, copper (II), iron (II), iron (III), lead (II) and zinc (formulae of complex ions are not required)
b) describe tests to identify the following anions: carbonate (by the addition of dilute acid and subsequent use of limewater); chloride (by reaction of an aqueous solution with nitric acid and aqueous silver nitrate); iodide (by reaction of an aqueous solution with nitric acid and aqueous lead (II) nitrate); nitrate (by reduction with aluminium and aqueous sodium hydroxide to ammonia and subsequent use of litmus paper) and sulfate (by reaction of an aqueous solution with nitric acid and aqueous barium nitrate)
c) describe tests to identify the following gases: ammonia (using damp red litmus paper); carbon dioxide (using limewater); chlorine (using damp litmus paper); hydrogen (using a burning splint); oxygen (using a glowing splint) and sulfur dioxide (using acidified potassium dichromate (VI))

Fig. 11.1 Learning objectives related to the identification of ions and gases as specified in the secondary chemistry syllabus (UCLES, 2011, p. 7)

there should not have been any reaction; the students could not figure out why there were no reactions, so they decided to create the observations for the tests. Other researchers have also commented that students seemed to be unaware of what they should be doing and thinking about during practical work and did not or could not use the theory that they had learnt to guide them in their practical tasks (Berry, Mulhall, Gunstone, & Loughran, 1999; Hodson, 1992). Cognitive overloading (Johnstone & Wham, 1982; Nakhleh & Krajcik, 1994) could be a cause of student difficulties in QA as students need to read instructions, carry out procedures using reagents and apparatus, observe and record reactions, interpret results and be mindful of the time left to complete the practical tasks such that they have little capacity left to think and reflect on what they are doing. In addition, the aims and requirements of QA practical work might not have been explicitly elaborated by teachers and understood by students, so students think that they just need to follow instructions and make the required observations; this reduces QA practical work to a mechanical level requiring very little thinking (Gunstone, 1991; Tasker & Freyberg, 1985).

Understanding of QA

To investigate students' understanding of the reactions and procedures involved in secondary QA, a two-tier multiple choice instrument (Treagust, 1995), the Qualitative Analysis Diagnostic Instrument (QADI), was developed (Tan, 2000; Tan, Goh, Chia, & Treagust, 2002b). An example of the items in the QADI is given in Fig. 11.2. This instrument was administered to 915 secondary 4 students, 360 junior college years 1 and 2 students and 181 graduate pre-service teachers (Tan, Goh, Chia, & Treagust, 2002a). The secondary 4 students (60 % female and 40 % male) were from 11 secondary schools whose teachers agreed to participate in the study. These 11 schools included seven co-educational schools, two girls' schools and two boys' schools. The junior college students (50 % females and

Question 11

A student is given a solution which may contain carbonate ions. She decides to add aqueous barium nitrate (V) ($Ba(NO_3)_2$) to the solution first, followed by dilute nitric (V) acid. Is it possible for her to determine whether carbonate ions are present when she carries out the above procedures followed by the identification of the gas involved (if any)?

A Yes B No

Reason/Justification

1. An insoluble carbonate would be formed leaving no free carbonate ions in solution to react with the acid.
2. Dilute nitric (V) acid will react with both a soluble and an insoluble carbonate.
3. The above procedure is strictly to test for the presence of sulfate (VI) (SO_4^{2-}) only.
4. The acid must be added directly to the solution to test for the presence of a carbonate.
5. The unknown compound will only react with the dilute nitric (V) acid and not with the barium nitrate (V).

Fig. 11.2 An example of the items in the qualitative analysis diagnostic instrument

50 % males) were from three junior colleges, one in the upper band of junior colleges, one in the middle band and one in the lower band. As a whole, the sample of junior college students could be considered to be representative of junior college students in Singapore, while the sample of Grade 10 students could be considered to be slightly biased towards the population of higher-achieving students. The demographic data of the pre-service teachers were not considered because many pre-service teachers did not supply their details.

The cross-age study showed that alternative conceptions in QA such as precipitation reactions in the tests for cations and anions, and addition of acids in the test for anions, were prevalent among all participants (see Table 11.1), though in general, the percentage of pre-service teachers (T) having alternative conceptions was the lowest, followed by the junior college (JC) and secondary (S) students. This trend was expected as the participants should have a greater understanding of the chemistry involved in QA with increasing number of years of chemistry education. However, it was a concern that a number of pre-service teachers still had alternative conceptions in a topic which they would be teaching in school.

Many cations and anions are identified by reactions involving the formation of precipitates. For example, copper (II) ions are initially identified by the formation of a blue precipitate in a reaction between copper (II) salt solutions and aqueous ammonia, with further reaction with excess aqueous ammonia resulting in a deep blue tetraamminecopper (II) solution. In the formation of the blue solid, copper (II) ions combine with hydroxide ions, and these oppositely charged ions aggregate to form insoluble copper (II) hydroxide which precipitates out. However, many students and pre-service teachers described the precipitation process as a displacement reaction in which 'a more reactive ion displaced a less reactive ion' (Tan et al., 2002a, 2002b). A displacement reaction is an example of a reduction-oxidation reaction in which a more reactive element donates electrons to the ions of a less reactive element; for example, when magnesium is introduced into a solution of copper (II) sulfate (VI), the metal will donate electrons to the copper (II) ions present in the solution to produce magnesium ions and copper metal. Displacement reactions were taught in the topics of '*reactivity of metals*' and '*periodicity*', and it seemed that the students and pre-service teachers could not differentiate between

Table 11.1 A sample of alternative conceptions of the secondary, junior college and pre-service teacher participants

		Percentage of students with the alternative conception (%)		
Alternative conception	Question	S ($n = 915$)	JC ($n = 360$)	T ($n = 181$)
Precipitation				
A more reactive ion displaces a less reactive ion in precipitation reaction	Q1 (A3)	25	12	17
	Q5 (A1, A4)	37	35	27
	Q13 (A3, A4)	15	11	12
	Q18 (A1, A2, A4 & A5)	29	21	17
Addition of acid				
Acid is needed to acidify the mixture so that reaction can proceed properly	Q6 (A5)	22	22	17
Carbonate ions cannot be determined if acid is added after the addition of barium nitrate (V)	Q17 (A4)	35	38	31
The procedure is strictly a test for sulfate (VI)	Q11 (B3)	20	24	11
The acid must be added directly to the unknown sample	Q11 (B4)	25	33	18
Test for oxygen				
Test for oxygen has to be carried out when a substance is heated because all gases have to be tested	Q10 (A1)	23	27	16
Oxygen gas need not be tested when a substance is heated because oxygen will not be liberated during heating; instead it will be used up	Q10 (B3)	27	19	32

the combination of oppositely charged ions which occur in a precipitation reaction and the reduction-oxidation processes involved in a displacement reaction. This could be due to the formation of solids in both precipitation and displacement reactions (even though the products are vastly different, e.g., blue copper (II) hydroxide compared to brown copper metal), as well as the use of sodium hydroxide solution in the tests for cations as they might think that 'more reactive' sodium ions displace 'less reactive' cations in the unknown solutions to form precipitates. The latter alternative conception could have originated from students memorising the reactivity of metal series and thought that sodium ions were just as reactive as the sodium metal and could take part in displacement reactions (Tan et al., 2002a).

The pre-service teachers and students also had difficulty understanding the use of acids in the test for anions (Tan et al., 2002a, 2002b). Chloride, sulfate (VI), sulfate (IV), carbonate and iodide anions can be determined by the formation of their insoluble salts and whether these salts react with specific dilute acids to liberate gases which can be subsequently identified. Their lack of understanding of the role of the dilute acids could be seen from Table 11.1. Many indicated that acids were needed so that reactions could proceed 'properly' without being able to elaborate what 'proceeding properly' meant (Tan et al., 2002a). In addition, they also seemed to think that acids must be added directly to the unknown samples in order to test for carbonates; they did not seem to realise that if barium nitrate (V) solution was added to unknown solutions containing carbonate anions, the barium carbonate precipitate formed could still react with dilute nitric (V) acid to produce carbon dioxide. Related to this difficulty, a number of students and pre-service teachers also thought that the use of barium nitrate (V) solution followed by dilute nitric (V) acid could only be used to test for sulfate (VI) ions; they did not realise that carbonate, sulfate (IV) and sulfate (VI) anions could be differentiated using this procedure.

Testing for gases liberated in QA experiments was generally difficult for students because they could not determine which gas to test for in a given situation, so they tested for all the gases that were included in the syllabus (Tan et al., 2001). This random testing wasted time as unnecessary tests were carried out; the students did not realise that the procedures and reagents used could give indications of the gases that could be liberated. For example, if dilute acid is added to a white solid, it is unnecessary to test for hydrogen and oxygen as an acid will not react with any inorganic compound given to students in basic QA experiments to liberate hydrogen or oxygen gas. If the dilute acid is added to a metallic substance and a gas is evolved, then a lighted splint must be used to test for hydrogen gas as some metals will react with dilute acids to give hydrogen gas. If an unknown solid is heated, then oxygen gas needs to be tested as many nitrates (V) will decompose on heating to give the oxide, nitrogen dioxide gas and oxygen gas. In item 10 of the QADI, many students and pre-service teachers indicated that all gases had to be tested for when an unknown solid was heated or that oxygen gas would not be liberated when an unknown solid was heated because oxygen gas would be used up (as in combustion/oxidation reactions). These findings were indications that the respondents did not understand or were not aware of the reactions and concepts underlying the procedures in QA experiments. Sweeney, Bula and Cornett (2001) reported similar findings that students 'preferred to follow prescribed, step-by-step instructions and struggled even then to understand the chemical concepts which provided a rationale explaining why these steps were being followed' (p. 418) in a practical work session in a US school where the students had to identify a series of unknown compounds.

Learning QA by Inquiry

One possible way to help students learn QA meaningfully and to address alternative conceptions is for students to learn QA by inquiry. Inquiry involves finding solutions to problems by asking questions that can help to solve the problems, designing and conducting experiments to answer the questions, collecting and analysing data, drawing conclusions and reporting and defending the findings of the inquiry (Chinn & Malhotra, 2002). This is the definition of inquiry adopted in this chapter, and it is in line with social constructivist views of learning that students construct their knowledge collaboratively when they grapple with the challenges of designing experiments and making sense of data (Crawford, 2000, 2012). Teachers play the key role of scaffolding the inquiry process through the use of 'questions, prompts, task selection and structured classroom interactions as cognitive supports for recall of appropriate knowledge and skills in a timely fashion to maintain the continuity of investigative work' (Flick, 2000, p. 109). Depending on the ability and readiness of the students (and the teachers), learning by inquiry can be more open-ended with students posing their own questions for investigation and deciding what data to collect and how to analyse them, to the other end of the continuum where inquiry by students is very guided with the activities determined by the teacher and procedures spelt out explicitly for the students to follow (National Research Council [NRC], 2000).

An instructional package on qualitative analysis, the Qualitative Analysis Teaching Package (QATP) (Tan, 2000; Tan, Goh, Chia, & Treagust, 2004, 2009), was developed to facilitate the learning of QA in line with the essential features of inquiry advocated by the National Research Council (2000) where students would learn the key procedures, reactions and concepts involved in QA through guided inquiry. An example of an activity for student to gain tacit experience (Woolnough & Allsop, 1985) of precipitation is given in Fig. 11.3 where students are instructed to add sodium hydroxide solution to sodium chloride solution and then to iron (III) chloride. They will observe that sodium chloride solution and iron (III) chloride solution interact differently with sodium hydroxide solution; in the case of sodium chloride solution, there is no visible change, whereas a reddish brown precipitate is formed in the case of iron (III) chloride solution. They are to explain their observations, compare them with their classmates' explanations, discuss and resolve any conflict and arrive at a consensus which they will compare with their teacher's explanation; the iron (III) chloride solution has reacted with the sodium hydroxide solution to form an insoluble compound, iron (III) hydroxide, which precipitates out, whereas sodium ions will not form any precipitate with hydroxide ions. The students will be encouraged to resolve any differences between their explanations and those of their teacher. The teacher will remind them that they have been introduced to double decomposition and precipitation reactions in the preparation of insoluble salts section of the topic of '*acids, bases and salts*' (see Fig. 11.4) to make explicit links between the theory that they learn in class and

Addition of sodium hydroxide to sodium chloride and iron (III) chloride

Procedure	What you see
Add aqueous sodium chloride into a test tube until it reaches a height of 1 cm. Add two drops of aqueous sodium hydroxide and shake the mixture.	
Add aqueous iron (III) chloride into a test tube until it reaches a height of 1 cm. Add two drops of aqueous sodium hydroxide and shake the mixture.	

1. Explain the differences in what happened above.
2. Compare your answer with your partner or another pair/group and try to reconcile differences (if any). Write down the agreed explanations if it is different from (1).
3. After listening to the explanation given by your teacher, in your own words, explain what is meant by: (you may discuss with your classmates)
   - double decomposition reaction
   - precipitation reaction
   - sodium chloride (common salt) is soluble in water
   - Iron (III) hydroxide is insoluble.
4. Now make your own precipitate
   Name an insoluble salt that you wish to make.
   Name two solutions that you would use to make the insoluble salt.
   Was your attempt successful? Explain why you managed to obtain the insoluble salt.

Fig. 11.3 Activity on precipitation

the experiments that they do in the laboratory, something which 'current common practice neglects' (Sere, 2002, p. 628).

To help students better understand the double decomposition and precipitation reactions, the students need to be able to visualise the processes at the submicroscopic level. Computer animations or magnets can be used to illustrate the different ions in the solutions before they are mixed. When they are added together, a precipitate forms when certain oppositely charged ions are attracted to each other and aggregate together. The water molecules are unable to pull these ions apart, so the salt formed is insoluble. The teacher can ask the students to explain why sodium chloride solution does not form a precipitate with sodium hydroxide solution (i.e., no insoluble compound can be formed) to help them clarify and integrate their knowledge of precipitation and solubility. To provide further opportunities for students to apply their understanding of double decomposition, precipitation and solubility, the students are asked to prepare their own insoluble salts. Students are required to memorise a list of soluble and insoluble salts, but this knowledge is seldom put into practice in the laboratory. Thus, students need to choose their two starting solutions carefully in order to obtain their chosen salts. The teacher can also challenge them to explain why soluble salts cannot be prepared by double decomposition reaction (a mixture of soluble salts will be obtained which are very difficult

a) describe the techniques used in the preparation, separation and purification of salts as examples of some of the techniques specified in Section 1.2(a) (methods for preparation should include precipitation and titration together with reactions of acids with metals, insoluble bases and insoluble carbonates)
b) describe the general rules of solubility for common salts to include nitrates, chlorides (including silver and lead), sulfates (including barium, calcium and lead), carbonates, hydroxides, Group I cations and ammonium salts
c) suggest a method of preparing a given salt from suitable starting materials, given appropriate information

Fig. 11.4 Learning objectives related to salts as specified in the secondary chemistry syllabus (UCLES, 2011, p. 11)

to separate) to help them make sense of the different methods used to prepare different types of salts.

How the procedures are carried out is very important to get accurate data to be used as evidence in identifying any given unknown substance, so this has to be highlighted to the students. For example, if an unknown solid is given to test for cations, students need to make a solution of the unknown solid and then add sodium hydroxide solution to it rather than adding sodium hydroxide solution directly to the solid as it is difficult to determine if any resultant solid is due to the formation of a precipitate or the original solid remaining undissolved. They also need to add the aqueous sodium hydroxide dropwise to the solution of the unknown solid rather than introducing a large volume at one time which may cause any precipitate formed to react immediately with excess alkali. They have to practise these techniques correctly and repeatedly and to understand why these procedures have to be carried out in a certain way as well as in a specific sequence; students need to master the required skills in order to conduct laboratory work successfully (Hodson, 1990). The importance of acquiring the manipulative skills to conduct QA experiments proficiently is acknowledged in the QATP, and exercises (Woolnough & Allsop, 1985) are included to help students gain the necessary competencies. For example, in the test for cations, students will practise adding reagents to the cations and learn the rationale for the specific steps that they need to take so that the procedures will make sense to them and help them to link theory with laboratory procedures (see Fig. 11.5).

The '*Salt-Acid Reaction Experience*' section (see Fig. 11.6) of the QATP helps students to learn about the role of acids in the tests for anions and gases. Students will find out that, in addition to adding a dilute acid directly to sodium carbonate solution, an appropriate dilute acid can also be added after the introduction of barium nitrate (V) solution to produce carbon dioxide. They will observe that the barium carbonate precipitate will still react with nitric (V) and hydrochloric acids but not sulfuric (VI) acid and have to be able to use their knowledge of insoluble salts to produce an explanation; insoluble barium sulfate (VI) is formed when sulfuric (VI) acid initially reacts with barium carbonate, and this will coat the barium carbonate particles, preventing further reaction between the hydrogen ions from the acid and the barium carbonate. The students also need to apply similar reasoning to explain why all three dilute acids can be added to sodium carbonate

Aims:
You will learn to:
Put the appropriate amount of the zinc solution to be tested into a test tube.
Add a small amount of aqueous sodium hydroxide to the zinc solution.
Add aqueous sodium hydroxide until it is in excess.
Shake the mixture after each addition of aqueous sodium hydroxide.

2. Skills to be practised

a. Solution to be tested
When you add the zinc solution into a test tube, you should add till it reaches a height of about 1 cm. Why?

b. A small amount of reagent
When you add a small amount of aqueous sodium hydroxide, you should add about two drops and shake the mixture thoroughly (by shaking the test tube or using a glass rod to stir). Why?

c. Excess reagent
When you add aqueous sodium hydroxide in excess, you should add 1 cm height portions at a time, shaking the mixture thoroughly each time until the mixture reaches about 2 cm from the top of the test tube. If there is a need to, pour away ¾ of the content of the test tube and continue adding as described.

Fig. 11.5 Skills required for the testing of cations

solution. Thus, students need to ponder over the observations obtained from the activities as well as what they learn in class to explain the reactions and concepts underlying the observations.

After a series of guided experiments, students will be given the opportunities to carry out more open inquiry investigations to determine the identity of an unknown compound without any instructions given. This requires them to plan a series of tests and determine if their plans are comprehensive enough. They have to record their observations accurately and analyse the positive and negative results to draw inferences. If there are conflicting data, the students need to repeat the relevant procedures to check if the data are correct or to scrutinise their plans to determine if there are loopholes in them. Further confirmatory tests may also be required. For example, lead (II) and aluminium ions give similar results with aqueous ammonia and sodium hydroxide, so an iodide solution has to be used to differentiate between the two cations. Another example is that both sulfur dioxide and carbon dioxide react with limewater to produce insoluble white calcium sulfate (VI) and calcium carbonate, respectively. Thus, the limewater test cannot differentiate between sulfur dioxide and carbon dioxide, so the students need to prepare acidified potassium dichromate (VI) to test for sulfur dioxide in addition to the limewater test. The more open inquiry tasks definitely require more extended reasoning and analytic thinking on the part of the students, unlike the 'mindless' following of procedures in traditional QA experiments, giving students additional impetus to try and understand the relevant reactions and concepts involved in the experiments (Schauble, Klopfer, & Raghavan, 1991)

Aqueous sodium carbonate + acid.

Procedure	Observations
Add aqueous sodium carbonate to 1 cm height in a test tube. Add twice amount of any dilute acid to the aqueous sodium carbonate. Observe.	

Did you expect to observe any reaction? Why?

Is the choice of acid to be added important in the above reaction? Why?

Aqueous sodium carbonate + aqueous barium nitrate (V) + dilute nitric acid

Procedure	Observations
Add aqueous sodium carbonate to 1 cm height in a test tube. Add an equal amount of aqueous barium nitrate (V) to the aqueous sodium carbonate. Observe. Divide the above mixture into 3 parts. To one part, add dilute nitric (V) acid until there is no further change. Observe. To the second part, add dilute sulfuric (VI) acid until there is no further change. Observe To the third part, add dilute hydrochloric acid until there is no further change. Observe	

1. Explain what happened when aqueous barium nitrate (V) was added.
2. Explain what happened when dilute nitric (V) acid was added.
3. Does it matter if dilute sulfuric (VI) acid or dilute hydrochloric acid is used in place of dilute nitric (V) acid? Why?
4. Can aqueous barium chloride and dilute hydrochloric acid be used instead of aqueous barium nitrate (V) and dilute nitric (V) acid? Why?

Fig. 11.6 Experiencing salt-acid reactions

Implementation of the QATP

The QATP was implemented in one secondary school in November 1999, during the end-of-school-year holidays, over 6 days, three-and-a-half hours daily (Tan, 2000; Tan, Goh, Chia, & Treagust, 2004). Two intact classes, each with 45 secondary 3 students (going on to secondary 4 in January 2000) taking chemistry as a single subject at the O-level, underwent the QATP: I taught one class while Mrs. Ang (pseudonym used), the original chemistry teacher of both classes, taught the other class. These two classes were the top two secondary 3 classes in the school, and the school wanted both classes to learn QA through the QATP. To prepare for the November implementation, Mrs. Ang and I met thrice in August/September 1999 to discuss the content of the teaching package, the preparation of reagents and apparatus and other administrative matters. Mrs. Ang reviewed the teaching package and contributed practical suggestions which were used to revise a few activities and worksheets in the package. She voiced her concerns that her students may not be able to participate actively in the discussions as they were not used to talking about concepts and expressing their ideas. This proved to be true; when questions were asked during the sessions, there usually was a murmur of voices but nobody would volunteer an answer, so the students had to be nominated by name to answer the questions. Mrs. Ang was also not confident that her students would be able to plan investigations on their own, saying, 'I hope they will be able to think...it is typical of them to only follow instructions'. However, this concern was unfounded as most of the students were able to plan investigations and carry them out satisfactorily. Her third concern was the time required to complete the QATP as she felt that in normal school days, she could not afford the time to complete all the activities in the QATP.

Four teachers from the school observed several sessions of the QATP; two teachers were present throughout the first 4 days, while the other two teachers joined in only when they were free from other duties in the school. The four teachers commented that the QATP was very structured and comprehensive and would help the more academically inclined students to gain a deeper understanding of the reactions and concepts involved in QA. 'If you have the time, it is very good. It is very detailed...actually gives the students a very good idea on what is happening behind the procedures, and they can appreciate them better' was a comment made by one teacher. They also said that the computer animations used could help students better understand the submicroscopic interactions during dissolving and precipitation. However, they had reservations using the whole QATP with the less academically inclined students as they believed that the package was too detailed and the reactions discussed might be too difficult for these students to understand; one teacher commented on her students, 'The more I tell them the more confused they get. Don't need to go into details because they will not understand'. Their comments were understandable as the QATP was designed for the more academically inclined students who were taking chemistry as a single subject rather than students taking chemistry as part of a combined science subject. Students taking chemistry as a single subject need to learn more topics and in greater depth compared to those taking chemistry as part of a combined science subject.

The four teachers shared Mrs. Ang's views about the time required to complete the whole teaching package. Mr. Ho (pseudonym used) commented that the seven hours required to complete the section on cations was too long and tedious and might bore the academically less inclined students. In addition, seven hours were approximately equivalent to 6 weeks' worth of normal weekly practical work sessions, so too much time was allocated to a small part of QA. Mrs. Khoo (pseudonym used) pointed out that the weekly practical work schedule might not

Table 11.2 Descriptive statistics for the two groups of secondary 4 students to whom the QADI was administered

	S4	S
No. of cases	90	915
No. of items	19	19
Alpha reliability	.75	.68
Mean	6.66	5.76
Standard deviation	3.72	3.28
Median	6.00	5.00
Mode	8	5
Minimum	0	0
Maximum	18	18

be suited for the implementation of QATP because students might forget what was discussed in the previous week's lesson. She believed that doing the QATP continuously over 6 days was much better as students would better remember what they did, but this would be difficult to implement in school. In addition, the less academically inclined students would not want to come back for lessons during their holidays compared to the students in the top two classes. In summary, the teachers in the school believed that QATP was very structured and detailed, but they had concerns over the time required to implement it and the ability of students, especially less academically inclined ones, to understand and discuss the reactions and concepts involved.

The QADI was administered to the 90 students 4 months later to provide sufficient time to minimise memory effects that could have occurred if the QADI was administered immediately after they completed the QATP. The descriptive statistics for the 90 students (denoted as S4) to whom the QADI was administered are given in Table 11.2 together with that of the 915 students (denoted as S) to whom the QADI was administered in 1999. As there were many confounding variables such as the profile of the students and their teachers, the different number of students in the two samples, the modes of instruction and the length of time given to qualitative analysis instruction to the students in the two groups, statistical tests for significance would not be highlighted. The mean, median and mode scores of the S4 group were higher than that of the S group (see Table 11.2), and the percentage of S4 students having stated alternative conceptions were lower than the students in the S group in six out of the ten items in Table 11.3, indicating the feasibility of using the QATP to facilitate students' understanding of QA. However, there was generally not much difference between the types and levels of alternative conceptions between the two groups of students. This seemed to indicate that the effect of the QATP might not be uniform, that is, some students in the S4 group might develop deeper understanding of the reactions and concepts involved in QA from the QATP and some might not. Though the students seemed to show that they understood what was going on during the QATP sessions, some of them somehow failed to apply their knowledge in answering the QADI. Unfortunately, constraints of time and funding prevented a more in-depth study on the cognitive, affective and psychomotor aspects of the QATP.

Table 11.3 Comparison of a sample of alternative conceptions of the S4 and S groups of students

Alternative conception	Choice combination	Percentage of students with the alternative conception	
		S4	S
Precipitation			
A more reactive ion displaces a less reactive ion in a precipitation reaction	Q1 (A3)	21	25
	Q5 (A1, A4)	21	37
	Q13 (A3, A4)	21	15
	Q18 (A1, A2, A4 & A5)	35	29
Addition of acid			
Acid is needed to acidify the mixture so that reaction can proceed properly	Q6 (A5)	11	22
Carbonate ions cannot be determined if acid is added after the addition of barium nitrate (V)	Q17 (A4)	31	35
The procedure is strictly a test for sulfate (VI)	Q11 (B3)	13	20
The acid must be added directly to the unknown sample	Q11 (B4)	30	25
Test for oxygen			
Test for oxygen has to be carried out when a substance is heated because all gases have to be tested	Q10 (A1)	24	23
Oxygen gas need not be tested when a substance is heated because oxygen will not be liberated during heating, instead it will be used up	Q10 (B3)	20	27

Conclusion

QA is difficult for students because they usually follow instructions in QA practical work with little understanding of the reactions behind the procedures and the results that they obtained. Learning QA by inquiry has the potential of making it more meaningful by requiring students to engage with the reactions and concepts underlying QA through guided activities and making sense of the results obtained. The study described in this chapter highlights the feasibility of using inquiry to facilitate students' understanding of the procedures and concepts involved in practical work, helping them to link what they have learnt in class with what they do in the laboratory.

References

Berry, A., Mulhall, P., Gunstone, R., & Loughran, J. (1999). Helping students learn from laboratory work. *Australian Science Teachers Journal, 45*, 27–31.

Chinn, C. A., & Malhotra, B. A. (2002). Epistemologically authentic inquiry in schools: A theoretical framework for evaluating inquiry tasks. *Science Education, 86*, 175–218.

Crawford, B. A. (2000). Embracing the essence of inquiry: New roles for science teachers. *Journal of Research in Science Teaching, 37*, 916–937.

Crawford, B. A. (2012). Moving the essence of inquiry into the classroom: Engaging teachers and students in authentic science. In K. C. D. Tan & M. Kim (Eds.), *Issues and challenges in science education research: Moving forward* (pp. 25–42). Dordrecht, The Netherlands: Springer.

Flick, L. B. (2000). Cognitive scaffolding that fosters scientific inquiry in middle level science. *Journal of Science Teacher Education, 11*, 109–129.

Gunstone, R. F. (1991). Reconstructing theory from practical experience. In B. E. Woolnough (Ed.), *Practical science: The role and reality of practical work in school science* (pp. 67–77). Milton Keynes, UK: Open University Press.

Hart, C., Mulhall, P., Berry, A., Loughran, J., & Gunstone, R. (2000). What is the purpose of this experiment? Or can students learn something from doing experiments? *Journal of Research in Science Teaching, 37*, 655–675.

Hodson, D. (1990). A critical look at practical work in school science. *School Science Review, 70*, 33–40.

Hodson, D. (1992). Redefining and reorienting practical work in school science. *School Science Review, 73*, 65–78.

Hofstein, A. (2004). The laboratory in chemistry education: Thirty years of experience with developments, implementation, and research. *Chemical Education Research and Practice, 5*, 247–264.

Johnstone, A. H., & Wham, A. J. B. (1982). The demands of practical work. *Education in Chemistry, 19*, 71–73.

Nakhleh, M. B., & Krajcik, J. S. (1994). Influence of levels of information as presented by different technologies on students' understanding of acid, base and pH concepts. *Journal of Research in Science Teaching, 31*, 1077–1096.

National Research Council (NRC). (2000). *Inquiry and the national science education standards: A guide for teaching and learning*. Washington, DC: National Academy Press.

Schauble, L., Klopfer, L. E., & Raghavan, K. (1991). Students' transition from an engineering model to a science model of experimentation. *Journal of Research in Science Teaching, 28*, 859–882.

Sere, M.-G. (2002). Towards renewed research questions from the outcomes of the European project 'Labwork in Science Education'. *Science Education, 86*, 624–644.

Sweeney, A. E., Bula, O. A., & Cornett, J. W. (2001). The role of personal practice theories in the professional development of a beginning high school chemistry teacher. *Journal of Research in Science Teaching, 38*, 408–441.

Tan, K. C. D. (2000). *Development and application of a diagnostic instrument to evaluate secondary students' conceptions of qualitative analysis*. Unpublished Ph.D. thesis, Curtin University of Technology, Western Australia.

Tan, K. C. D., Goh, N. K., Chia, L. S., & Treagust, D. F. (2001). Secondary students' perceptions about learning qualitative analysis in inorganic chemistry. *Research in Science & Technological Education, 19*, 223–234.

Tan, K. C. D., Goh, N. K., Chia, L. S., & Treagust, D. F. (2002a). A cross-age study on the understanding of the reactions involved in basic inorganic chemistry qualitative analysis. *Journal of Science and Mathematics Education in Southeast Asia, 25*(2), 1–14.

Tan, K. C. D., Goh, N. K., Chia, L. S., & Treagust, D. F. (2002b). Development and application of a two-tier multiple choice diagnostic instrument to assess high school students' understanding of inorganic chemistry qualitative analysis. *Journal of Research in Science Teaching, 39*, 283–301.

Tan, K. C. D., Goh, N. K., Chia, L. S., & Treagust, D. F. (2004). Qualitative analysis practical work. *School Science Review, 85*, 97–102.

Tan, K. C. D., Goh, N. K., Chia, L. S., & Treagust, D. F. (2009). Linking the macroscopic, sub-microscopic and symbolic levels: The case of inorganic qualitative analysis. In J. K. Gilbert & D. F. Treagust (Eds.), *Multiple representations in chemical education* (pp. 137–150). Dordrecht, The Netherlands: Springer.

Tasker, R., & Freyberg, P. (1985). Facing the mismatches in the classroom. In R. Osborne & P. Freyberg (Eds.), *Learning in science: The implications of children's science* (pp. 66–80). Auckland, New Zealand: Heinemann.

Treagust, D. F. (1995). Diagnostic assessment of students' science knowledge. In S. M. Glynn & R. Duit (Eds.), *Learning science in the schools: Research reforming practice* (pp. 327–346). Mahwah, NJ: Lawrence Erlbaum Associates.

University of Cambridge Local Examinations Syndicate (UCLES). (1997). *Chemistry: Report on the November 1996 examinations*. Cambridge, UK: UCLES.

University of Cambridge Local Examinations Syndicate. (2011). *Chemistry: GCE Ordinary Level (Syllabus 5072)*. Cambridge, UK: UCLES.

Woolnough, B., & Allsop, T. (1985). *Practical work in science*. Cambridge, UK: Cambridge University Press.

Chapter 12
Bridging Research, Policy and Practice of Inquiry in Singaporean Classrooms

Barbara A. Crawford

Keywords Critical thinking • Inquiry • Nature of science • Inquiry learning • Inquiry teaching • Investigation • Model and modeling • Argumentation • Professional development • Scientific literacy • Problem-based learning

The focus of these four chapters—inquiry teaching and learning in Singaporean science classrooms—is an area of utmost importance in preparing children to be scientifically literate in the twenty-first century. Most science educators believe strongly that the time for change in the way science is taught has arrived. First, there is virtually no way anyone can learn all the exponentially increasing volume of new science facts by simply memorising this information. Second, there is no evidence that memorisation leads children to be critical thinkers. Third, in today's rapidly changing world and with advances in scientific discoveries, it is important for children to understand what science is, and what science is not, and aspects of the nature of science. Research suggests that inquiry-based science instruction has the potential to move diverse children towards independent and critical thinking (see Blanchard et al., 2010; Cuevas, Lee, Hart & Deaktor, 2005). Yet, the reality is that little change is taking place in many classrooms in the United States and in other countries (see Capps & Crawford, 2013). Understanding how we can support teachers in changing the way science is taught to children is a worldwide issue. Internationally, there is a distinct shift from expecting teachers to use primarily lectures and fact-based instruction to use reform-based inquiry pedagogy in their science classrooms. In Singapore, the national science curriculum was updated in 2008 to include inquiry as the core and guiding philosophy. In Europe there is increased attention to the question of how to effectively support teachers in changing their practice to involve students in asking and answering questions,

B.A. Crawford (✉)
Mathematics and Science Education Department, The University of Georgia, Athens, GA, USA
e-mail: barbarac@uga.edu

A.-L. Tan et al. (eds.), *Inquiry into the Singapore Science Classroom*, Education Innovation Series, DOI 10.1007/978-981-4585-78-1_12,

using data as evidence and justifying their explanations (i.e., Osborne & Dillon, 2008).

One problem involves communication. Researchers, curriculum developers and practitioners often use different definitions of inquiry teaching. A critical issue in researching inquiry in science classrooms relates to *which* particular view of teaching science as inquiry is being used, and these views can vary greatly. Inquiry can be viewed as a pedagogy or as the learning outcome of instruction; pedagogy being the method of engaging students in designing and carrying out investigations, and the learning outcome refers to learning science subject matter by engaging in these investigations. In addition, a third view of inquiry relates to learning 'about' the nature of scientific inquiry. Learning 'about' the nature of scientific inquiry overlaps in some ways with developing understandings of nature of science. Three uses of the term 'inquiry' in science classrooms include (1) *scientific inquiry* (the various ways in which scientists study the natural world), (2) *inquiry learning* (a process by which children acquire knowledge of science concepts and learn about nature of science) and (3) *inquiry teaching* (broadly defined as the pedagogy by which teachers engage students in inquiry) (modified from Anderson, 2002).

In the latest version of the *Handbook of Research on Science Education*, Crawford (in press) utilises the following definition of inquiry-based teaching, based on a synthesis of definitions from the National Research Council [NRC] (1996, 2000, 2012):

> Teaching science as inquiry involves engaging students in using critical thinking skills, that includes asking questions, designing and carrying out investigations, interpreting data as evidence, creating arguments, building models, and communicating findings, in the pursuit of deepening their understanding about the natural world by using logic and evidence.

Not all researchers use the exact same definition. But, it is important that each author clarifies his or her definition.

From a close reading of the chapters, it appears that these authors have slightly different views of teaching science as inquiry. Chew-Leng Poon's and Shirley Lim's definition seems that of a pedagogical approach, *inquiry teaching*, by which teachers engage students in inquiry to help children learn core science concepts. They provide a table comparing documented pedagogical frameworks. Not readily evident in their table is the presence of a *question* that drives a science investigation. Yew-Jin Lee's view of inquiry teaching appears to involve children practising inquiry as equivalent to carrying out process skills. Further, Lee views inquiry as an activity involving critical reasoning. Seng-Chee Tan and Jennifer Yeo reference the National Research Council's view of a process approach for learning science through investigation of a natural phenomenon, which engages learners in questioning, conducting scientific investigation, collecting evidence, interpreting results, constructing explanations for the findings, reporting results, asserting claims and presenting findings. They also view children learning in a social collaborative setting by creating knowledge artefacts. Daniel Tan's definition of inquiry teaching aligns with social constructivist views and involves children finding solutions to problems by asking questions that can help to collaboratively solve problems, designing and conducting experiments, collecting and analysing data, drawing conclusions and reporting and defending the findings of the inquiry.

In addition to some confusion about what it means to teach science as inquiry, policy documents may advocate for reform-based teaching practices. Few examples are given. As might be predicted, many teachers find it difficult to make actual change a reality. As stated shortly after publication of policy documents in the United States (see *National Science Education Standards*, NRC, 1996, 2000), 'Everyday events are often left to the imagination of the classroom teacher ending in frustration from attempting inquiry-based strategies' (Crawford, 2000, p. 918). We need to know more about what it takes to elicit change in science classroom practice. Thus, practical research situated in the classroom, like the kind presented in these four chapters, is needed to form a bridge between what is asked by policy documents and what teachers can reasonably strive for and effectively accomplish.

In the United States, there exists a new challenge on the horizon that of carrying out science instruction that focuses on Science and Engineering Practices (see K-12 Framework, NRC, 2012, and Next Generation Science Standards [NGSS], Achieve, Inc., 2013). The new US standards documents involve a shift from the essential features of inquiry (NRC, 2000) to a greater emphasis on children creating evidence-based explanations and using modelling and argumentation to answer scientific questions. Effectively preparing new teachers and supporting experienced teachers in carrying out this kind of inquiry-based instruction requires targeted and sustained professional development programmes and resources. A recent review of the literature (Capps, Crawford, & Constas, 2012) revealed few empirical studies related specifically to science inquiry teacher professional development, published in major peer-reviewed journals. The reality is that worldwide, many teachers either do not attempt inquiry-based pedagogy or they struggle in trying to enact it in their classrooms. Strategic and concerted efforts are needed to support teachers in engaging children in science practices and in scaffolding children in classroom discussions focused on using data as evidence, a central component of inquiry-based instruction.

In Chew-Leng Poon's and Shirley Lim's chapter, they tackle an important and challenging problem—that of documenting actual classroom practice over time in elementary level classrooms. They cite Hogan et al. (2006) who depicted inquiry practice as not widespread in Singapore classrooms. Poon and Lim noted that in Singapore, there are large classes (averaging about 40 students) and great pressure on teachers to prepare students for national examinations. It makes sense to explore what happens in Singapore classrooms, as there are many obstacles present in Singapore classrooms evident in other settings. To be perfectly honest, the more one looks into science classrooms around the world, the more one realises that inquiry-oriented practices are not widespread in many countries and that includes the United States.

Poon and Lim's research goal focused on efforts to support teachers in engaging students in inquiry science practices. The researchers carefully examined the science instruction in four elementary classrooms. In selecting this sample the authors reported these teachers had attended various workshops on inquiry. The teachers had different educational backgrounds. Two of the four teachers, Dorothy and Isa, had virtually no formal science background. However, Dorothy had 40 years of experience teaching science. One of the strengths of this chapter includes using a strong theoretical framework of how people learn science as a lens for their research questions and analyses.

Using a case study approach, Poon and Lim videotaped not just one or two lessons of each teacher. In the first phase of their study, 13 inquiry science lessons were observed and video-recorded over a period of two school terms (6 months). In the second phase, they observed another 49 inquiry lessons, this time over a 6-month time frame. This kind of intense data collection is rare. Thus, these authors should be commended for gathering data over many school months, which is imperative for capturing the teacher-student and student-student interactions that occur in a classroom over time and for understanding the evolution of a teacher's views and practice.

A close look into these elementary teachers' classrooms revealed that various aspects of teaching science as inquiry were indeed present, at least to some extent. However, the researchers did note that all these four classrooms reflected a more teacher-directed than student-directed inquiry environment. It was the classroom teacher who mainly took charge of posing the questions and problems. In all fairness, it is a myth that inquiry only involves students in generating their own questions. The critical aspect of inquiry teaching is that there *is* a scientific question, and that children collect and analyse data and then grapple with this data and use it as evidence to develop explanations and justify their claims. Whether or not it is the answer to the question, who is doing the thinking during the inquiry?

In Poon and Lim's study, two of the four teachers, Dorothy and Isa, had little or no science background. Dorothy appeared to be successful in carrying out certain aspects of inquiry instruction—as least in the nature of the group work. The authors noted that in Dorothy's class, they observed students censuring group members who threatened to disrupt their activities with off-task behaviours. Isa, in contrast, recognised her lack of science content knowledge and cited it as a factor contributing to her lack of confidence. It is likely that for teachers to adopt and carry out the practice of teaching science as inquiry, longer, sustained programmes will likely be needed for those teachers who lack science background knowledge and familiarity with inquiry. Having in-depth science subject matter is a huge advantage when responding to children's questions. In a recent study by Capps and Crawford (2012), the researchers observed change in a range of teacher cases, demonstrating that effective professional development can impact a teacher's views and practice related to inquiry instruction, even in a teacher who initially has very limited understandings. The differential influence of the professional development observed across the cases suggested that change in knowledge, views and teaching practice can occur quite rapidly for some teachers. Yet, for others, growth and change may require much more time. In Poon and Lim's study, the level of direction and guidance given by each teacher varied with the ages and grades of their children. Not simply age and grade but also the developmental ability of children is a key consideration in the amount of teacher guidance needed. The amount of guidance is necessarily part of a continuum; an ebb and flow is orchestrated during this kind of sophisticated teaching practice.

In Yew-Jin Lee's chapter, he hits the 'nail on the head' in identifying some of the problems inherent in implementing innovative curricula, some of which exist in spite of what he termed 'unbridled support from the government'. As stated by Lee, Singaporean students are among the best in the world in performing on international benchmarking tests, and the national science curricula have been long praised for their coherency and conceptual focus (Ministry of Education [MOE], 2012). The Singapore government recognises the need to work towards children acquiring critical thinking skills, despite students' high performance on international assessments. Therefore, the fact that Singapore is striving to change the nature of science instruction in schools is particularly noteworthy. Lee analysed the data from both a perspective of a former high school science teacher in Singapore and an academic teacher educator whose beliefs include the importance of children developing scientific literacy, instead of merely regurgitating more and more science facts.

The study site was a school in the western part of Singapore known as a 'neighbourhood school'. Over 2 years, Lee and his team amassed a great deal of data, including field notes, video recordings of hours of lessons, teacher and student interviews and various documents. Lee provided insight into the workings of curriculum implementation. Using the powerful metaphor of a 'straightjacket', Lee described the constraints these Singapore teachers experienced while enacting the inquiry-based *SciencePlus* curriculum.

The story is not one of total success. It takes courage and intellectual honesty to admit that things did not go as planned. This kind of story is quite rare in published literature, as researchers and curriculum developers would rather tell stories of success. Yet, sometimes we can learn as much, if not more, from events that do not work out as planned. It is very important to ask the question, 'why?' Why did things not work out? Lee identified one failing of the *SciencePlus* curriculum that involved a lack of philosophical underpinning apparent to the teachers. Understanding the underlying philosophy of a set of instructional materials such as these is critical for successful implementation. Following the frenzy of post-Sputnik, a lack of teacher understanding of the underlying philosophy was identified as one of the failings of the US 1960s innovative curricula initiative (Yager, 2000).

Lee identified a critical issue of implementation—that of teachers having difficulties in managing groups of students pursuing different inquiries, thus being pulled in many directions. This same challenge was highlighted in the Crawford (1999) study, in which a new biology teacher attempted to involve her students in open inquiry projects during her student teaching semester. Lee recognised that although 10 weeks were available in the Singapore school, the teachers still experienced a great deal of pressure when supervising 13 groups of students. Orchestrating classroom discussions to support students in developing and defending arguments was an important, but often missing, element of inquiry instruction. Further, Lee reported teachers viewed their training as hurried and reported minimal professional development. The power of professional development for teachers should not be underestimated.

Personally, I admire the courage of Lee to highlight aspects of a state-supported inquiry initiative that did not fully succeed. I believe this chapter can be very useful to curriculum developers, professional developers and teachers in other countries. Teaching science as inquiry is complex; the best curriculum in the world will fail unless there is adequate external and internal support of teachers, who are the key agents of change in carrying out the instruction. In addition, it is *imperative* that well-designed assessments match the innovative curricular goals. Perhaps one undervalued benefit of the *SciencePlus* programme was that many children reported they enjoyed it. In addition, children reported acquiring certain process skills, and they felt an increased relevance of science in their lives. This affective result is to be celebrated!

In their chapter, Seng-Chee Tan and Jennifer Yeo report on two intervention studies conducted in Singapore classrooms, and they critiqued how these interventions were carried out related to knowledge creation among the students. They described two kinds of inquiry science learning environments, in which they used a theoretical knowledge creation metaphor of learning. First, Tan and Yeo used a knowledge building framework to analyse what was happening in two primary 4 classrooms. Second, they used a knowledge creation problem-based learning (KC-PBL) framework to determine what was going on in a junior college physics classroom. In the primary 4 classroom, there was evidence of students working as a community of learners, in developing an evidence-based explanation of an unknown prehistoric animal. As is true of real scientists, these primary 4 school children engaged in a social collaborative process of knowledge building. Another aspect of this teacher's inquiry instruction was the use of a learning technology, the Knowledge Forum, a web-based forum supporting asynchronous online discussion. Although technology is not necessarily an essential part of inquiry teaching, the use of this kind of technology can support students in ways that is not possible in a non-technology-based traditional classroom. Thus, an analysis of this learning technology is an exciting and valuable aspect of Tan and Yeo's research study.

The second study reported by Tan and Yeo was vastly different from the first, especially related to the grade level of the students. The second study took place in a Singaporean junior college. The approach was that of Problem-based Learning (PBL) in a physics classroom. PBL, a variation of inquiry-based teaching, engages students in solving complex real-world questions. A key aspect of this instruction is that students carry out meaningful investigations as a vehicle for learning science facts and principles (Krajcik, McNeill, & Reiser, 2008). Tan and Yeo reported that the problem posed was a real world, but fictitious scenario—one involving a roller coaster accident. Since the problem was fictitious, it could be viewed as real world, but not authentic. The teacher guided students in an iterative process of investigation and evidence gathering. Tan and Yeo described how design principles were used to modify the instruction. One of the intriguing aspects of the instruction was the use of the software called, Knowledge Constructor.

In analysing the classroom interactions, Tan and Yeo used a framework similar to the 5 E Learning Cycle approach to teaching inquiry (Bybee et al., 2006). In both cases, there was evidence of students using epistemic discourse, evaluating claims and creating knowledge artefacts and using technology, important components of inquiry-oriented collaborative discussion. In each case, students engaged in the kind of academic classroom discussions that could lead to deep understanding of science concepts. This is exactly the type of classroom environment envisioned in the latest NRC 2012 K-12 framework proposed in the United States. In their analyses, the authors highlighted the role of the teacher in only one of the components—the *Prepare to Investigate* phase, in which scaffolding was identified. It is very likely that the classroom teacher was a critical player in several, if not all, other components of the instruction. I encourage the authors to re-evaluate the role of teacher in an inquiry-based learning environment. Misguidedly, some researchers view the highly sophisticated approach of teaching science as inquiry, as an unguided, unchartered, random discovery approach taken on by students in an often aimless unproductive way (see Furtak, Siedel, Iverson, & Briggs, 2012). Driver (1989) acknowledged that discovery learning with minimal teacher guidance does not work with children, nor should we expect it to work in the twenty-first century.

In Daniel Tan's chapter, he focused on what is ultimately important in inquiry-based instruction; that is, what do children learn? The other important feature of his work is that it centres on the domain of chemistry and, in particular, children understanding, integrating and applying chemical concepts, rather than mindlessly running through laboratory procedures. In previous research by Tan and colleagues (Tan, Goh, Chia, & Treagust, 2001), children have admitted they merely followed instructions during practical work with little idea about what they were doing. Assessments have uncovered alternative understandings. Additionally, Tan noted that prospective teachers might hold some of the same alternative conceptions as their students, making the problem even worse. A similar intellectual situation was found in a study of prospective biology teachers unveiling their alternative understandings of evolution by natural selection (Crawford, Zembal-Saul, Munford, & Friedrichsen, 2005). It is hoped that changing the nature of classroom instruction will lead to more robust conceptual understandings and a more critically thoughtful self-directed learner.

As part of the teaching strategy, chemistry teachers in the Tan study were expected to have students explain their observations, compare them with their classmates' explanations, discuss and resolve any conflict and arrive at a consensus which they will compare with their teacher's explanation. One of the important aspects of this kind of instruction is that the students are doing most of the work, yet the teacher is an active player as a co-constructor of knowledge. The emphasis is on explanation, not regurgitation.

In the Tan study, one chemistry teacher taught one intact class of 45 secondary 3 students, and the author taught another intact class of 45 students. Interestingly, 45 students is a high number as compared with average numbers of students per class in other countries, thus adding to the challenge of carrying out this kind of instruction. Both classes appeared to have students of high ability. Other teachers observing these classes expressed their reservations that this kind of instruction might not work with the less academically inclined students. Another concern that emerged was that this kind of intense teaching took too much time. Assessment of these students who engaged in the inquiry-based practical work suggested that these students learned the concepts more deeply and they had fewer alternative conceptions. However, the author noted some issues of confounding variables, which prevented use of statistical tests for significance. Still, the results reported in this chapter are promising.

In conclusion, the issues raised in these studies about the challenges of teaching science as inquiry, and the advantages of engaging students in carrying out investigations and thinking about data, are very important contributions to the literature. Each of these authors is to be commended for tackling important questions surrounding what actually happens in real classrooms when teachers attempt to carry out the practice of teaching science as inquiry. Education frameworks in many countries advocate teaching children not only core science facts and principles through inquiry but also about nature of science, including the tentativeness, cultural and social aspects of scientific knowledge. In these chapters, each author contributed important insights into some problems, as well as the promises, of teaching science as inquiry. More studies like these should be undertaken in other countries. We are just beginning to appreciate the realities of what happens in classrooms when change is attempted—in other words, when the 'rubber hits the road'.

Finally, it is absolutely critical not to be pejorative against classroom teachers. Anyone who has attempted to change his or her classroom instruction to align with this kind of sophisticated inquiry-based pedagogy understands the enormous complexity of the situation. In many countries that have high-stakes standardised testing and varied resources depending on the school, there are added challenges. Teachers need to be assured of tangible benefits when changing the nature of their teaching and be given a true safety net when engaging children in asking and answering questions.

As highlighted in these studies, it often takes additional class time when helping children to build and test models and develop and defend arguments. It requires skill and in-depth science knowledge to help children change direction when they come up with wrong answers, not by telling them the answer but by asking them the helpful follow-up questions. All these aspects of teaching science as inquiry point to the sophistication of this kind of teaching (Crawford, 2000). The reality is that it takes more time for a learner to deeply grasp the meaning of complex and abstract science concepts and principles than it takes to memorise them. Ultimately, what counts most in the classroom is a teacher's expert pedagogical content knowledge to effectively provide experiences for students and scaffold them in using logic and evidence and critical thinking to find answers to scientific questions (Crawford, 2012). As a student learns how to build his or her personal understanding of a concept or issue, he or she is also constructing an intellectual mechanism, which can be used to address all future questions. The inquiry method of teaching science (teaching science as inquiry) may seem to take more time initially than simple memorisation, but the experience steadily builds information for future learning and thinking.

References

Achieve, Inc. (2013). *Next generation science standards: For states, by states*. Washington, DC: The National Academies Press.

Anderson, R. D. (2002). Reforming science teaching: What research says about inquiry. *Journal of Science Teacher Education, 13*(1), 1–12.

Blanchard, M. R., Southerland, S. A., Osborne, J. W., Sampson, V. D., Annetta, L. A., & Granger, E. M. (2010). Is inquiry possible in light of accountability? A quantitative comparison of the relative effectiveness of guided inquiry and verification laboratory instruction. *Science Education, 94*, 577–616.

Bybee, R., Taylor, J., Gardner, A., Van Scotter, P., Powell, J., Westbrook, A., et al. (2006). *The BSCS 5E instructional model: Origins and effectiveness. A report prepared for the office of science education National Institutes of Health*. Colorado Springs, CO: BSCS.

Capps, D. K., & Crawford, B. A. (2012, March). *Tracking teachers' change in teaching science as inquiry: Different teachers, different journeys*. Paper presented at the annual meeting of the National Association for Research in Science Teaching, Indianapolis, IN.

Capps, D. K., & Crawford, B. A. (2013). Inquiry-based instruction and teaching about nature of science: Are they happening? *Journal of Science Teacher Education, 24*(3), 497–526. doi:10.1007/s10972-012-9314-z.

Capps, D. K., Crawford, B. A., & Constas, M. (2012). A review of empirical literature on inquiry professional development: Alignment with best practices and a critique of the findings. *Journal of Science Teacher Education, 23*, 291–318.

Crawford, B. A. (1999). Is it realistic to expect a preservice teacher to create an inquiry-based classroom? *Journal of Science Teacher Education, 10*, 175–194.

Crawford, B. A. (2000). Embracing the essence of inquiry: New roles for science teachers. *Journal of Research in Science Teaching, 37*, 916–937.

Crawford, B. A. (2012). Moving the essence of inquiry into the classroom: Engaging teachers and students in authentic science. In K. C. D. Tan & M. Kim (Eds.), *Issues and challenges in science education research: Moving forward* (pp. 25–42). Dordrecht, The Netherlands: Springer.

Crawford, B. A. (in press). From inquiry to scientific practices in the science classroom. In N. G. Lederman & S. K. Abell (Eds.), *Handbook of research on science education* (Vol. II) New York: Rutledge.

Crawford, B. A., Zembal-Saul, C., Munford, D., & Friedrichsen, P. (2005). Confronting prospective teachers' ideas of evolution and scientific inquiry using technology and inquiry-based tasks. *Journal of Research in Science Teaching, 42*(6), 613–637.

Cuevas, P., Lee, O., Hart, J., & Deaktor, R. (2005). Improving science inquiry with elementary students of diverse backgrounds. *Journal of Research in Science Teaching, 42*, 337–357.

Driver, R. (1989). The construction of scientific knowledge in school classrooms. In R. Miller (Ed.), *Doing science: Images of science in education* (pp. 83–107). New York: Routledge.

Furtak, R., Siedel, T., Iverson, H., & Briggs, D. (2012). Experimental and quasi-experimental studies of inquiry-based science teaching: A meta-analysis. *Review of Educational Research, 82*(3), 300–329.

Hogan, D. J., Luke, A., Kramer-Dahl, A., Lau, S., Liau, A., & Koh, K. (2006). *Core research program: Year two progress report* Unpublished CRPP Technical Report, National Institute of Education, Nanyang Technological University, Singapore.

Krajcik, J., McNeill, K., & Reiser, B. (2008). Learning-goals-driven design model: Developing curriculum materials that align with national standards and incorporate project-based pedagogy. *Science Education, 92*(1), 1–32. doi:10.1002/sce.20240.

MOE. (2012). *International studies affirm Singapore students' strengths in reading, mathematics & science*. Retrieved December 21, 2012, from http://www.moe.gov.sg/media/press/2012/12/international-studies-affirm-s.php

National Research Council. (1996). *National science education standards*. Washington, DC: National Academy Press.

National Research Council. (2000). *Inquiry and the national science education standards*. Washington, DC: The National Academies Press.

National Research Council. (2012). *A framework for K-12 science education: Practices, crosscutting concepts, and core ideas*. Washington, DC: The National Academies Press.

Osborne, J., & Dillon, J. (2008). *Science education in Europe: Critical reflections*. London: Nuffield Foundation.

Tan, K. C. D., Goh, N. K., Chia, L. S., & Treagust, D. F. (2001). Secondary students' perceptions about learning qualitative analysis in inorganic chemistry. *Research in Science & Technological Education, 19*, 223–234.

Yager, R. (2000). The history and future of science education reform. *The Clearing House, 71*(4), 51–54.

Chapter 13
Knowledge Building as a Boundary Object in Formal/Informal Learning

Jennifer Yeo

Abstract The literature on informal science learning indicates that many benefits can be derived from structured/nonstructured enrichment programmes or hobbies, but school science learning tends not to tap on the potential of these informal platforms. This disjoint in learning between formal and informal environments creates a boundary around each context that ignores the fluid and relational set of practices which transforms learning. The focus of this chapter is to propose knowledge building as a bridge between formal and informal learning environments to engender the use of resources from both contexts for science learning. Using the notion of boundary object and boundary crossing from the third-generation activity theory, this chapter describes a case study of how a "homework" task assigned to students set in the context of knowledge building could result in a hybrid space in which resources from formal and informal settings mediate the process of knowledge advancement.

Keywords Informal and formal science learning • Knowledge building • Boundary object • Activity theory • Boundary crossing • Primary science learning

Introduction

School science learning tends to be confined within the four walls of the classroom, disjointed from learning that might take place in other contexts such as *structured/nonstructured* enrichment or hobbies. Yet, literature (e.g., National Research Council [NRC], 2009; Tal, 2012) on informal science learning indicates that many benefits can be derived from such contexts. Informal science learning includes all forms of learning that take place outside the traditional, formal science

J. Yeo (✉)
National Institute of Education, Nanyang Technological University, Singapore, Singapore
e-mail: jennifer.yeo@nie.edu.sg

A.-L. Tan et al. (eds.), *Inquiry into the Singapore Science Classroom*, Education Innovation Series, DOI 10.1007/978-981-4585-78-1_13,

classroom. These include all forms of learning arising from field trips, visits to the museum, surfing the Internet and engagement in hobby. They can be organised by teachers or community groups, or left to students' own discretion. Learning in these environments can take place over a short or long period. According to Aubusson, Griffin and Kearney (2012), students learning in such environments are often motivated and driven by a clear sense of purpose and have some control over what they want to learn and how they go about it. In such learning environments, students have the opportunity to explore and inquire their own questions and about the natural phenomenon, instead of being constrained by a set of predetermined curricula objectives and restrictive teaching strategies. Studies looking at learning outcomes arising from such informal settings found that the development of knowledge and skills extends beyond classroom learning, interaction and socialising skills are improved, and self-concept, self-esteem and physical and mental health are increased. Specific to science, Orion and Hofstein (1994) found that learning from outdoor activities could bring about better understanding of the abstract ideas of science encountered in the classroom. Such learning outcomes derived from informal learning are similar to those desired in formal science learning but continue to be a challenge to achieve.

Despite the benefits of learning in informal settings, such platforms are still relatively untapped by schools. While schools organise field trips to national parks and museums or cocurricular activities related to their hobbies, learning from these informal contexts seldom gets transformed in the science classrooms. This could perhaps be explained by the mismatch between features of informal learning and formal schooling (Aubusson, Griffin, & Kearney, 2012). For example, the goal of acquiring a set of prescribed science content in school science learning is often regarded as violating the principle of free choice associated with informal learning. Teachers planning outdoor activities may be well intentioned in organising a set of structured activities to meet a predetermined set of learning objectives or to avoid students wandering aimlessly. This organised structure may rob students of the freedom to choose what they really enjoy and want to learn and how to learn. These out-of-classroom activities are often considered by teachers and students as enrichment, an add-on to what is learnt in the classroom and not crucial to what is stipulated in the science curriculum. The converse is also true. Bamberger and Tal (2007) found that while students generally enjoyed the experience in informal learning settings, these environments might not necessarily provide much support for school-based content learning.

Another possible problem could arise from coordinating learning in and out of classroom. Tal (2004) found that successful learning in informal setting requires the employment of sociocultural pedagogies and involves multiple parties such as parents, community and students. These conditions require special attention in coordinating time and activities with different groups of people, which could prove a challenge to teachers designing such complex learning environments.

To tap the advantages of learning achieved in informal settings, the problems of the focus of the learning activity and its pedagogical structure and organisation need to be addressed. This requires the boundary of each context of learning to be renegotiated and a new object of activity created. This will enable the set of fluid and relational practices in informal learning environments to transform the way science is learnt in the classroom, while the structuredness of formal schooling provides direction for purposeful inquiry. Teachers could perhaps explore the possibilities of extending their classroom learning to that of informal space. This is already done in schools in the form of homework. While the term "homework" generally refers to all tasks assigned for students to do outside the classroom, it tends to conjure images of pages and pages of worksheets that students had to plough through after school. Such out-of-school activity might not seem motivating and interesting enough for students to want to inquire deeper into the scientific knowledge. There is potential to look beyond such take-home activities and explore other kinds of "homework" activities that could better engage the vast resources outside of the classroom. To engender the crossing between the formal and informal space and to motivate students to further inquiry, the purpose of "learning" needs to be re-examined.

The goal of this chapter is to describe a case study of how a "homework" task assigned to students set in the context of knowledge building could result in a hybrid space in which resources from formal and informal settings mediate the process of knowledge advancement. In this case study, a class of primary 4 students was tasked to understand the factors that affected the germination of seeds. Over a period of five months, the students were instructed to investigate their hypotheses at home, while they reported their findings to their classmates back in school, in which ideas were critiqued and synthesised. This take-home activity got the students involved in familiar activities happening at home, such as gardening, playtime with pets and casual chat. These activities generated knowledge that became important resources in the knowledge advancement of the class.

To help understand the knowledge advancement that took place between settings, I introduce the concepts of boundary objects and boundary crossing, taken from third-generation activity theory, as a lens to understand learning that cuts across boundaries of well-contained activity systems. I then illustrate how working on the common object of knowledge building created a hybrid space in which practices and resources from school and home were being used to advance students' knowledge about germination of seeds. This case study provides teachers some possibility of the use of a common classroom activity (homework in this case) to engender this boundary crossing. For researchers, the case study suggests the need to extend the unit of analysis beyond that of the classroom to the out-of-classroom activities that might provide insights into students' learning.

Boundary Objects and Boundary Crossing

Learning has been traditionally conceived to happen within the confines of a school. Students' development in scientific knowledge, skills and attitudes is often considered in terms of what is being done in the classroom. Events that students encounter outside the classroom, such as family outings, surfing of Internet and engagement in hobbies, are often not utilised in the science classroom by the teacher or considered by researchers when analysing students' learning. Instead, the learning students derive from these settings was often considered as prior knowledge, which are sometimes considered a nuisance as they are thought to contribute to misconceptions. Such is the perspective of learning taken by the earlier versions of activity theory, in which learning as an object of an activity is often thought of in terms of a fixed activity system such as schooling. In contrast to this traditional view, the third-generation activity theory conceives learning to be boundary-free; rather it emerges relationally between multiple parties and multiple perspectives and has the potential to be utilised in a range of domains and sites based upon participation in multiple communities of practices (Edwards, 2005).

Designing boundary-free learning environment creates new challenges to the instructional designer or teacher. According to Akkerman and Bakker (2011), this calls for the need to reconceptualise one's identity in different contexts (e.g., school versus home), coordinate communication and translation of artefacts between different contexts (e.g., knowledge learnt in formal and informal learning environments), and reflect on practices in other activities that might facilitate crossing of boundaries and transformation of practices that might have long been entrenched in the culture. In regard to these mechanisms, implications of third-generation activity theory to learning goals and practices are discussed in the paragraphs that follow.

In the context of science learning, the implication of third-generation activity theory is the reconceptualisation of science learning as events taking place not only within the confines of a school but also how it can take place across different settings, involving different people and diverse perspectives. It is clear that schoolgoing children do not stop learning science once they step out of the school gate. Instead, when students attend enrichment classes organised by the community centre or school, help their parents with gardening or read popular science magazines, students participate in activities that embody knowledge, skills and attitudes related to science. Learning from such out-of-school activities is seldom considered in school science learning or when analysing students' scientific development. Instead of separating science learning according to settings, the third-generation activity theory also beckons science educators to consider how science learning can take place across different settings in our analysis and also how the advantages of informal learning can be tapped to improve students' learning in school, and vice versa.

To consider learning as cutting across boundaries of school and out-of-school settings, we can draw on two concepts from the third-generation activity theory – boundary object and boundary crossing – to refer to ongoing two-sided actions and interactions between contexts. The concept of boundary crossing originally denotes how professionals at work may need to cross territory with which they might be unfamiliar and/or unqualified (Suchman, 1994) and to negotiate and combine ingredients from different contexts to achieve hybrid situations (Engestrom, Engestrom, & Karkkainen, 1995). In the context of this study, the concept of boundary crossing can be conceived of in two ways. First, it refers to the ways in which teachers familiar with formal learning environment cross boundaries to create hybrid situations which blurs the formal/informal boundaries. This means that teachers could explore ways to bring resources from informal learning environments into the classroom, while the structuredness of schooling could better organise the free-form learning that characterises learning in informal settings. Second, it can also refer to the ways in which students learning within these hybrid situations move between the formal and informal activity systems. This second conception of boundary crossing suggests the need for a focal learning activity that can motivate students to be engaged in the activity within and outside the classroom. The concept of boundary objects is thus used to refer to the artefact that can fulfil a specific function in bridging between intersecting practices in the formal and informal learning environments (Star & Griesemer, 1989).

The third-generation activity theory emphasises the notion of an object shared between activity systems instead of an object within a single activity system (Paavola, Engeström, & Hakkarainen, 2012). Taking the third-generation activity theory as its theoretical perspective generates two implications. The first implication relates to the design of the learning environment. To engender crossing of boundary of formal and informal learning, the object of activity needs to be redefined. Schooling is often associated with acquisition of knowledge as the object, while participation in hobby, community activities and online activities associated with science are considered to engage one in the practices of science-related communities. The objects in these two learning environments were thought to be at odds. Homework, done outside the confines of the classroom, is at best an extension of the teacher's intention to enhance acquisition of knowledge. It might not motivate students to go beyond the assigned task or to inquire deeper into the knowledge learnt in the classroom. Participating in a hobby activity, such as gardening, within the familiar surroundings of home might be more fun and engaging than homework, but it focuses more on the doing, which often does not bring to one's consciousness the related scientific knowledge.

Paavola and Hakkarainen (2005) came up with a third metaphor of learning, knowledge creation, to describe another form of learning environment that focuses not only on either content or social practices alone but both. Unlike individual knowledge acquisition common in schooling, knowledge creation emphasises on collective advancement of knowledge. Instead of enculturation into the practices of the community, the practice being emphasised in knowledge creation is advancing the knowledge of the community. An example of a knowledge creation environment is knowledge building (Scardamalia & Bereiter, 2003). Knowledge building refers to the process of creating new cognitive artefacts through collective efforts of a community and support of technology (Scardamalia & Bereiter). By focusing the activity on knowledge advancement of the community, it extends beyond the goal of individual acquisition of knowledge in a formal science classroom (Paavola & Hakkarainen). With scientific ideas at the centre of the activity, the emphasis on theory-building overcomes the problem of situatedness of knowledge associated with interacting within specific contexts in informal learning environments. With pervasiveness as one of the principles of knowledge building (Scardamalia, 2002), the implication is that the knowledge creation activity should not be confined within the four walls of the classroom but pervades the mental life of the students, in and out of school. In this respect, knowledge creation produces a hybrid space whereby activities can take place within or outside the confines of the classroom as students go about contributing towards the knowledge advancement of the community. In the case example that I will describe in the next section, classroom activities were used to provide structures for activities that took place outside the classroom as homework tasks, while knowledge generated from these out-of-classroom activities became resources for the advancement of ideas facilitated by the teacher in the classroom.

A second implication of taking the third-generation activity theory as the theoretical stance is that analysis needs to be extended to hybrid spaces in which boundaries are crossed. An activity theory perspective considers learning to be mediated by social practices and objects (Edwards, 2005). Using the hybrid activity as the unit of analysis, activity theory was used as a lens to trace the shift in the knowledge artefact generated as well as the mediating artefacts. It is hypothesised that with advancement of knowledge as the boundary object, practices and resources from different contexts may be combined to form something new and unfamiliar. A problem space is thus created in which boundaries may be crossed and new practices are created. In the case example to be described in the next section, tracking of students' activities and development of ideas took into account what was done within and outside the classroom. Within the classroom, the students' sharing of their scientific ideas during the (almost) weekly show-and-tell was videotaped. As observation of students' activity outside the classroom was not easily accessible, data relied solely on students' account of what was done outside the classroom. In short, analysis of data took into account the activities that took place in the hybrid space, that is, activities related to knowledge creation.

Crossing the Boundary of Formal and Informal Settings: A Case Example

The case example is taken from a larger study that looked at how knowledge building can be orchestrated in primary school science classrooms. Knowledge building was introduced to a class of primary four high-achieving students to hone their inquiry skills and to extend their learning beyond the textbook. Being the top-achieving class of the grade level, the teacher had wanted the class to be engaged in more than content acquisition or mere participation in experimental activities. She saw knowledge building as an opportunity to stretch the potential of the best students in the level. As she was new to the concept of knowledge building, she collaborated with the research group of the larger research project to design and co-facilitate the knowledge building sessions.

The knowledge building activity in this study was centred on the topic of germination of seeds. Students were expected to advance their understanding about the factors affecting the germination of seeds. They had learnt that seeds required three basic conditions (warmth, air and water) to germinate and were expected to go beyond the basic recall and comprehension of these conditions to develop a deeper understanding of germination. While the learning outcomes might not be part of the national primary science syllabus, it is part of the curriculum for the high-achieving students in the school. The process of the knowledge activity essentially involved students sharing their ideas/problems and/or consolidating their knowledge in the classroom and conducting investigations into their problem at home. Activities were jointly designed and facilitated by the teacher and the researchers.

The activity started out as a mundane activity of planting seeds. Students were given seeds of different varieties, shapes and sizes to germinate at home. What started out as a homework task resulted in activities that involved a larger community other than the teacher and their classmates as students got themselves engaged in pursuing questions about germination of seeds. The knowledge building activity was triggered by students' observation that the three conditions alone did not seem sufficient to ensure the germination of seeds. The surprise observations led to puzzling questions being asked as students shared their findings and challenged one another's ideas. One such example of a puzzling question was the effect of the size of seeds on its rate of germination. Some claimed that small seeds germinate faster than bigger ones, while others claimed otherwise. Another puzzling question was the effect of the size of seeds on their probability of germination, which arose from one student's observation that her pets ate up the seeds.

As there was limited school curriculum time for students to explore the hypotheses they constructed, the researcher and the teacher decided that students could carry out their investigation at home and share their findings in the next session. More puzzling questions that arose from the activity resulted in further investigations or research into the problem. The knowledge building activity concluded with a rise above (i.e., a consolidation of ideas that is accepted by the community of students) in which ideas were synthesised. Table 13.1 summarises the sequence of activities and their settings in which knowledge building activities were taking place over a period of five months. This extended period was a result of time needed for students to conduct the investigations, share their findings in class and reconduct their investigations, as well as school term breaks and examination.

As Table 13.1 shows, boundaries were crossed as students went about advancing their knowledge on seed germination. Two examples will be given to illustrate the boundary crossing arising from the knowledge work done.

In the first instance, one student became involved in her mother's hobby of growing wheatgrass as she sought to find out whether the size of seeds affected germination of seeds. Using wheatgrass seeds from her mother's garden, and others like black-eyed pea and green beans, she grew them together with her mother to test her hypothesis. Instructions received in school about the need for a large sample size helped her and her mother to structure their activity at home better. Their

Table 13.1 Sequence of knowledge building activities and their settings

Activity	Setting	Activity description	Actions
1	Classroom	Introducing phenomenon	Students were asked to list down what they knew about seed germination. They were given seeds to plant and told to observe the germination process.
2	Home	Seed planting	Students planted their seeds at home.
3	Classroom	Show-and-tell	Students brought their seeds or plants back to school and reported what they saw. Each group also reported any phenomenon that they were puzzled about. Some conclusions contradicted with others. Students were instructed to find evidence to better support their hypotheses.
4	Classroom	Investigation planning	Students planned investigations to test their hypotheses with their group members and facilitators (teacher and researchers).
5	Home	Carrying out investigation	Students carried out investigations outside the curriculum time.
6	Classroom	Show-and-tell	Students reported their findings.
7	Home	Investigation/ seeking explanation	Students sought explanation for their findings.
8.	Classroom	Rise above	Teacher consolidated students' advancement of knowledge.

findings showed that there was no discernible relationship between size of seeds and rate of germination. Her results helped the class to conclude that size of seeds did not affect the rate of germination. As the class continued to wonder why some seeds seemed to germinate faster than others, the child sought advice from her grandmother, who also enjoyed gardening. She found out from her grandmother that the thickness of the seed coat might affect the rate at which water penetrated into the seed leaves. Terminologies introduced to the child by her grandmother were subsequently shared with the class, resulting in a new hypothesis being generated about how the thickness of the seed coat might affect germination of seeds. However, as it was close to the end of the school year, it was not possible to find out how the new puzzling question was addressed.

In the first instance, we see the desire to resolve the contradicting conclusions made about the effect of the size of seeds led to the crossing of boundary between schooling and engagement in a hobby. The result was that resources from the two settings mediated the advancement of knowledge about seed germination. Without the knowledge building activity, the student would not be involved in her mother's hobby. Without the instruction from school, they might not be conducting a reliable test at home. Further puzzling questions led the student to seek another expert's opinion – her grandmother. The result was a set of new terminologies and knowledge introduced to the class. Figure 13.1 shows how the resources from each setting mediated the students' advancement of knowledge.

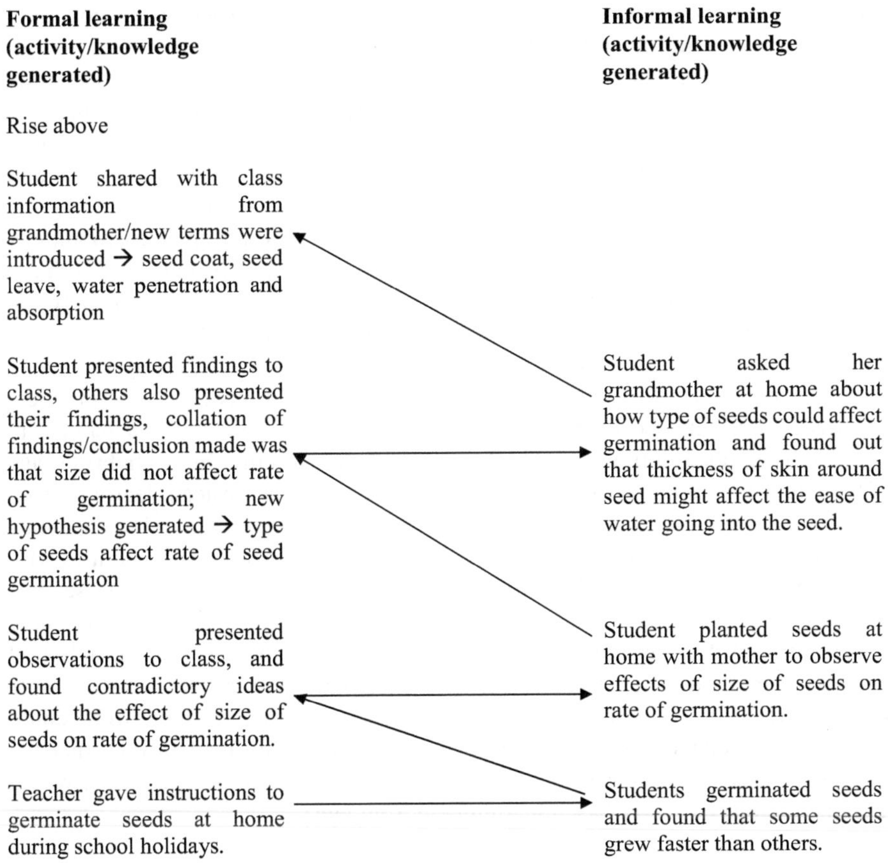

Fig. 13.1 Resources from formal and informal settings contributing to the advancement of knowledge

In the second instance, one student was surprised that her pets ate up the seeds she was trying to germinate. In relation to the puzzling problem about the effect of the size of seeds on germination, she wanted to find out how, in the presence of consumers in the environment, the size of seeds might affect the chance of germination. The knowledge building activity provided her with the platform to further investigate the behaviour of her pets as well as to resolve her puzzling question she had about germination of seeds. Using a variety of seeds of different sizes, she placed them in her garden before releasing her pet dogs and birds into the garden. She found that the bigger seeds were eaten up by her pets, while the smaller ones were relatively untouched. Her sharing of her findings with her class led to a different perspective taken about germination of seeds. Like the first instance, the common object of knowledge building provided her with the opportunity to extend her observation at home to a more structured inquiry into the relationship between her pets' behaviour and germination of seeds, and hence crossing the boundary between school learning and informal learning.

The two instances showed that knowledge building as a boundary object created a new hybrid space as practices and knowledge derived from the formal and informal learning spaces were crossed. In the first instance, practices and knowledge gained from engagement on hobby get transformed into the classroom and inquiry practices learnt in the classroom get translated in gardening. Similarly, the learning resulting from interacting with pets, when it gets into the classroom, can trigger interesting inquiry questions that result in greater understanding of both the behaviour of pets and germination of seeds. In both cases, the divide between learning in informal settings and formal settings disappeared as practices and knowledge from each setting were transformed in the other. Learning science is no longer demarcated by settings, activity or identity. Instead, the new practice, knowledge building, becomes a pervasive practice that does not just happen at home or in school, but transcends both aspects of a child's life. In this hybrid space of knowledge creation, learning is no longer discontinuous; instead, learning from each setting intersects and a new culture of learning is created in which students learn not only from the teacher and peers but also from the other communities with whom their peers interact. Similarly, knowledge and practices from the science classrooms get translated into informal communities. Besides illustrating how knowledge building as the boundary object mediated the creation of a hybrid space for the interaction between formal and informal resources, this case example also illustrates how a common out-of-school activity such as homework could potentially engender this hybrid space. However, it is pertinent to note that the homework tasks in this case study did not involve doing worksheets as one would imagine in an acquisition-based learning environment.

Implications to School Science Learning

Two implications can be drawn from this case example for teachers and researchers studying students' learning. The first relates to the design of learning activities which allows boundary crossing and creates hybrid spaces. According to Akkerman and Bakker (2011), four mechanisms that constitute boundary crossing were identified – identification, coordination, reflection and transformation. In this case study, learning points arising from how these mechanisms mediated boundary crossing were drawn.

Identity. One of the challenges of boundary crossing is the issue of identity. Practices in school (e.g., acquisition of knowledge) are often not considered by students and teachers to be aligned with practices at home (e.g., play and hobby). Instead of perceiving learning and play to be discontinuous, knowledge creation reconstructs the practices between the boundaries and a renewed identity of a knowledge builder helps to overcome the disparity between the two identities often associated with different settings.

Coordination. Akkerman and Bakker (2011) found that several studies showed that learning at the boundary requires careful coordination of communication and translation between boundaries. According to Landa (2008), as cited in Akkerman and Bakker, communication is crucial for multiple perspectives to be shared. In this case study, show-and-tell provided the platform for students to share new information with the class. The opportunity to discuss and argue about the new ideas put forth brought about new questions that required the engagement of participants outside the classroom.

An intersubjective ground and diversity of ideas are needed to mediate translation of practices. For boundaries to be crossed, the different communities must be willing to allow movement between different settings. In the case study, strong support from both the teacher and family members (and pets) facilitated knowledge and practices from one setting to be translated in the other. In this sense, the pervasiveness of knowledge building, embraced not only by the students and teachers but also family members, made this translation of practices possible. In this respect, the homework task that is taken seriously by students and parents in Singapore provided the structure for this translation.

Reflection. Reflection of one's perspective of practices in other activity systems would facilitate boundary crossing. In this case study, the teacher's dissatisfaction with engaging her best students in mere acquisition of knowledge led her to reflect on how she could better stretch her students. Her participation in this research allowed her to see how school science learning can extend beyond acquisition of knowledge in the classroom by engaging students in knowledge creation, which is a more pervasive form of learning.

Transformation. Hybridisation requires a profound transformation of teaching and learning practices. Learning science no longer took place only in the science classroom during a fixed time frame in this case study. Instead, the knowledge building activity took almost half a year long as time was needed for the iterative process of sharing and investigations amid other school activities. Time and space must be given for this transformation to take place. Students' motivation and interest also need to be sustained for such prolonged inquiry into one area of their study.

A second implication relates more to researchers analysing students' learning in the knowledge creation paradigm. As knowledge building pervades the mental and social milieu of the students, analysis need to go beyond the classroom context.

In conclusion, this chapter illustrates how science learning can be enhanced by tapping resources from both formal and informal settings. The case example showed that reconceptualising learning as knowledge creation provided the impetus for students to be engaged in activities within and outside the classroom. The students' generated problems of understanding brought about the pervasiveness of knowledge building, whereby students worked towards solving their problems as well as to increase the knowledge of the class. In this way, knowledge building created the hybrid space for formal and informal learning resources to interact. On this note, common schooling activities such as homework and show-and-tell provided the structure for resources from both formal and informal settings to be cross-fertilised. For example, the homework tasks brought to the fore knowledge embodied in familiar home activities such as gardening, playing with pets and chit-chat sessions with grandmother, which might miss the consciousness of the students. The show-and-tell activity provided the platform for scientific ideas to be shared and critiqued by the class, bringing about both an advancement in knowledge about germination of seeds and generation of more puzzling questions for further inquiry. It is in this respect that knowledge building as a boundary object mediated the possibility of boundary crossing.

As a case study, the intention in this chapter is not to claim generalisability in the ways in which boundary crossing can be orchestrated. Instead, the purpose is to provide a case example whereby boundary crossing was made possible by knowledge building as a boundary object and homework as a mechanism that engendered the involvement in activities in the informal settings. Further research is necessary to better understand the potential of knowledge building as a boundary object in terms of the sociocultural factors that mediated the development of the hybrid space as well as its generalisability in a greater variety of topics in science and across more informal activities.

References

Akkerman, S. F., & Bakker, A. (2011). Boundary crossing and boundary objects. *Review of Educational Research, 81*(2), 132–169.

Aubusson, P., Griffin, J., & Kearney, M. (2012). Learning beyond the classroom: Implications for school science. In B. J. Fraser, K. G. Tobin, & C. J. McRobbie (Eds.), *The international handbook of science education* (2nd ed., Vol. 2, pp. 1123–1134). Dordrecht, The Netherlands: Springer.

Bamberger, Y., & Tal, T. (2007). Learning in a personal context: Levels of choice in a free choice learning environment in science and natural history museums. *Science Education, 91*, 75–95.

Edwards, R. (2005, July 5–7). *Contexts, boundary objects and hybrid spaces: theorizing learning in lifelong learning*. Paper presented at 35th annual SCUTREA conference, University of Sussex, England, UK.

Engestrom, Y., Engestrom, R., & Karkkainen, M. (1995). Polycontextuality and boundary crossing in expert cognition: Learning and problem solving in complex work activities. *Learning and Instruction, 5*, 319–336.

Landa, M. S. H. (2008). *Crossing the divide: A phenomenological study of early childhood literacy teachers who choose to work with children in high poverty schools*. Unpublished doctoral dissertation, University of Maryland, College Park.

National Research Council. (2009). *Learning science in informal environments: People, places, and pursuits*. Committee on Learning Science in Informal Environments. In P. Bell, B. Lewenstein, A. W. Shouse, & M. A. Feder (Eds.), Board on Science Education, Center for Education. Division of Behavioral and Social Sciences and Education. Washington, DC: The National Academies Press.

Orion, N., & Hofstein, A. (1994). Factors that influence learning during a scientific field trip in a natural environment. *Journal of Research in Science Teaching, 31*, 1097–1119.

Paavola, S., Engeström, R., & Hakkarainen, K. (2012). Trialogical approach as a new form of mediation. In A. Moen, A. Morch, & S. Paavola (Eds.), *Collaborative knowledge creation: Practices, tools and concepts* (pp. 1–14). Rotterdam, The Netherlands: Sense Publishers.

Paavola, S., & Hakkarainen, K. (2005). The knowledge creation metaphor – An emergent epistemological approach to learning. *Science & Education, 14*, 535–557.

Scardamalia, M. (2002). Collective cognitive responsibility for the advancement of knowledge. In B. Smith (Ed.), *Liberal education in a knowledge society* (pp. 67–98). Chicago: Open Court.

Scardamalia, M., & Bereiter, C. (2003). Knowledge building. In J. Guthrie (Ed.), *Encyclopedia of education* (pp. 1370–1373). New York: Macmillan Reference, USA.

Star, S. L., & Griesemer, J. R. (1989). Institutional ecology, 'translations', and boundary objects: amateurs and professionals in Berkeley's Museum of Vertebrate Zoology 1907–1939. *Social Studies of Science, 19*, 387–420.

Suchman, L. (1994). Working relations of technology production and use. *Computer Supported Cooperative Work, 2*, 21–39.

Tal, T. (2004). Community-based environmental education: A case study of teacher-parent collaboration. *Environmental Education Research, 10*, 523–543.

Tal, T. (2012). Out-of-school: Learning experiences, teaching and students' learning. In B. J. Fraser, K. G. Tobin, & C. J. McRobbie (Eds.), *The international handbook of science education* (2nd ed., Vol. 2, pp. 1109–1122). Dordrecht, The Netherlands: Springer.

Chapter 14
Science Centre Singapore as an Alternate Classroom

Anne Dairianathan and Tit Meng Lim

Abstract In the realm of science education beyond the classroom, science centres are important in motivating learners of science, both young and old. In this chapter, we examine the role(s) that science centres play as a complement to formal science education in schools in Singapore. Besides static and special feature exhibits, the Science Centre Singapore also runs customised programmes, including events and competitions, to complement the diverse forms of inquiry and learning activities offered by schools.

Keywords Informal science learning • Out-of-school learning • Science interest • STEM • Engagement • Science-based events • Science centres • Teacher resources

Introduction

Science centres and science museums are considered leisure attractions or edutainment centres that provide support for students, teachers and families with opportunities to experience and understand science in an enjoyable setting. More than 70 % of such attractions in the United States offer programmes catering to schools (Phillips, Finkelstein, & Wever-Frerichs, 2007). The types of programmes range from field trips, complementary enrichment laboratory programmes, camps, community or science-based service learning opportunities for students to professional development for teachers and building resources for use in classrooms. Such programmes offer an alternative (and usually) complementary way of learning science and can spark curiosity, inquiry and student interest (Pell & Jarvis, 2001).

The science centre learning experiences, variously called informal learning, non-formal learning, free-choice learning or out-of-school learning, have been

A. Dairianathan (✉) • T.M. Lim
Science Centre Singapore, Singapore, Singapore
e-mail: anne_dhanaraj@science.edu.sg

A.-L. Tan et al. (eds.), *Inquiry into the Singapore Science Classroom*, Education Innovation Series, DOI 10.1007/978-981-4585-78-1_14,

compared against classroom experiences. Such experiences have been viewed as less structured, non-assessed, learner-centred, learner-directed and open-ended with many possible outcomes and with a strong focus on the social aspect of learning, in relation to a classroom lesson (Dierking & Martin, 1997; Wellington, 1990). One example of a definition comes from Crane, Nicholson, Chen and Bitgood (1994):

> Informal science learning refers to activities that occur outside the school setting, are not developed primarily for school use, are not developed to be part of an ongoing school curriculum and are characterized as voluntary as opposed to mandatory participation as part of a credited school experience. (p. 3)

We believe that learning is not restricted to time or place as cognitive processes do not differ between classroom and out-of-classroom learning (Anderson & Lucas, 1997). Rather, the environments and contexts of places such as the science centre or other educational science institutions draw upon their unique characteristics to enhance this learning process.

Science Centre-School Partnerships: Bridging Formal and Informal Learning

Generally, school curricula are structured such that important education and critical competencies in children and youth are developed. The content is developed and carefully structured to match cognitive preparedness, spiralling upwards to more complex and abstract concepts. Classrooms are particularly well suited to developing sustained engagement with subjects and working on projects that span over a period of time as in months or even a year. Because of the long exposure over the school year, teachers and students are able to build a relationship with each other as well. Such features are usually not present in edutainment attractions such as science centres.

Although interaction with students is short and possibly sporadic, science centres can offer unique experiences such as authentic, unusual and large-scale exhibits (Leinhardt & Crowley, 2002), access to facilities and equipment which schools do not have, provide research level programmes and up-close encounters with scientists and science personalities. Science centres also have an advantage in being able to provide cultural and historical context for specific themes. This helps to give a better understanding of the place of science in society. Through concrete experiences, a personal and participatory learning style that builds on social aspects, science centres can make abstract and difficult ideas comprehensible, give a macro view of a topic, make connections between disparate science information or disciplines, impact views and attitudes and stimulate curiosity, interest as well as desire for further learning (Falk, Dierking, & Holland, 1995). Bell, Lewenstein, Shouse and Feder (2009, p. 11) found that informal science learning experiences "lead to further inquiry, enjoyment, and a sense that science learning can be personally relevant and rewarding".

Learning outside of school has always had strong support from the science education community as they are seen to be important and effective complements to the formal curriculum (Hofstein & Rosenfeld, 1996; Semper, 1990) and could serve as a bridge between formal and free-choice settings (Bybee & Legro, 1997). Learning through play was the inspiration for the formation of science centres and the ubiquitous participatory exhibit that it pioneered (Oppenheimer, 1972). The development of such interactive experiences involving visible phenomena was influenced by work of thinkers such as Piaget and Dewey. Both emphasised the ways information was learnt through active, experiential investigations and in asking questions. Vygotsky's (1978) sociocultural theory has had a major impact in shaping science centre experiences as it posits that learning takes place through interaction with the environment and importantly through language and facilitated by people such as teachers, parents and facilitators (more knowledgeable others), through a zone of proximal development. These constructivist views of learning science apply to the experiences in science centres where programmes build on prior experiences of students and make connections to everyday life. These programmes give opportunities for students to apply skills and information to solve or debate real-world issues. This is aligned with the twenty-first-century education where schools are like a nerve centre connecting teachers, students and the community to the wealth of knowledge that exists beyond the school building.

There are now extensive data to show that programmes in informal learning institutions such as science centres help increase the level of science content knowledge in students (Bell et al., 2009; Moore, 2001). The benefits are not only in the cognitive domain but in the affective areas as well as in the development of specific laboratory and research-based skills. Manipulating real objects and phenomena in science centres and the stimulation of curiosity make science centres unique learning environments for students that complement the learning process within the classroom (Ramey-Gassert, 1997).

Science Centre Singapore

Science Centre Singapore (SCS) positions itself as a place where science befriends and transforms the minds of millions. It is the country's only science centre with the mission to promote interest, learning and creativity in science and technology, through imaginative and enjoyable experience, and to contribute to the nation's development of its human resource. SCS works in close partnership with schools to inspire youths to take up science and technology-related careers.

The SCS receives more than a million visitors annually of which more than 250,000 are students who come to learn science through informal, hands-on and inquiry-based activities. It strives to be a centre of excellence for informal education and a desired partner on everything relating to science and technology.

Government agencies such as the Agency for Science, Technology and Research (A*STAR), Defence Science Organisation (DSO), Economic Development Board (EDB), the Singapore National Academy of Science (SNAS) and institutes of higher learning collaborate with the SCS on a regular basis to promote science, technology, engineering and mathematics (STEM).

The SCS houses more than 1,000 exhibits, many of which are interactive. These exhibits cover a wide range of topics related to STEM and take up a total of about 20,000 m^2 of indoor and outdoor space. A series of specialised laboratories and workshops provide for a wide range of enrichment programmes for students to complement the schools' formal science education. In the course of the year, the SCS organises more than a hundred events, competitions and outreach activities for participants of a very wide range of age and interest annually.

Types of Informal Learning Programmes That Can Foster Interest and Inquiry: Exhibitions Provide Immersive Learning Environments

The SCS is known for the unique exhibitions it develops and hosts. Dinosaur animatronics and fossils have always been highly popular with schools whenever they are exhibited, and the SCS exhibitions with highest visitor attendances have all been linked to unusual exhibits such as *Dinosaurs, CSI exhibits* and *Body Worlds* that feature human bodies in artistic and tasteful settings. Unique learning experiences such as high tension and liquid nitrogen demonstrations and customised exhibits such as 3D visualisations and virtual reality are experiences that expand students' thinking and understanding. Showcasing sophisticated equipment used in science such as telescopes, electron scanning microscopes and cloud chambers can also inspire students and help them to appreciate "big" science (Braund & Reiss, 2006, p. 1379). Unique exhibits at the SCS include the *Fire Tornado* exhibit, real-time data streaming in *Earth: Our Untamed Planet* and live interactions with scientists in *Scientist for a Day*.

High-technology-based exhibits have high potential for inspiring students and providing a different dimension to the learning experience. Examples of such exhibits that used to be or are still present at the science centre are: the *Cave-Automated Virtual Environment* (CAVE) which is a supercomputer-based multimedia system, *Digistar* which is a computer graphics-based projection system and the *Virtual Cell*, a three-dimensional (3D) programme that takes viewers on a journey through a human cell increasingly magnified till the components of the DNA molecule are reached. The cognitive and affective potential of such exhibits was evaluated by Anthony (2008) who showed a positive correlation between the use of high-technology-based exhibits and better understanding of topics showcased in the systems, like states of transformation of water, the solar system and the cell. Such high-technology exhibits are expensive to purchase, install and maintain, and schools are not likely to invest in these exhibits. As such, the exhibits at SCS complement the school experiences of students through a different means. Further, *Exhibit Resource Guides* developed by teachers on attachment help make links between exhibits and science concepts taught in the school curriculum. They provide information on the various exhibits so that teachers are equipped to guide their students at the exhibits and in follow-up activities.

Similarly, *Gallery Pathways* are designed to enhance students' understanding and application of science concepts taught in the primary, secondary and junior college school science curriculum. These pathways also complement the exhibits and are designed to guide students as they learn how to better visualise abstract ideas such as force, infection and transmission of infections, genetics, nanotechnology, environmental impact, etc. Such pathways tend to be more structured – they focus attention on specific exhibits that link to topics covered in school and involve a written guide or worksheet. Such resources have been shown to increase learning but not enjoyment (Flexer & Borun, 1984; Stronck, 1983).

Example 14.1: Problem-Based Gallery Resource

The SCS offers a suite of *problem-based gallery pathways* to cater to different levels and science topics. Students are presented with a problem which they attempt to research and solve, with learning taking place in the process. This approach is premised that the experience begins and continues back in school and the SCS visit is part of the learning journey. The problem-based learning (PBL) programme consists of three parts:

(a) Previsit (starts in school)
A problem is presented as a stimulus for the class and followed by discussions on the nature of the problem and kinds of information needed to solve the problem.

(b) SCS visit
The visit is to experience and collect information about the specific question that needs to be answered for the problem. This helps to put the fact-finding component in context.

(c) Post-visit (back in school)
The information collected is revisited in class as the student groups refine and present their information to the class. Extension projects could also be possible if there is sufficient interest or time.

As an illustration, a template for a topic on energy is provided in Table 14.1. The stimulus is energy related and leads to fact-finding and question generation, after which students come to the SCS to look for answers at exhibits. The teacher is provided with a set of resources that will help the brainstorming activity, definition of terms and website resources that are relevant and useful for the activity. In

Table 14.1 Template for topic of energy

Title of problem: Electromagnetic induction and alternate energy sources

Target group students in year 5–6

Links with the school topic of energy

Duration of problem: 1 h

1. Outcomes

Knowledge	Process
Students should be able to:	Observe the outcomes from the exhibits
Comprehend the principles of electromagnetic induction	Predict the different possibilities
State at least 3 alternative energy sources and their benefits	Carry out an investigation
	Compare their findings
	Draw conclusion of a backup system and alternate energy sources
	Work in teams

2. Prior knowledge and skills

Students should be able to:

State the different forms of energy

Observe the conservation of energy in electrical appliances

Compare the benefits gained from different energy sources

Access websites

3. Problem format: Stimulus

Ah Hock owns a small factory in Singapore. Sometimes he encounters blackouts which disrupt his manufacturing line. Even though he has a backup generator, it does not seem to provide enough power for all his machines.

(a) What can he do to make his generator more effective?

A few months later, Ah Hock sends you a "thank-you" note for solving his "blackout" problem. However now he has a new problem – his PUB (electricity) bills are getting higher! He now wants to explore other ways to get electricity.

(b) Can you recommend the best alternative way?

4. Pilot-test problem

Facts	Ideas	Learning issues	Resources needed (specific energy exhibits)
(a) Ah Hock owns a small factory	How does a generator produce the electricity?	Power generator	Energy exhibition (letters denote specific exhibits)
He encounters blackout	What are the compartments needed to simulate the similar system?	Workings/ mechanics	AT-08 grand shuttle
He owns a backup generator but it does not provide sufficient power	What are the factors to affect the performance of this system?	Principles behind it	EG-12 jumping ring
(b) Ah Hock's PUB bill is getting higher	What are the different forms of energy?	Necessary inputs	EG-21 electric motor

(continued)

Table 14.1 (continued)

He is asking for suggestions on alternative way to get electricity	What are the alternate energy sources?	Possible variables	EG-22 electric generator
	What are the benefits of these energy sources?	Different forms of energy sources	(EG-80) natural gas
		The methods to harness energy from these sources	(EC-03) solar cells
		Compressed natural gas	(EG-26) solar powered discs
		Solar energy	(EG-46) photovoltaic cell
		Fuel cell	(EC-07) fuel cells
		Hydroelectricity	(EG-47) hydroelectric power
		Tidal/wave power	(EG-48) tidal and wave power
		Wind power	(EG-49) windmill
		Geothermal energy	(EG-83) geothermal power

addition, a follow-on team activity in class is also provided, in this case, the process and resources to make a rubber band machine.

This pathway programme incorporates the findings that pre- and post-visit activities enhance learning (Anderson, Lucas, Ginns, & Dierking, 2000). It also allows question generation by learners who then look for the answers to their questions. Although no data has been collected on this programme, feedback from teachers who use it has been very positive. One drawback of this and similar programmes is that it requires extensive planning of resources and commitment by the teacher to devote time before and after the visit to extend the content that has been learnt in the visit (DeWitt & Storksdieck, 2008; Griffin, 2004).

A variation of this pathway that has also been implemented is the mobile tech trail which offers an activity with a moderate degree of structure and the students have some freedom to explore. Students decide what they would like to learn about and feature on a specified theme and then go about creating a video clip on the topic of interest. This allows the student to ask the questions and research for the answers. They take charge of the learning which results in greater interest and motivation.

Laboratories Providing Authentic Experiences

Specialised laboratories, spanning the physical and life sciences, offer schools opportunities for authentic science and access to unique experiences. Such programmes allow students to use scientific equipment and to experience procedures carried out by

Table 14.2 Summary of laboratories in SCS

Labs	Participation level	Student numbers reached
Robotics Learning Centre	Preprimary-secondary levels	1,000/year
CRADLΣ	Upper secondary and above	More than 600 upper secondary and tertiary level students (15–17 years) since it opened in 2011
Kitchen science lab	Upper primary and above	More than 4,000/year
DNA learning labs	Upper primary and above	More than 25,000/year
Observatory and star lab	Preschool and above	About 2,000/year and about 6,000 public visitors annually
Ecogarden and Ecolab	Preschool and above	About 50,000/year including tours
Digital design studio	Primary and above	About 2,000/year
Movie making lab	Primary and above	Nearly 1,000/year

scientists (Luehmann, 2009). Exposure to real-world science investigations can have positive influence on students' decision to pursue a career related to science (Knox, Moynihan, & Markowitz, 2003; Markowitz, 2004). Examples of such facilities are the *Robotics Learning Centre*, *CRADLΣ* (a physics and engineering-based facility), *Movie Making Lab*, *Digital Design Studio* and the *DNA Learning Lab* (see Table 14.2).

Example 14.2: DNA Learning Lab

This facility allows primary to junior college level students to have hands-on experience in conducting genetics experiments. The programmes range from the study of cells and bacteria, gene transfer of fluorescent protein from jellyfish to bacteria and DNA sequence inspection using computer bioinformatics. The lab programmes are very popular with students who enjoy the opportunity to conduct the same type of experiments (e.g., extraction of DNA) that scientists would be conducting in their labs. In the 10 years it has been running, more than 200,000 students have experienced a DNA programme in one of two laboratories at the SCS.

Feedback was solicited from 274 teachers attending the classes with their students in 2011/2012. All (100 %) found the lab and hands-on activities being helpful and overall programme good or very good. 99 % of the teachers found that the DNA programme their students attended stimulated their interest in science. More than 60 % of the free responses written by teachers mentioned hands-on activities and experiments to be most helpful for their students. Other components they found helpful were clear instructions, interesting story (in some lessons) that captured

Table 14.3 Comments by teachers on programme

Category	Comments
Connection to curriculum	Lab: Genetic diseases
	"It covers quite a lot of bio chapters on genetics"
	Lab: DNA in forensic science
	"Excellent for the IB curriculum"
Real-world context	Lesson: DNA in forensic science
	"Real case study for the pupil's attention"
	Lab: Outbreak
	"Interesting hands-on activity conducted with a real-world context"
Teaching approach	Lab: Diversity of cells
	"Very interesting way of teaching"
	Exhibits and discussion: Bioethics
	"Informative and provides opportunity for discussion"

attention and passionate educators. Further, the teachers were asked to present their feedback to the question: "*Would you recommend this programme to others in the future? Why*?" All teachers in the survey (100 %) would recommend the programme to others and the reasons given can be grouped into three categories: (1) connection to curriculum, (2) real-world context and (3) teaching approach. Table 14.3 provides some examples of comments given by teachers on the programme.

Feedback was also collected from 2,834 participating students who filled up a short questionnaire at the end of the class. The questionnaire included multiple choice questions as well as components for open-ended responses. Ninety-seven percentage of students found the lesson interesting, with 94 % indicating that the hands-on activities were fun and interesting. With regard to the question "*One thing I enjoyed most in the lesson*", students were varied in their responses. This suggests that the activities appealed to the learners in different dimensions and the learning experiences were generally positive. Some learning experiences that the students enjoyed include "*The hands-on experiment with the lab tools*", "*Experience what a scientist would experience*" and "*The experiment made me feel like CSI*".

A study carried out at the SCS to investigate primary students' learning through participation in a DNA Lab programme which focused on DNA and genes showed statistically significant gains in understanding of inheritance (Dairianathan & Subramaniam, 2011). The sample size was 245 primary 5 (11 years) students. The students were tested on knowledge, comprehension and application of information learnt in the lesson, through pre-, post- and delayed post-tests. A survey form was also administered to measure student feedback. Learning gains were still statistically significant when tested 2 months after the initial programme, and high levels of interest were recorded from the participants with 94 % saying they had enjoyed the experience and 92 % that they were able to understand DNA better. 86 % said they could relate DNA to their daily life. Students found the activities interesting (96 %) and educational (92 %).

Example 14.3: Centre for Research and Applied Learning in Science (CRADLΣ)

CRADLΣ, a new research and applied learning hub, was set up in 2011 at SCS, for the purpose of nurturing students with the attitude and aptitude in STEM, particularly in the area of integrated science and engineering. CRADLΣ is an initiative supported by the Agency for Science, Technology and Research (A*STAR), the Defence Science Organisation (DSO) and the Ministry of Education (MOE).

The motivation behind CRADLΣ is the belief that understanding science requires more than memorising textbook knowledge. Hands-on experimentation results in experiences that lead to intuitive understanding and develop the practical skills needed to do research. CRADLΣ also administers a student research project grant to encourage students to develop independent thinking and scientific inquiry. Students with an idea to explore, with a question to answer, with a problem to solve or with a hypothesis to test can apply to CRADLΣ for research project funding. Students will also be mentored by researchers and teachers who can provide guidance and advice as well as referrals to other experts and resources. About 600 students have already benefitted from the programmes and facilities. Participants found the opportunity to experiment and do hands-on activities most enjoyable.

Fostering Inquiry and Interest Through Events and Competitions

As out-of-school programmes or enrichment generally evoke positive affective responses, they are very useful as complements to formal instruction in facilitating learning and stimulating interest. Exposure to extracurricular activities in science at upper primary or lower secondary levels (13–14 years) contributed to students having increased interest in science activities compared to students exposed to regular science in school (Maoz & Rishpon, 1990). There is a direct relationship between informal learning environments, teaching procedures and scientific reasoning ability (Gerber, Cavallo, & Marek, 2001). Students who participate in clubs, societies and hobbies and visit to museums, zoos, cultural centres, etc., had significantly higher reasoning abilities compared to those with impoverished informal learning environments.

There is great potential in science centres creating positive attitudes towards science and research (Salmi, 2003). Participation in informal science activities leads to greater interest in science and uptake of science courses at undergraduate level (Salmi, 2003). Involvement in youth programmes in science centres has also been shown to help participants to pursue careers in science teaching (Siegel, 1998). Student learning and engagement in STEM has also been shown to be enhanced through participation in informal learning activities such as excursions, visits by scientists and competitions such as mathematics and science Olympiads (Tytler et al., 2008).

The key component for organising many of the events and competitions is that there is an innovative task setting. Students need to become involved in a process of scientific investigation and active knowledge construction or manipulation of variables. In working on a physical project, students need to have mental models which may have to be modified with new information learnt in the course of the project. An engaging project can help students learn a lot more about science concepts than listening to the theoretical aspects in a classroom, especially for kinaesthetic learners who learn best through other modalities. Harel and Papert (1991) suggest that the best way for children to construct their ideas about the world is to build a working model.

Besides the deeper understanding of science concepts and the acquisition of new skills, the longer and deeper immersion in a project imparts other values such as patience, integrity, teamwork, discipline and project planning. SCS works with dedicated partners who share the same vision of providing youth programmes that challenge participants to be creative and resourceful. These partners help promote STEM through organising science and engineering-related events and competitions all year round, and many are highlights in the school calendar. A sample of the major events is presented in Table 14.4. Each competition has a following with schools who adopt competitions as part of their programmes if they see a beneficial impact on their students' learning and interest in science.

Example 14.4 Singapore Science and Engineering Fair (SSEF)

The SSEF is a national competition organised by the MOE, A*STAR and SCS. The SSEF is the pinnacle of achievement for science fair projects in Singapore as only the best nationwide are selected for the final judging rounds. The SSEF is affiliated to the prestigious Intel International Science and Engineering Fair (Intel ISEF), which is regarded as the Olympics of science fair competitions. SSEF is open to students between 15 and 21 years of age who are required to submit research projects from the Science Research Programme (SRP), Technology and Engineering Research Programme (TERP) and Science Mentorship Programme (SMP) or projects done at school or cluster level. Projects cover all areas of science and engineering.

In schools, these students are already selected on the basis of aptitude and academic performance to participate in mentorship or attachment programmes. The students are given opportunities to work on a project with a mentor. The scheme involves many scientists from the various research institutions in Singapore who willingly take such students under their wing to guide them in research. The students have to work within a timeline (although they can extend to the next year if there is great personal motivation and potential in the research work) to complete

Table 14.4 Competitions and events organised by SCS

Participation level	Event	Participant numbers/feedback
Upper Sec and JC students aged 15–21 years	Singapore Science and Engineering Fair (SSEF)	SSEF representatives at Intel International Science and Engineering Fair (ISEF) have won awards for every year of participation in ISEF ISEF, the world's largest precollege science fair, provides an opportunity for students from different parts of the world to showcase their science projects
Upper Sec and JC students aged 15–21 years	A*STAR Talent Search (ATS)	*1171* since it started in 2002. The number of participants has increased from 51 in 2002 to 189 in 2012 The Chief Judge for this event is usually a Nobel Laureate
For schools Primary-tertiary levels	National Junior Robotics Competition (NJRC)	*600 teams (3,000 students)/year* Over the past 15 years, more than 30,000 participants and 200,000 supporters have gathered for this premier robotics competition
For schools Primary level and above	The Amazing Science-X Challenge (ASXC)	*Increased from 177 in 2009 to 453 in 2012* *Feedback from participants:* The ASXC increased my interest to know more about science and technology – 93 % agree Through ASXC I have greater understanding about physical science phenomena – 99 % agree
For schools Upper primary and above	Singapore Amazing Flying Machine Competition (SAFMC)	*Increased from 200 participants in 2009 to 333 in 2013*
For schools Primary 1–6 students (7–12 years)	The Sony Creative Science Award (SCSA)	Since it started, number of submissions has doubled to more than *4,000 toy entries/year* More than *40,000 toys* have been submitted since the competition started in 1997 These students do not lose the passion for creating – many continue their inventiveness by joining national invention competitions such as the Tan Kah Kee Young Inventors Competition
Primary level and above Ages 7–70	Science Buskers Festival	To date *914 participants* since the event started in 2008 Feedback indicates 100 % agreement from participants that taking part in the Festival made them realise science is fun and the activities helped them understand science better
Primary school to lower sec, families	Science in the Mall (SITM)	*283,511 (2003–2012)* Dec 2012 SITM: 188 responses

(continued)

Table 14.4 (continued)

Participation level	Event	Participant numbers/feedback
		Helps my children realise that science is fun – 91 % agree
		SITM is an exciting way to learn science – 91 % agree
		SITM increased interest in science – 89 % agree
		Would recommend the event to my friends – 89 % agree
Students, general public and professionals	Singapore Science Festival	*761,113, since 2002. Average of 60,000/year*
		An example is the STAR lecture which featured the speaker from Royal Institution lectures
		Feedback from nearly 1,600 responses:
		The lecture was interesting – 97 % agree
		I learnt new things from the lecture; speaker was inspiring – 95 % agree
		I am more interested in science and technology after the lecture – 88 % agree

their research projects. Such students are often cited as co-authors in publications with their mentors. MOE provides very strong support for the students through the various research programmes mentioned. There are also opportunities for school-based research to be exhibited and these may be mentored by teacher scientists.

Feedback from SSEF participants revealed that they learnt about the qualities of a good researcher, exposure to the work of a researcher and use of research equipment that was different from school labs, forming a hypothesis and designing the experiment to test the hypothesis. Such exposure to a scientist mentor or the research laboratory often initiates the interest in a science career. In 2012, 79 % of respondents indicated that they planned to pursue a career in science research, medicine or engineering as opposed to business or non-science careers.

ISEF, the world's largest precollege science fair, is held in the United States annually and provides an opportunity for students from different parts of the world to showcase their science projects. The 1,500 finalists are chosen from more than 500 regional ISEF-affiliated science fairs from over 40 countries. A select group of students who win at SSEF are chosen to participate at ISEF – between two and three students in individual and team categories. From about 80 projects and 210 participants at the inaugural SSEF in 2002, there has been a sevenfold increase to 605 projects involving 1,227 participants in 2012. Both SSEF and ISEF help to provide a platform for bright students to move to the next level, which is the A*STAR Talent Search (ATS). The talent scheme has served to inspire talented youth who show depth of scientific understanding beyond their age, independent thinking, scientific curiosity and creativity as well as deep motivation and passion for research. Over the years, many ATS participants and finalists have gone on to pursue careers in science, engineering and medical research.

One example among many, Dr. Loh Xian Jun who presented his research project as a 17-year-old ATS finalist went on to read Bioengineering. He is currently a research scientist studying general polymer characterisation techniques with the potential for such novel materials to be used for the delivery of drugs or as scaffolds for living cells. Many ex-SSEF participants have continued to pursue their passion in science in further studies and careers in science and engineering. Ms. Zhao Ye, who won a top prize in ISEF 2008, went on to read Physics at Harvard University and is continuing graduate studies at Stanford University. A participant from ISEF 2005, Mr. Hong Wenxian, a President's Scholar, pursued a Masters in Mechanical Engineering at Massachusetts Institute of Technology (MIT). He attributed his interest in engineering and technology to his positive research experience in SSEF.

Example 14.5: The National Science Challenge (NSC)

Launched in 2004, the National Science Challenge (NSC) aims to promote science among students and the local public. It is jointly organised by the A*STAR and SCS, with support from the MOE. This competition is a rigorous series of quiz rounds with many problem-solving challenges, culminating in a nationally televised finals that garners a viewership of about 387,000.

Reflections of participants from winning teams of the NSC indicate that participation in such competitions have a personal impact, including stimulating interest in a career in science:

> During the competition, we visited one of the agencies and learnt about blood diseases. It was the visit that introduced the subject matter and sparked my interest in human diseases. Now I'm working with A*STAR (research attachment) at Genome Institute of Singapore. (Wei Qiang, NSC 2005)
>
> The demo round (in NSC) really made me realise that science should be for everyone to understand and appreciate. And that connecting with people and helping them to understand the science around them was just as important as the hard science. (Nigel, NSC 2007)
>
> During NSC, we went to the Institute of Bioengineering (IBN) for the competition and since then I've wanted to apply the knowledge I learnt in science and that is why I decided to study Engineering (Stanford) because I want to learn how to use science to solve problems people face. (Gloria, NSC 2009)

School-Based Activities That Sustain Interest in STEM

Activities that engage young people in self-directed science activities from an early age are important, as interest in science develops before 11 years for the majority of youth who decide to pursue a science-related career (Bayer Facts IV, 1998; Bayer Facts XIV, 2010). As such, SCS works with schools in planning programmes that can potentially encourage self-directed learning among students.

Example 14.6: Young Scientist Badge Scheme

The objectives of this scheme are to stimulate interest in science activities among primary school students aged 7–12, enabling them to carry out self-directed activities in areas of interest in science and to provide opportunities for students to develop initiative and creativity. There are at least 17 different *Young Scientist* badges. To earn a badge, they have to get a corresponding activity card and carry out the activities listed in the card. For each activity completed, students earn a number of stars. After they have earned the required number of stars, they qualify for a badge.

The badge scheme was launched in 1983 and more than a million badges have been awarded. There is high awareness of this scheme and many adults and students recall working on the activities in primary school, especially with family members. In its 30th year, the card activities have been revamped to be still relevant and incorporate more inquiry-based activities, through a partnership with academic scientific bodies such as the Singapore Association for the Advancement of Science, the Science Teachers Association of Singapore, Singapore National Academy of Sciences and the Academy of Singapore Teachers. Other partners include institutions such as the Public Utilities Board (PUB) and the Housing and Development Board (HDB) as there are other places where students may carry out activities. These include the zoo, nature parks, waterways, homes and schools.

The badges are awarded in an annual ceremony, and scientists who had earned their badges when they were in primary school have been invited to speak about how working on the badges got them interested in doing science and continuing to stay in a science-related career. One of the scientists, for example, was a multiple award winning scientist who had started his passion in science with working on 12 different badges. One of Singapore's atmospheric scientists also professed in the media that his interest in atmospheric science started with working on badges for astronomy, meteorology and geology.

Deepening Science Interest Through Attachments and Volunteer Programme

Like many other science centres around the world, the SCS also accepts students to work alongside our science centre educators in an area of their interest. Students (must be at least 14 years of age) may interact with the visitors, explain the principles behind each exhibit or show them how a test kit or resource works. They could also help in the specialised labs as facilitators or engage in public stargazing sessions at the SCS Observatory. Students earn credits for community service after they fulfil a minimum of 20 h of public service. This mandated minimum period of service ensures students commit to the programme and plan their time appropriately. Student volunteers have been increasing and student attachments spanning 2–6 weeks to various labs doubled from 59 in 2011 to 107 in 2013.

More than half of the 3,000 volunteers who participate annually stay on after completing a mandated 20 h of service. Our record shows that the percentage of youth who continued to volunteer after completing 40 h increased over a 3-year period, from 10 % in 2010 to 16 % in 2012. This is most apparent in the astronomy club where students continued with the group, serving hundreds of voluntary hours. Such youth programmes engage teens and young adults in a science programme that fuels their interest. Although astronomy is not part of the science curriculum, astronomy clubs have started in many schools, and higher institutions and undergraduate modules on astrophysics in the two universities (National University of Singapore and Nanyang Technological University) have a high take-up rate. The youth volunteer programme is part of the overall SCS youth science movement to engage and build on their energy and idealism to drive self-initiated projects.

Besides learning skills, such as getting along with others and learning science content, and communication skills, volunteering helps to stimulate interest in science, based on student reflection logs and exit survey feedback. The holistic development of a 14-year-old volunteer is evident from what she wrote in her reflection – "*I had a wonderful experience at Science Centre. I came here to be a volunteer and learnt how to be a professional person while teaching others. At the same time I became more interested in science. This voluntary experience had enhanced my learning experience in science. In addition the staff are friendly and caring too. Thank you so much* ☺".

Research on out-of-school learning shows that after-school community-based science work also helped to increase students' interest in science by allowing them to actively use what they had learnt and make connections to the community at large (Fusco, 2001). An exit survey involving 100 student volunteers was conducted in 2012. The survey showed an overall positive experience about student volunteers' engagement at the SCS. Of the 100 student volunteers, all agreed or strongly agreed that serving as a SCS volunteer enhanced their interest in science. They also indicated that their volunteer experience made them think about pursuing a career in science, for example:

List two aspects of the programme that you find most relevant and useful to you:

I understand science better. I would like to explore more science knowledge.
I understand science better. I will think about relevant science knowledge when I see some interesting phenomenon.
I learnt how to cope with customers, little children, time management.
I have learnt how to explain experiments.

Besides student volunteers, the SCS depends heavily on a dedicated and passionate group of scientists, educators and other science professionals, who commit many hours of their time and share expertise as judges at the many competitions organised, sit on advisory committees, develop and conduct some programmes and give talks and write guidebooks or articles. At least 500 such scientists and educators donate their time each year. Their commitment and partnership are critical for the SCS to tap on a broad range of expertise.

Nurturing Interest Through Youth-Initiated Projects: Singapore Academy of Young Engineers and Scientists (SAYES)

This newly launched initiative aims to create a youth science movement in Singapore. It is a student-centric youth club, with its headquarters at the SCS. The SAYES is empowered through facilitative help and professional networks to nurture aspiring young scientists and engineers. An executive committee formed by the students is entrusted to decide and generate relevant science community and research projects for involvement.

The youth club took part in a project called Science Centres Engagement and the Rio Summit (SCEnaRioS) which was a pilot project by the Association of Science-Technology Centres (ASTC) in preparation for the United Nations Conference on Sustainable Development 2012 (Rio +20). The aim was to engage youth to think about the topic of sustainability and produce online resources that would be shared and featured at the Planet Under Pressure conference in London and the Rio +20 Earth Summit in Rio de Janeiro.

The SCS collaborated with Questacon in Australia and Guangdong Science Centre in Guangzhou, China, to bring youth from Singapore together with youth from Australia and China to explore the issue of Water and Life (Chew, 2013). The activities involved a video conference with students from Australia and China, online survey, water testing activity at World Water Day and a video for *Planet under Pressure*. The SAYES group of youth carried out the various components of their project including creating videos on NEWater production, which can be viewed at http://www.science.edu.sg/schoolprogrammes/Pages/SCEnaRioS.aspx.

Working with Educators Creates Multiplier Effect in Informal Science Education

Teacher development is also a very important component of the collaboration and partnership between the SCS, the MOE (Sciences Branch) and the Academy of Singapore Teachers (AST). Teacher workshops are developed together with the Academy to ensure that teachers are able to use the resources at the SCS in ways that bring maximum benefit to their students.

For example, Master Teachers from AST worked with SCS staff to plan and conduct a workshop for teachers to use the genetics exhibits in the teaching of a rather abstract topic. The workshop involved using the teacher resource guide, the genetics exhibits and activities developed by the SCS staff. However, critically, Master Teachers created awareness of the resources, elicited feedback if they were helpful and used collaborative learning techniques to engage teachers' subject matter knowledge (SMK) and pedagogical content knowledge (PCK). This approach helped to make links to the topic taught in schools and helped teachers to develop customised lesson plans. Master Teachers also refined the format, created a lesson plan template exemplar to help participants get started and loaded all resources on a website immediately for peer feedback. The outcomes of such workshops enabled more teachers to understand how exhibits and resources at the SCS can be used to aid teaching and learning in class.

Teacher Attachments at SCS

The SCS also takes in teachers on 2-year or shorter-term attachments, and many will return to formal teaching with new methods or innovative approaches adopted from an informal education experience. These teachers are a valuable bridge between schools and the education department at SCS, as they advise on needs for different students and also initiate development of resources that would benefit students and teachers. An example is the teacher resource guide meant for teachers to equip them with the information and skills to optimise a class visit to the SCS exhibitions.

Developing Teaching Resource Kits for Classroom Use

The SCS works closely with the Curriculum Planning & Development Division (CPDD) in the Ministry of Education to develop a series of classroom learning tools called STaR Kits® (Science Teaching and Resource Kits). These kits are designed to support the teaching and learning of science through inquiry-based activities. Targeted at primary and secondary levels, the kits can be used to develop inquiry process skills in students and to stimulate student talk as they construct understanding about science concepts or ideas. We develop the prototype with input from educators before mass producing them and selling to schools for classroom use. Some examples of kits developed are shown in Table 14.5.

Informal Teaching Workshops and Pre-service Teacher Programmes

Teachers' workshops are conducted on a regular basis to propagate the teaching of science using exhibition galleries and science show tricks. The SCS is the summer

Table 14.5 Learning kits developed

Types of kits	Objectives
Light kit	To help students develop an understanding of light and shadows
Black box	To help students develop an understanding of the nature of light by observing and experimenting with some of its properties
Mystery sponge	To help students develop an understanding of the nature of science through a collaborative process of creating knowledge based on observations and inferences
Heat kit	To help students explore the concept of heat conduction and learn more about the type of materials that make good conductors
Density kit	To help students explore the concept of density by comparing different materials and their ability to sink/float in water
Chemistry ICT tool kit	To reinvent the means to learn and enhance students' knowledge of chemistry symbols, ionic compounds and formulae through animation and playing a time-based game

course venue for the Masters of Science in Science Communication Program jointly offered by the Australian National University and the National University of Singapore. The 2-week intensive course requires participants, mainly teachers, to learn and share creative ways of communicating and teaching science while demystifying certain topics and correcting common misconceptions in science.

Before graduating from the National Institute of Education, all pre-service teachers attend a compulsory orientation at the science centre to familiarise themselves with resources useful to them and to understand the partnership role of the SCS. Language and humanities teachers are included in the orientation programme because they can use the content and technology we offer to enhance their teaching, especially when science and technology transcend all disciplines. Close to 1,000 pre-service teachers undergo this orientation each year.

In-service teacher workshops that focus on science content and how teachers can make science teaching interesting and facilitate understanding of concepts are also conducted. In 2012, a total of 3,293 in-service teachers attended teachers' workshops, learning journey, teachers' previews, feedback sessions as well as symposium and seminars organised by SCS.

Feedback on Enrichment Programmes from Teachers

As it is important for SCS to ensure that its enrichment programmes are relevant to students and teachers, a nationwide survey was conducted in 2005 (Anthony, 2008) and feedback was gathered from 130 primary school teachers and 108 secondary level teachers. The teachers, who were represented equally in terms of years of teaching experience, indicated that visits to the SCS were planned to complement science lessons. Secondary level teachers brought their students to excite them about science and when the schools did not have the facilities available at SCS. The majority of the teachers reported that they made references to the visit of the students between 1 week and 3 months later, and the survey revealed that 100 % of primary level teachers and 98 % of secondary level teachers indicated they would bring another group of students to the SCS.

The International Science Centre Impact Study (ISCIS)

SCS has taken part in a 2013 survey involving 17 science centres in 13 countries to analyse the impact of science centres on adults' and youth's knowledge and understanding of science, interest in and engagement with science (in school and out of school), likelihood of pursuing careers or hobbies related to science, creativity and problem-solving skills. Preliminary results have shown that science centres do bring about a difference to their visitors. For example, more than half of all 14- to 15-year-old youth (60 % of 5,792) in the combined sample had visited one of the 17 science centres at least once during their lifetime. The data also supported the argument that visiting a science centre, even once, was empirically related to enhanced science literacy and engagement. Science centre visits are significantly related to enhanced science understanding, interest, curiosity, and engagement with science-related activities, in school and out of school. (Falk et al., personal communication, 2013).

Conclusion

The SCS as an out-of-school learning environment has features that allow learning to take place differently, as learning is not assessed, not time sensitive and generally voluntary unless it is a bused school group. The positive affective responses of student and participant feedback from enrichment programmes, competitions and events at SCS indicate that youth and adults can enjoy learning something new at their own pace. The partnership and collaborations between formal and informal educational institutions should be seen as leveraging each other's strengths. For one, school structure is necessary for students to systematically develop deep conceptual understanding in progressing to advanced content. At the same time, science centres not only bring a concrete realisation of abstract phenomena but, because of their structure, contribute a contextualised, sociocultural and arguably more accessible approach to science and its place in society (Bevan et al., 2010).

Singapore places a lot of emphasis on education, and being the only science centre in a small island state with easy access to all schools and agencies, we are privileged to play a vital role in local science education. We believe in synergy and strategic alliances with formal and informal education agencies and tapping into industrial and academic R&D networks. It is through such an "education ecosystem" that bridging the gap between formal and informal science education can be engaging, effective and empowering, making SCS a worthy alternate classroom for both young and old.

References

Anderson, D., & Lucas, K. B. (1997). The effectiveness of orienting students to the physical features of a science museum prior to visitation. *Research in Science Education, 27*, 485–495.

Anderson, D., Lucas, K. B., Ginns, I. S., & Dierking, L. D. (2000). Development of knowledge about electricity and magnetism during a visit to a science museum and related post-visit activities. *Science Education, 84*, 658–679.

Anthony, S. X. (2008). *Use of technology-based exhibits in science centres to foster affective and cognitive outcomes among students*. Unpublished Masters thesis, National Institute of Education, Nanyang Technological University, Singapore.

Bayer Facts of Science Education Survey. (1998). *The Bayer facts of science education IV: Scientists on science for the 21st century*. Retrieved January 31, 2013, from http://www.bayerus.com/MSMS/Survey/survey98.aspx

Bayer Facts of Science Education Survey. (2010). *The Bayer facts of science education XIV: Female and minority chemists and chemical engineers speak about diversity and underrepresentation in STEM*. Retrieved January 31, 2013, from http://www.womenscolleges.org/files/u1/BayerFactsofScienceEducation.pdf

Bell, P., Lewenstein, B., Shouse, A. W., & Feder, M. A. (Eds.). (2009). *Learning science in informal environments: People, places, and pursuits*. Washington, DC: The National Academy Press.

Bevan, B., Dillon, J., Hein, G. E., Macdonald, M., Michalchik, V., Miller, D., et al. (2010). *Making science matter: Collaborations between informal science education organizations and schools* (A CAISE Inquiry Group Report). Washington, DC: Center for Advancement of Informal Science Education (CAISE).

Braund, M., & Reiss, M. (2006). Towards a more authentic science curriculum: The contribution of out-of-school learning. *International Journal of Science Education, 28*, 1373–1388.

Bybee, R. W., & Legro, P. L. (1997). Finding synergy with science museums: Introduction to the National Science Education Standards. *ASTC Newsletter, 25*(2), 6–7.

Chew, L. L. (2013, January/February). Bringing youth's ideas about water to Rio+20. *Dimensions, 15*(1), 44–47.

Crane, V., Nicholson, H., Chen, M., & Bitgood, S. (Eds.). (1994). *Informal science learning: What the research says about television, science museums, and community-based projects*. Dedham, MA: Research Communications Ltd.

Dairianathan, A., & Subramaniam, R. (2011). Learning about inheritance in an out-of-school setting. *International Journal of Science Education, 33*, 1079–1108.

DeWitt, J., & Storksdieck, M. (2008). A short review of school field trips: Key findings from the past and implications for the future. *Visitor Studies, 2*, 181–197.

Dierking, L. D., & Martin, L. (1997). Guest editorial: Introduction. *Science Education, 81*, 629–631.

Falk, J. H., Dierking, L. D., & Holland, D. G. (1995). What do we think people learn in museums? In J. H. Falk & L. D. Dierking (Eds.), *Public institutions for personal learning: Establishing a research agenda* (pp. 17–22). Washington, DC: American Association of Museums, Technical Information Service.

Falk, J. H., Dierking, L. D., Needham, M. D., & Prendergast, L. (2013, November/December). Science centers make a difference: Results from the International Science Center impact study. *Dimensions, 15*(6), 27–30.

Flexer, B. K., & Borun, M. (1984). The impact of a class visit to a participatory science museum exhibit and a classroom science lesson. *Journal of Research in Science Teaching, 21*, 863–873.

Fusco, D. (2001). Creating relevant science through urban planning and gardening. *Journal of Research in Science Teaching, 38*, 860–877.

Gerber, B. L., Cavallo, A. M. L., & Marek, E. A. (2001). Relationships among informal learning environments, teaching procedures and scientific reasoning ability. *International Journal of Science Education, 23*, 535–549.

Griffin, J. (2004). Research on students and museums: Looking more closely at the students in school groups. *Science Education, 88*(Suppl. 1), S59–S70.
Harel, I., & Papert, S. (Eds.). (1991). *Constructionism*. Norwood, NJ: Ablex.
Hofstein, A., & Rosenfeld, S. (1996). Bridging the gap between formal and informal science learning. *Studies in Science Education, 28*, 87–112.
Knox, K., Moynihan, J., & Markowitz, D. (2003). Evaluation of short-term impact of a high school summer science program on students' perceived knowledge and skills. *Journal of Science Education and Technology, 12*, 471–478.
Leinhardt, G., & Crowley, K. (2002). Objects of learning, objects of talk, changing minds in museums. In S. Paris (Ed.), *Perspectives on object-centered learning in museums*. Mahwah, NJ: Lawrence Erlbaum Associates.
Luehmann, A. L. (2009). Students' perspectives of a science enrichment programme: Out-of-school inquiry as access. *International Journal of Science Education, 14*(64), 52–89.
Maoz, N., & Rishpon, M. (1990). Attitudes towards school science: A comparison of participants and nonparticipants in extracurricular science activities. *School Science and Mathematics, 90*, 13–22.
Markowitz, D. G. (2004). Evaluation of the long-term impact of a university high school summer science program on students' interest and perceived abilities in science. *Journal of Science Education and Technology, 13*, 395–407.
Moore, J. M. (2001). *The effects of inquiry-based summer enrichment activities on raising eighth-graders' knowledge of science processes, attitude toward science, and perception of scientists.* Unpublished doctoral dissertation, University of Mississippi, Oxford.
Oppenheimer, F. (1972). The exploratorium: A playful museum combines perception and art in science education. *American Journal of Physics, 40*(7), 978–983.
Pell, T., & Jarvis, T. (2001). Developing attitude to science scales for use with children of ages from five to eleven years. *International Journal of Science Education, 23*, 847–862.
Phillips, M., Finkelstein, D., & Wever-Frerichs, S. (2007). School site to museum floor: How informal science institutions work with schools. *International Journal of Science Education, 29*, 1489–1507.
Ramey-Gassert, L. (1997). Learning science beyond the classroom. *The Elementary School Journal, 97*(4), 43–50.
Salmi, H. (2003). Science centres as learning laboratories: Experience of Heureka, the Finnish science center. *International Journal of Technology Management, 25*, 460–475.
Semper, R. J. (1990, November). Science museums as environments for learning. *Physics Today*, pp. 2–8.
Siegel, E. (1998). The science career ladder at the New York Hall of Science. *Curator, 41*(4), 246–290.
Stronck, D. R. (1983). The comparative effect of different museum tours on children's attitudes and learning. *Journal of Research in Science Teaching, 20*, 283–290.
Tytler, R., Osborne, J. F., Williams, G., Tytler, K., Clark, J. C., Tomei, A., et al. (2008). *Opening up pathways: Engagement in STEM across the primary-secondary school transition. A review of the literature concerning supports and barriers to Science, Technology, Engineering and Mathematics engagement at primary-secondary transition.* Melbourne, Australia: Deakin University.
Vygotsky, L. S. (1978). *Mind in society: The development of higher psychological processes.* Cambridge, MA: Harvard University Press.
Wellington, J. J. (1990). Formal and informal learning in science: The role of the interactive science centres. *Physics Education, 25*, 247–257.

Chapter 15
Public Education About Science in Singapore: The Role of Science Journalism via Newspapers

R. Subramaniam

Abstract This chapter explores the role of science journalism in contributing towards public education about science in Singapore, an aspect which is under-explored in the literature. Though the key platforms for promoting public understanding of science are spearheaded mainly by institutions that promote informal science learning as well as by learned societies, the contribution from newspapers is also important. Content analysis of one continuous week of a national newspaper was undertaken, and the results show that coverage of science and technology was modest and predominantly news driven. For news stories set locally, there was a pronounced emphasis on the use of institutional actors in the text as well as quotations. Articles on health/medicine and earth science elicited the most coverage in the press. Some suggestions for using newspaper science to introduce inquiry in the science classroom are discussed.

Keywords Science centres • Scientific societies • Singapore • Science journalism • Content analysis • Straits Times • Science and technology • Inquiry • Nature of science

Introduction

Developments in science and technology are impacting society to such an extent that it is increasingly important for the public to keep abreast of these developments to some extent. They would then be in a position to make informed decisions on matters related to science and technology that affect them. For example, high cholesterol foods are increasingly being partaken by people in today's urban societies – are they aware that cholesterol can be deposited on the walls of blood

R. Subramaniam (✉)
Natural Sciences & Science Education Academic Group, National Institute of Education, Singapore, Singapore
e-mail: subramaniam.r@nie.edu.sg

A.-L. Tan et al. (eds.), *Inquiry into the Singapore Science Classroom*, Education Innovation Series, DOI 10.1007/978-981-4585-78-1_15,

vessels and later give rise to heart problems? Furthermore, with the pervasive influence of globalisation, borders between countries have become more porous, and developments in one country can affect another. Air travel has shrunk distances between countries, and infectious diseases can spread quite rapidly from its source, which could be in another distant part of the world; also, forest fires in one country can affect air quality in neighbouring countries. As such, people do need to know what steps can be taken in order to mitigate such risks. More such examples can be given, but the point to note is that people need to have some basic science literacy in order to manage risks as well as to be able to make prudent decisions on matters related to science and technology that affect them. This is where public education about science can play a useful role.

Science and technology are also recognised as instruments for socio-economic development. Many countries in the West have used this approach to provide their people with a good standard of living. Such an approach for development is possible when people are receptive to the potential of science and technology to promote economic development and their well-being. Again, public education about science is essential. A scientifically literate citizenry can help governments to realise socio-economic objectives (Tan & Subramaniam, 1998).

Public Education About Science in Singapore

Various mechanisms exist for the dissemination of science and technology to people in Singapore, a country located in South East Asia. Singapore has made great strides in providing its people with a good standard of living, mainly through science and technology-driven economic development as well as good governance. Science and technology impact on many aspects of societal endeavours in this tiny island state.

The role of schools in imbuing students with basic science literacy is very important in their early years. There are also a number of institutions that promote informal science learning to people, and these efforts give rise to useful tributaries for the cause of public education on science. In this context, the roles of the Science Centre Singapore (Caleon & Subramaniam, 2005, 2007; Dairianathan & Subramaniam, 2011; Tan & Subramaniam, 1998, 2003) and the Singapore National Academy of Science as well as its constituent societies (Tan & Subramaniam, 1999, 2009) are well documented. Destinations such as the zoo, bird park and botanic gardens have also played a useful role in public education about science. However, some aspects of public education about science in Singapore have been under-explored (or not done so systematically) in the literature. One of these is the role of print journalism. In reaching out to the public through the print media, there are two key avenues in Singapore – articles in the best-selling local science magazine, the *Singapore Scientist*, and science-based articles in the national daily, *The Straits Times*.

The *Singapore Scientist* is a popular science magazine published by the Science Centre Singapore. It has been in publication since 1977 and has a readership of about 60,000. Its readers comprise mainly students, but copies of the magazines are also sent to the local institutes of higher learning as well as various government departments and statutory boards. The quarterly magazine is available for sale, mainly through subscriptions for students and direct sales through bookshops. Articles in the magazine come in various forms – for example, articles written by science centre staff and local academics on contemporary topics in science, quizzes, competitions (with prizes) and information about upcoming attractions at the science centre. As the magazine is printed in colour and on glossy paper, it is attractive from a design standpoint. Typically, it is of 50 pages in length. The efforts of popular science magazines also play a useful role in helping to enthuse the school-going public about the multifaceted splendours of science and technology. While somewhat different from other established popular science magazines, such as *Popular Science* and *New Scientist*, as the latter two are managed by full-time staff, inclusive of editorial, production and marketing teams, as well as having an international readership, the *Singapore Scientist* magazine relies almost totally on staff from the science centre to contribute most of the articles and take care of the editorial work and artwork. Profit is not the motive for publishing the magazine, popularising science is. Its circulation is also mainly local. Since the writing in the magazine is generally pitched at the level of school students, often difficult topics can be communicated in a way that they can reasonably understand. More importantly, the fact that a popular science magazine has been in production for over 35 years in Singapore without a full-time editorial and management team and with minimal reliance on advertisements is a testimony to its effectiveness in contributing to public science education in Singapore.

The Straits Times, a national newspaper which has a history of over 150 years, regularly carries articles related to science and technology as well as those related to socio-scientific issues, besides others. The articles comprise those written by its own journalists as well as those provided by international wire services. The importance given to science can also be seen from the fact that a few reporters cover the local science and technology beat. Given the high circulation of the national newspaper, such articles do reach out to a good number of people in the country and contribute in some way to sensitising them about issues related to science and technology. It is, however, of concern that no study has so far explored the role of newspapers in science journalism in the Singapore context.

As it is beyond the scope of this chapter to explore both newspaper coverage of science and how a popular science magazine reach out to the public, this study focuses only on the former. The latter is worthy of a separate study.

Literature Review on Coverage of Science in Newspapers

Examination of the science communication or science education literature shows that newspapers have been the subject of quite a number of studies. Some have explored science coverage in the press, while others have focused on how newspapers could be used in the classroom for teaching, promoting critical thinking skills and examining controversial issues in science.

Science coverage in the various national presses has been the subject of a number of studies – for example, Pellenchia (1997) and Logan, Peng and Wilson (2000) studied it in the US context; Bucchi and Mazzolini (2003) explored it in the Italian context; Wellington (1991) surveyed it for the case of Britain; Metcalfe and Gascoigne (1995) probed it in the Australian context; and Dimopoulos and Koulaidis (2003) presented it in respect of Greece. The general impression that emerges from these studies is that science is gaining prominence in the press, but its share of the total coverage is still rather low. Topics which cornered the predominant proportion of the column space allocated to science in these papers are related to health and medicine.

Very few newspapers have dedicated science sections, and for those that have, only one has been the subject of study. For example, *The New York Times* has a Science Times section, and this has been the subject of studies by Fursich and Lester (1996) and Clark and Illman (2006). It was found that a range of science topics was covered in this section and that this section has been growing in importance over the years.

A number of authors have explored the coverage of selected topics in the sciences in newspapers. For example, Morvillo and Brooks (1995) focused on articles in biology in the press to stimulate students' interest in the subject, while Stamm, Clark and Eblacas (2000) explored how the public understood the issue of global warming.

Some authors have explored the coverage of certain events in the press. For example, Friedman, Gorney and Egolf (1992) studied how the US media covered the Chernobyl nuclear accident. Massey (2000) explored how three newspapers in South East Asia depicted the haze which enveloped the region in 1997–1998; one of the newspapers studied was *The Straits Times* from Singapore.

The literature review suggests that almost all studies of science coverage in newspapers originate from Western countries. No comprehensive study of science coverage in the press in Asian countries has been located. There is thus a gap in the literature that would be of interest to address.

Methodology

Content Analysis

Studying the coverage of science in newspapers is a challenging task. The sheer diversity of the articles and the content therein necessitate the use of appropriate methodologies in order to uncover answers to research questions. Content analysis is one such technique. Riffe, Lacy and Fico (2005) define content analysis as 'the systematic and replicable examination of symbols of communication, which have been assigned numeric values according to valid measurement rules and the analysis of relationships involving these values using statistical methods'. It is an effective tool that can be used to analyse written text in books, newspapers and other printed documents (Krippendorff, 1980). The technique is regarded as labour intensive and time consuming since extensive coding schemes have to be developed for the analysis. It is also non-obtrusive in nature since the approach does not entail dealing with human subjects.

In the literature, content analysis has been one of the common approaches that have been used to analyse newspaper coverage of articles. For example, Granner, Sharpe, Burroughs, Fields, and Hallenbeck (2010) have used this technique to evaluate newspaper articles related to physical activity. Some studies that have used content analysis to explore science coverage in newspapers include those by Dimopoulos and Koulaidis (2003); Logan, Peng and Wilson (2000); and Clark and Illman (2006).

The framing of an article in the press is an important consideration in its appeal to readers. The term 'framing' refers to the 'modes of presentation that journalists and other communicators use to present information in a way that resonates with existing underlying schemas among their audience' (Scheufele & Tewksbury, 2007, p. 12). These authors also noted that the way an issue is depicted in a press article has a bearing on how the public would interpret it and form their opinions. Sometimes issues are complex to understand, and the reporters have to resort to particular perspectives in order to get the message across to readers. By giving a certain slant to an article, it is possible for a particular message to be conveyed to readers. The notion of 'frame sponsors' (Deprez & Raeymaeckers, 2010) or the participants selected to partake in the process of building the frame is also closely tied to legitimising the views of an article. It helps if the participants are in positions of authority so that they can lend some credibility to the news report (Wilkins, 1993). Content analysis can be used to study framing in a news article.

Using content analysis, the principal objectives of this exploratory study were to:

1. Explore the scope of coverage of science and technology in *The Straits Times*
2. Study the domains into which science and technology coverage can be mapped
3. Examine how articles with a local flavour are framed when reported in the press
4. Suggest some implications of newspaper science for science teaching based on this study

Table 15.1 Categories and examples

Health/medicine:	Dengue, cancer, surgery, H7N9, etc.
Earth science:	Clogged drains, forest fires, air quality, recycling, earthquake, environment, etc.
ICT/technology:	Copper-based phone line, digital map, iPhone, app, 4G, etc.
Chemistry:	Corrosive chemicals, toxic gas, names of chemicals, etc.
Biology:	Otter, tulip, etc.
Energy:	Waste-to-energy plant, LNG terminal, energy security, etc.
Astronomy/space science:	Planet, satellite, etc.
General:	Interviews, opinions, general news, etc.

Choice of Sample

A major English language daily from Singapore (*The Straits Times*) was selected as the newspaper of choice for this study. A week of newspapers, running from 20 April 2013 (Sunday) to 26 April 2013 (Saturday), was chosen for analysis. This represents a convenient sample. It was not possible to procure 'two constructed weeks' (Hansen, Cottle, Negrine, & Newbold, 1998) of newspapers for a year as not all the articles are stored in online databases and also the university library stores hard copies of the newspapers for only a few months owing to space constraints.

Operationalising of Categories

The literature was consulted for the choice of categories to classify the articles. For classifying an article, its thrust needs to be predominantly in line with a particular category. Some examples for the relevant categories are now given (Table 15.1):

Coding of Articles

Each article was read and coded for 15 variables by the author (Table 15.2).

Prominence Index

The term 'Prominence Index' has been used in the literature (Caburnay, Kreuter, & Luke, 2003; Granner et al., 2010; Schooler, Sundar, & Flora, 1996) as a numerical measure of the extent of coverage an article has been given in the press. The

Table 15.2 Coding scheme used to classify the articles

Variable	Description
Date	The selected period runs from 20 to 26 April 2013
Day	The choice of values runs from Sunday to Saturday
Title of article	This refers to the headline
Column space	This refers to the space occupied by an article and is expressed in cm^2
Column length	This refers to the length of the column and is expressed in cm
Type	This refers to the positioning of an article, and the possible choices are: News article, Letters to the Editor, Editorial, Opinion Piece, Interview
Visuals	This refers to whether there is a picture or graphic accompanying the article. The possible choices are yes/no
Headline font size	This refers to the size of the headline and is measured in cm
Location	This refers to where an article appears in the newspaper; the possible positions are left, above fold; right, above fold; left, below fold; and right, below fold
Section	This refers to whether the article appears in the main section (i.e., in the section attached to the front page or in one of the other sections)
Category	This refers to the subject matter covered in the article and is one of the following: health/medicine, earth sciences, ICT/technology, chemistry, biology, energy, astronomy/space science and general (interviews, policy, opinion, etc.)
Institutional actor(s)	This refers to the institution or industry which is mentioned in the article and is restricted to local articles
Byline	This refers to whether the article is written by a staff reporter or obtained from wire services
Quote	This is restricted to articles generated in the local context, and the responses are yes/no
Science term	This refers to the first occurrence of word(s) of a scientific nature in the article

theoretical range of the Prominence Index is 7–18. Articles with a high Prominence Index are more likely to be read than those with a low Prominence Index.

We shall define the Prominence Index (I) as follows:

$$I = X_1 + X_2 + X_3 + X_4 + X_5 + X_6$$

where

X_1 = whether article appeared in the front page (3 if yes and 2 if no)
X_2 = whether article appeared in main section (3 if yes and 2 if no)
X_3 = location of article on page (3 if left, above fold; 2 if right, above fold; 1 if left, below fold; and 0 if right, below fold)
X_4 = size of headline in cm (3 if > 1.91; 2 if > 1.27 and ≤ 1.91; 1 if > 0.64 and ≤ 1.27; and 0 if ≤ 0.64)
X_5 = presence of image or graphic (3 if yes and 2 if no)
X_6 = length of column in cm (3 if > 91.4; 2 if > 34.8 and ≤ 91.4; and 1 if ≤ 34.8).

Table 15.3 Distribution of science articles during 20–26 April 2013 (Saturday to Sunday)

Day	Number of articles	Percentage (%)
Saturday	17	34.0
Sunday	14	26.0
Monday	5	10.0
Tuesday	3	6.0
Wednesday	4	8.0
Thursday	3	6.0
Friday	5	10.0
Total	51	100.0

Inter-rater Agreement

As the number of items in each category was relatively few (2–16), only % agreement was used to quantify the level of agreement between the two raters for the classification of the articles. A 75 % agreement is generally regarded as being acceptable (Skinner, Rhymer, & McDaniel, 2000).

Results

The spread of science articles in the week of 20–26 April 2013 is shown in Table 15.3.

It is clear that articles on science appear on a daily basis throughout the week. A higher percentage (60 %) of the articles appears over the weekend, with more appearing on Saturday than on Sunday. The number of articles appearing on weekdays runs into the single digits, and, taken together, it is still less than the numbers appearing on the weekend.

Table 15.4 shows the distribution of articles according to the eight categories. Clearly, the articles explore content in multiple areas of science and technology but the coverage is not uniform.

In terms of numbers, articles on health/medicine predominate, followed by those on earth science and then ICT/technology. Articles related to the natural sciences have rather low representations – five articles related to chemistry followed by four on biology. Surprisingly, there were no articles related to physics.

In relation to inter-rater agreement, except for the categories of health/medicine, earth science and ICT/Technology, a value of 100 % was achieved for the remaining categories. For the former three categories, the % agreement was respectively 81.3 %, 77.7 % and 87.5 %. Overall, the levels of agreement are considered acceptable. It needs to be noted that a factor contributing to the maximum % agreement in some categories was the relatively small number of items in these categories. The distribution of articles, when measured by space allocation, is shown in Table 15.5.

Table 15.4 Coverage of science in *The Straits Times* according to categories for the period 20–26 April 2013

Category	Number of articles	Percentage (%)
Health/medicine	16	31.4
Earth science	9	17.6
ICT/technology	8	15.7
Chemistry	5	8.2
General (interviews, policy, opinion)	5	8.2
Biology	4	7.8
Energy	4	7.8
Astronomy/space science	2	3.9
Total	51	100.6

Note: Percentages do not add up to 100 because of rounding off

Table 15.5 Column space areas of articles on science in *The Straits Times* during 20–26 April 2013

Column space (cm^2)	Number of articles	Category
>1000	3	General (2), earth science (1)
800–999	2	General (1), earth science (1)
600–799	6	Earth science (2), health/medicine (2), biology (1), ICT/technology (1)
400–599	14	Health/medicine (5), energy (1), astronomy/space science (1), chemistry (2), biology (1), earth science (2), general (1)
100–399	18	Health/medicine (5), energy (1), astronomy/space science (1), chemistry (3), biology (1), earth science (2), general (1), ICT/technology (5)
<100	7	Health/medicine (2), biology (2), earth science (2), ICT/technology (1)

A surprising trend was observed when the articles were analysed on the basis of space allocation. There were two articles related to the general category and one on the earth science category which garnered the most column space (>1,000 cm^2) in the week surveyed. In the general category, both articles were opinion pieces written by experts. The article on 'Riding the Internet wave' on 20 April was by a university academic who explored the sensitivities of online and offline discourse and what can be done to promote responsible behaviour and accountability on the Internet. The other article on 'Ensuring healthy outcomes for all' on 22 April was by a doctor in the private sector who explored the increasing cost of health care in Singapore and how existing insurance and other policy schemes can be tweaked to promote better outcomes for the population. The article published on 21 April is related to 'Powerful quake rocks Sichuan again', an event attributed to a quirk of nature and which garnered the highest column space among the three articles. Articles on earth science were represented in all divisions of the column space. It seems that articles related to the natural sciences are given less column space, with

Table 15.6 Prominence indices of science articles covered in *The Straits Times*

Prominence index	Number of articles	Category
16	6	Earth science (4), health/medicine (2)
15	6	ICT/technology (2), chemistry (2), general (2)
14	9	Health/medicine (3), astronomy/space science (2), earth science (1), general (2), energy (1)
13	6	Health/medicine (2), earth science (2), ICT/technology (1); biology (1)
12	6	Health/medicine (2), earth science (1), ICT/technology (1), biology (1), chemistry (1)
11	5	Health/medicine (2), ICT/technology (1), energy (1), chemistry (1)
10	4	Chemistry (2), health/medicine (1), ICT/technology (1)
9	2	Earth science (1), general (1)
8	7	Health/medicine (2), ICT/technology (2), biology (2), earth science (1)

Note: In using the value of column length to calculate Prominence Index, an average column width of about 5.5 cm applies for most of the above articles in *The Straits Times*. In a few articles, the column width is slightly greater than this norm, and for such articles, the appropriate normalisation factor was applied to ensure consistency in calculating the column lengths. The size of the headline font size refers to the first letter in the headline

physics not featuring at all in the period surveyed. The Prominent Indices of the articles surveyed are depicted in Table 15.6.

There were no articles which attained the highest (18), second highest (17) or lowest (7) Prominence Index. A Prominence Index of 16, which was the maximum noted in this study, was achieved by six articles: four in the earth science category and two in the health/medicine category. A Prominence Index of 7, which was the minimum noted in this study, was achieved by seven articles: two in health/medicine, two in ICT/technology, two in biology and one in earth science categories. The distribution of scientific terms and their frequencies in the newspaper articles are presented in Table 15.7.

The nature of the science terms appearing in the articles can provide some indication of the extent to which an article can connect with the public. The scientific terms were coded into nine categories. As expected, a greater proportion of the terms belonged to the health/medicine category followed by the earth science category and ICT/technology category. Among the natural sciences, biology elicited more terms than chemistry, while physics terms were the least in number. Within each category, a few terms occurred with more regularity in the articles surveyed. For example, there were four mentions of each of the terms dengue, virus and bird flu under the health/medicine category. Under the earth science category, recycling had two mentions. Among the natural sciences, the subject of biology had four mentions of the term 'mosquito'. Overall, the articles in the health/medicine category contained quite a number of terms which the public may have some difficulty understanding or appreciating – for example, Chikungunya, cardiopulmonary resuscitation, epidemiological studies, psychotherapy, cognitive

Table 15.7 Occurrence of scientific terms in newspaper articles for the period 20–26 April 2013

Category	Scientific term
Health/medicine	Dengue (4), virus (4), bird flu (4), H7N9 virus (3), Aedes mosquito (3), coughing (3), vaccine (2), diabetes (2), Aedes albopictus (2), fever (2), flu (2), Aedes aegypti (2), respiratory failure (2), infection, disease-bearing vectors, Chikungunya, first aid, cardiopulmonary resuscitation, cardiac arrests, defibrillator, infectious diseases, human airways, epidemiological studies, resistance to drugs, strain, mental health, psychotherapy, cognitive behavioural therapy, transcranial magnetic stimulation dementia, Alzheimer's disease, hallucination, Parkinson's disease, neurodegenerative disease, stiffness of muscles, poor balance, second-degree burns, nerve damage, hand hygiene, drug screening, health-monitoring apps, debilitating diseases, cancer, nicotine, anti-viral drugs, pandemic, immunity, strain, coronavirus, mammal host, drugs, respiratory infection, pathogens, surgery, quarantine, epidemic, heart surgery, cardiac, kidney and brain procedures, heart valves, smoking, obesity, hypertension, kidney failure, breast cancer, respiratory and heart problems, genomic analysis, genes, sequencing of genome, stuffy noses, sinuses, prenatal exposure, autism, rashes, skin irritation, headache, sneezing, runny nose, sweating, fatigue, pneumonia, H5N1, hypertension, hepatitis B carrier
Earth science	Recycling (2), biomass, power, steam, landfill, fossil fuel, municipal and commercial waste, wind power, carbon emissions, pollution, forest fires, air quality, haze, Pollution Standards Index (PSI) reading, waste of resources, air quality, smog, 2.5 μm particles, 10 μm particles, epicentre, 7.8 magnitude, landslides, tectonic plates, tsunami, climate change, global warming, drought, flooding, cyclones, polluted water, super storms, 2.1 magnitude earthquake, water management, sewage, drought, 7.0 earthquake, landslide, pollutants, coal-fired plant, automobile emission, air pollution, green issues, solar cookers, air-filtration systems
ICT/technology	Internet (2), copper-based phone lines (2), iPhone, app, parallel processing, computer, fibre-optic technologies, next-generation nationwide broadband network, modem, fax machines, wall phone socket, digital maps, cartography, telecommunication networks, Goggle, electric car, man–machine interface, 4G, mobile data, smartphones, streaming videos, wireless sensor
Chemistry	Ammonium nitrate (2), fertiliser (2), toxic gas, transparent liquid, metal, corrosive chemical, hazardous chemicals, volatile chemicals, steel, mercury, hydroquinone, salicylic acid, tretinoin
Biology	Mosquito (4), ants, pandas, fish, vertebrates, spine, torso, fossils, evolution, traits, Australopithecus sediba, otters, tulips, growing vegetables, eggs (of mosquitos), cells, mosquito larvae, breeding
Energy	Waste-to-energy plant, clean energy, energy security
Astronomy/space science	Solar storms, satellite, radiation, space debris, cosmos, light years, constellation, orbit, faster than light, galaxy
Physics	Electricity (2), lightning current, momentum, electricity, LEDs, charge, batteries, recharging, power
Others	Human computer, mental calculator

Note: *Numbers* in *parenthesis* indicate frequencies; where there is no number indicated, it means singular occurrence of the term behavioural therapy, transcranial magnetic stimulation, Parkinson's disease, neurodegenerative disease, debilitating diseases, coronavirus, pathogens and sequencing of genome. Most of the terms occurring in the ICT/technology category should not present difficulty to the ICT-savvy younger generation in Singapore except perhaps for man–machine interface and cartography; for the latter, a definition, however, was included in the article. As for the terms in the natural sciences, most are within the appreciation of those who have gone through the school science system. Terms with origins in physics appeared in articles in other categories; as mentioned earlier, there were no articles related to physics for the period surveyed.

In relation to the source of the articles, 22 out of the 51 articles were from wire services. This represents 43.1 % of the articles published in *The Straits Times* for the week surveyed. The remaining articles were generated by staff reporters based locally and overseas as well as from the public and other invited experts.

Of the 19 articles that have a local flavour and were generated by staff reporters, all made mention of some academic institution(s) or industry in the write-up. Often, there is a quote from the relevant sources.

Of the 51 articles surveyed, 37 took a news angle in the reporting; five were on advances in science and technology, five were letters to the editor, two were invited pieces, one was an editorial, and one was an interview with a scientist.

Discussion

Promotion of public education about science in today's society is a complex undertaking. No one organisation or platform can undertake this task in totality and neither can any one approach cater to this effectively. A multiplicity of approaches is essential. Institutions for informal science learning, such as science centres, science museums, zoos, bird parks, botanic gardens and other destinations, however, continue to be the mainstay of such dissemination efforts as these are dedicated for specific purposes within the ambit of public science education efforts and are also part of their mission objectives. Increasingly, owing to the growing importance of science and technology in this age of globalisation and the pace at which new developments in science and technology are occurring, other approaches which are able to reach out to the masses are also needed to complement such efforts. Print journalism in the form of newspapers is one such medium.

Even with the proliferation of online news platforms, print versions of newspapers still have a role to play in informing the public about issues related to science and technology. As newspapers are published practically every day, people are likely to use this as the principal source to apprise themselves of contemporary matters related to science and technology that can affect them – for example, an outbreak of dengue, bird flu or mad cow disease in the country or occurrence of high levels of haze in a potential holiday destination. The prominence as well as frequency and extent of coverage that newspapers give to such matters can also help to shape public perceptions of these issues as well as set the agenda for action where necessary.

Content analysis of the articles appearing in one week indicates that there is reasonable coverage of science. Articles on science appear every day in *The Straits Times*, more so over the weekend, thus indicating a pronounced commitment on the part of the editors to feature science-related content. A total of 51 articles related to science were identified for that week. These articles cover a spectrum of topics in health/medicine, earth science, ICT/technology, energy, biology and chemistry as well as those of a general nature. Their role in contributing towards the cause of raising public awareness of science and technology cannot be underestimated. The proportion of science coverage as a function of the total number of pages in the newspaper is estimated to be less than 1 %.

As expected, articles related to health/medicine dominate in the press. In the period surveyed, about 31.4 % of the articles published in the press can be linked to this category. People do care for their health and often rely on the press to keep them informed of epidemics and other relevant issues of importance. The dominance chalked up by this category is also reflected in a number of other studies which explored science coverage in newspapers – for example, Bucchi and Mazzolini (2003) and Pellenchia (1997).

A notable finding from this study is the lack of substantive coverage of the natural sciences in the press. While the subjects of biology and chemistry were able to elicit at least a few articles in the period surveyed, there was none whatsoever for physics. A few reasons can account for this. Firstly, it could be due to the limited time period surveyed. Secondly, there has been a marked decline in interest in physics among students at the secondary and postsecondary levels (Oon & Subramaniam, 2010), leading to a decline in undergraduate enrolment in physics at the local universities (Oon & Subramaniam, 2011). The lack of reporting of issues related to physics could well be a manifestation of this trend, that is, it is being perceived as being minimally newsworthy. Thirdly, and which is the most plausible reason in my opinion, is that physics has become very much embedded in many other disciplines, including technology, that its purveying per se is not really necessary to reinforce its importance. There is no technology in the medical, biological, chemical, computing and engineering fields that does not make use of some fundamental concepts grounded in physics. That is, there is an implicit, though not explicitly stated, manifestation of the influence of physics in the sciences. Notwithstanding the foregoing, there is a need for more advocacy efforts by university staff in physics as well as members of learned societies in physics to raise the profile of physics in the newspapers.

Judicious use of the editorial page to reinforce an important point of view is of significance. In the editorial on 'Joining hands on Earth Day', published on 22 April, a compelling stand was made for promoting greater individual action towards the cause of the environment. Using the auspices of this page to relay an important message is not only responsible journalism but is also of significance as, besides the front page, the editorial column is also considered to be prime space that attracts attention. The pronounced importance of this column in contributing towards the cause of agenda-setting has been noted in the media literature (Wallack, Woodruff, Dorfman, & Diaz, 1999).

Two articles that garnered the most column space were in the general category. In both these articles, advocacy for particular points of view was evident. For quite some time, there has been some tension on where the new 'normal' in online discourse, which permits anonymity and little accountability, should be positioned. The provision of a large column space in a national newspaper for a leading legal academic to air his views about the Internet has the effect of raising the visibility of the issue for the various stakeholders. Likewise in the other article on how healthy outcomes can be promoted for people in this age of increasing health costs, the views of a health professional in the private sector were given valuable column space to generate publicity for the issue. Distanced from the public health sector, his views provide a perspective from the private sector in the ongoing health-care debate. Such advocacy efforts by stakeholders through the media are well recognised in the literature for furthering the cause of an issue through its pivoting on not only the public agenda but also the agenda of policy makers as well (Davidson & Wallack, 2004; Dearing & Everest, 1996).

Citizen-generated science content in a newspaper is a welcome development. It promotes valuable insights from the ground on issues that may not have been covered in depth in the newspapers or which profess to offer refreshing perspectives. During the period surveyed, there were five persons who penned letters to the editor on science-related issues, mainly on health and ICT issues. Though the column space devoted to these particular letters is either low or modest, the auspices of the forum page can have the effect of elevating it to a higher level of 'visibility' for the views professed in these letters. For example, the writer of the letter on 'Clogged drains: a dengue risk' demonstrates commendable civic activism as he recounted how he took photographs of stagnant water in parks and drains, which are potential sites for mosquito breeding, and sent these to the relevant national agency for the necessary action, which they duly took. As Singapore was encountering a dengue epidemic at that time, such actions show how other members of the public can contribute in similar ways to the cause of public health. It is well known that letters to the editor must pass through a few layers of scrutiny before they are accepted for publication in the newspaper. Nevertheless, even within the aegis of this 'mediated arena' (Zamith, Pinto, & Villar, 2013), it is a welcome development that the public raises issues related to science and technology and that the editors see value in publishing such perspectives for furthering the cause of public discourse.

There is a liberal occurrence of science terms in the press for the period surveyed. Most of the terms with origins in chemistry and physics as well as some in biology should not present difficulties for upper secondary and postsecondary students – for example, terms such as corrosive chemical, hazardous chemicals, volatile chemicals, fertiliser, toxic gas and ammonium nitrate are well within the scope of the school curriculum in chemistry. Likewise for terms such as vertebrates, spine, torso, fossils, traits and larva in biology, as well as terms such as current, momentum, electricity, LEDs, charge, batteries and power in physics.

However, students are likely to find difficulties with more terms in biology than in chemistry or physics (Table 15.7). It is pertinent to note that one of the prime attributes of science literacy is the ability to understand scientific terms (Miller, 1983; Thomas & Durant, 1987). The baseline knowledge in science that schools provide students is adequate, and the onus is really more on students to increase their repertoire of scientific vocabulary and understanding of science topics through informal means in order to make greater sense of the science content presented in newspapers.

It is fortuitous that less than half of the number of articles related to science is from wire services. A good number of the articles were generated by staff reporters, with only a few by letter writers and invited experts. Cooper (2008) has noted that the bylines of articles can provide useful information about the source of the news and that a greater number of articles generated by staff reporters as compared to those from wire services indicate a higher level of professionalism in the newspapers. It ought to be recognised that all newspapers rely on a mixture of staff-generated content and those from wire services. The analysis of the data presented in this study indicates that science is getting the desired attention in the local newspapers.

With regard to the framing of the articles generated locally, an overwhelming majority took a news angle for the reporting. With the mountains of garbage in most countries being either directly incinerated or dumped into landfills, the article on 20 April that a Singapore company is building a waste-to-energy plant in Britain is a reinforcement of the point that Singapore is one of the very few countries that have adopted environmentally friendly practices in waste disposal and that the experience in this initiative has led the company to land a contract to replicate the Singapore experience in the UK. Waste-to-energy plants are a new development in the waste industry – their principal advantage is that the enormous amounts of heat generated during incineration is captured and fed to the country's electricity grid rather than being allowed to go to waste, as is the practice in many countries (Tan & Subramaniam, 2012). The 20 April article on 'Indonesia fires bring haze to Singapore' is a timely reminder to Singaporeans of the environmental consequences that the action of one country can have on another and of the need to stay tuned to the three-hourly broadcasts of the Pollution Standards Index (PSI) readings. Even most of the articles from overseas also took on a news angle for their reporting. The article on 'Taiwan confirms first case of H7N9 outside mainland China', published on 25 April, is a poignant reminder (without explicitly stating it) to citizens to exercise caution and discretion when travelling to affected countries and reinforces the point that in this age of international travel, people and local health authorities need to stay vigilant. In the article on 'A can of soft drink a day ups diabetes risk by 22 %: Study', published on 25 April, a subliminal message is reinforced to people to take greater responsibility for their own health. Such education of the public about issues in science and technology would seem to be more effective if the message is relayed as a news story rather than as a direct exhortation couched in purely scientific terms with scarce regard for context. It may be pertinent to recall the work of Hijmans, Pleijter and Wester (2003), who have noted that the use of news items allows for relevant scientific knowledge to be slanted into a story for public consumption. This holds true for the articles in *The Straits Times* for the period surveyed.

In relation to other aspects of framing, it is of interest to explore the nature of the institutional actors that characterise science articles generated locally. As mentioned earlier, of the 19 such articles, all made mention of some academic institution(s) or industry in the text. The positioning of an article in this manner may admit of a reportorial acknowledgement that greater authenticity could be endowed on the content if institutional actors are given some, albeit brief, column space. Thus, for example, in the article on 'Residents more worried about dengue', which appeared on 20 April, the Healthway Medical Group and the National Environment Agency earned mentions. It may not be prudent to feature such an article depicting people's concerns if major players in the big picture are left out and their views not sought. In another article on 'Boy burnt by liquid on seat of bus stop', which appeared on the same day, the name of the boy's school as well as the National University Hospital was mentioned. In the article on 'More sightings of wild otters in Singapore', which appeared on 22 April, there were mentions of the National University of Singapore, International Union for Conservation of Nature, the National Parks Board and the name of a wildlife consultant. The article on 'Mad about tulips', published on 24 April and which publicised the planting of the largest number of these flowers (20,000) in the country, had mentions of Temasek Junior College, KLM Royal Dutch Airlines and the Netherlands Charity Association. In the article on 'LNG terminal will boost energy security', there was mention of a minister's name, the name of the international conference he graced and KPMG Global Energy Institute for Asia Pacific. Most of such articles have also a quote from an authoritative figure or the concerned actor. Quotations can be a proxy for depth of coverage of an issue – if the issue reported on is perceived to be important, the journalist is likely to solicit opinions from credible sources (Davidson & Wallack, 2004). It would seem that adherence to such 'norms' – that is, mention of institution and use of quotations – is generally a characteristic feature of science articles generated in the local context. It may be of interest to note that Dimopoulos and Koulaidis (2003) have found that the linkages of science and technology to the corresponding 'social actors' are a characteristic feature of science portrayal in the Greek press. In their study, however, the principal actors were mainly those related to the economy and politics, while in the present study, these were mainly national institutions. The difference could be due to a number of reasons – for example, journalistic inclination to include such actors in order to endow greater credibility for their write-up or greater integration of the institutional 'spheres of influence' (Dimopoulos & Koulaidis) into the wider science and technology landscape in Singapore as compared to Greece.

Coverage in newspapers is also dictated by the linkages of potential items to recent events, contemporary trends and socio-scientific implications (Miller, 1999; Nelkin, 1995). For example, in the period under study, Singapore was facing a dengue outbreak, and this was reflected by the presence of articles related to dengue on three different days, including one letter to the editor. Internationally, there was an outbreak of H7N9 flu, and this elicited three articles. The earthquake in Sichuan was the subject of two articles on different days.

There may be an unintended tendency to equate high column space (inclusive of photographs) of articles with greater visibility for the issues purveyed therein. For example, of the articles that generated the most column space, two were in the general category and one was in the earth sciences category. When the visibility of the articles was analysed on the basis of Prominence Index, four were in the earth sciences category and two were in the health/medicine category. Surprisingly, the two articles that generated the most column space in this study did not generate the two highest Prominence Indices. In the literature, the column space area seems to be the de facto measure for endowing articles with visibility. The use of Prominence Index, which is of relatively recent origins, has been the subject of only very few studies. The results of the current study attest to the fact that more research is needed to explore the impact of an article through the use of Prominence Index. Use of multiple measures can help to address the weaknesses inherent in any one measure and help to portray a more realistic picture of the perceived impact of an article in the press.

Science journalism via newspapers has probably a greater reach as compared to institutions such as by science centres, mainly as a result of the print medium being able to reach out daily to their large readership base, which runs into the hundreds of thousands for newspapers. Especially, newspapers are ideally positioned to communicate new developments in science and technology on a timescale that is not possible by science centres – for example, it requires time, resources and funding for science centres to mount exhibitions and other programmes on these developments, and some of these developments many not even be amenable to treatment via exhibitry, the principal tool used by science centres to communicate to the public. Since newspapers are published every day, they are in a position to educate the public about new developments in a time-effective manner. However, it has to be noted that given the breadth and scope of scientific disciplines as well as the rate at which new developments in science and technology are occurring, newspapers can cover only a minuscule fraction of what is happening. The fraction that is covered must pass several tests before it can muster editorial acceptance. For example, is the item newsworthy? Is it likely to appeal to a good number of readers? Are there other pressing news items competing for column space at the same time? Is the level of language used in the item amenable to comprehension by the lay public? Is advertising space, which brings in revenue, affected in the process? and so on. Thus, it is not surprising that only a limited number of topics can be covered in newspapers. The lack of a daily science section in *The Straits Times* is one reason why the scope of coverage of topics is still limited as compared to other newspapers such as *The New York Times* (Clark & Illman, 2006) and USA Today. *The Straits Times*, however, has a section devoted specifically to science and which appears on Sundays; both locally generated articles and those from wire services are featured in this section.

Implications for Teaching

The way science is presented in school and the way it is portrayed in the press admits of a dichotomy that is not easy to bridge. Science in schools follows a curricular framework, while that in the media is more complex and very much issues driven. Because of the latter, students, depending on their educational level, may sometimes lack the necessary scientific vocabulary or critical thinking to make sense of the content presented in some of the articles though they may be able to follow these to some extent. This is a challenge for science teachers to address as after all, one of the aims of science education is to make students become more discerning in their approach when reading news in the media, a skill that would be especially needed as they transition to adulthood. McClune and Jarman (2010) call this 'critical engagement with science in the media'. The invoking of relevant articles in the press when teaching particular science topics in the class and coming up with appropriate activities that can provide meaningful learning experiences to students would be a possible approach. These activities can leverage on inquiry in the broad sense. It has to be reiterated that in the literature, there is no consensus on what exactly constitutes inquiry (e.g., Barrow, 2006). True inquiry is generally regarded as being promoted when students engage in tasks that mirror the work of scientists – for example, doing investigations in the laboratory. However, given the need to complete the curriculum in time and to allocate time for revision activities, there is a limit to how much time teachers can devote to inquiry activities for students in the true context of the word. The laboratory need not necessarily be the place, although it is highly desirable, to promote inquiry. In fact, Chiappetta (1997) notes that 'Scientific inquiry goes beyond constructing knowledge through hands-on activities. Much of the inquiry that scientists and engineers engage in involves reading and communicating with other people' (p. 25).

The classroom can thus be used to promote elements of inquiry to students – for example, use of questioning and crafting of activities that promote thinking without recourse to the laboratory environment. Questioning and thinking are two attributes that are used by scientists in their work, and these can be made the basis for promoting inquiry when using newspaper articles in the teaching of science. Some examples will now be given using the articles covered in *The Straits Times*. In using the article on 'LNG terminal will boost energy security', published on 26 April, for the teaching of chemistry, students can be given some tasks to do – for example, writing the chemical equation for the combustion of liquefied natural gas and comparing it with when petroleum products are combusted, and asking them to explain from the products formed why the former reaction is in line with clean energy initiatives; calculating the heat change for these reactions to see which reaction produces more energy; following the carbon trail for these fuels to study the effects of pollution; and doing simple cost-benefit analyses in respect of the fuels. Another article which appeared on the same day is 'Banned chemicals found in 3 cosmetic products'. In using this article for the teaching of chemistry, students can be asked to look into the properties of mercury, hydroquinone, salicylic acid and tretinoin from the point of view of its effect on human physiological systems and find out what is the critical concentration beyond which each of these chemicals can pose a risk to human health; by correlating these concentrations with those found in the cosmetic products, they can see why the authorities have banned these products. In line with the recommendations of Jarman and McClune (2007, 2010), further activities can be promoted when using press articles for the teaching of science in class – for example, engaging students in a brief question-and-answer session based on the article, promoting general discussion of the article and asking students questions such as what aspects that would allow them to make greater sense of the text are missing from the article; if given an opportunity, what would they want to ask the reporter who wrote the article; and what topics in the syllabus have allowed them to reasonably understand the article. Also, students could be given opportunities to write up a science topic taught in class (and which is covered in the newspaper) as a press article with catchy headlines and word limit (Jarman & McClune, 2002).

The interdisciplinary nature of contemporary science and technology is a theme that resonates in most articles. In schools, however, science is taught as separate subjects – for example, physics, chemistry and biology. Where combined science is taught at the secondary level, its division into the respective sciences is still made apparent in the syllabus or in the class during teaching – for example, subject combinations include physics/chemistry, chemistry/biology or physics/biology. Students need to realise that the natural world is interdisciplinary in nature and that these artificial divisions are made for pedagogical convenience and administrative reasons. Science-based articles in newspaper offer tremendous potential to reiterate this point. For example, in the article on 'Indonesia fires bring haze to Singapore' which appeared on 20 April, a number of subject domains will be apparent – the slight changes in the composition of the atmosphere due to the presence of colloidal particles as well as the reduced visibility arising from these can be linked to chemistry; forest clearing via burning to make a way for development can be linked to earth science; deforestation contributing to reduced transpiration and photosynthesis has implications in biology; and the body parts affected as a result of such pollution can be related to biology or medicine. In the article on 'Banned chemicals found in 3 cosmetic products', linkages can be made to not only chemistry but also biology. Such articles can be used to infuse not only real-world contexts of a contemporary nature in the course of teaching traditional science but also reinforce to students the cross-disciplinary nature of modern science.

There is good scope for infusing nature of science (Lederman, 1999) elements when appropriate press articles related to science are used for teaching. Articles such as 'H7N9 flu outbreak: Ducks may be the source' and 'Experts stumped by how H7N9 virus is spread', both of which appeared on 20 April, indicate that even though scientists have a reasonably good understanding of the H7N9 flu, their understanding is not complete and, in relation to certain aspects of the flu, their knowledge is still in a state of flux. The article on 'Golden age of citizen cartography', published on 21 April, is a good example of what Shapin (1992) terms 'science-in-the-making' (p. 29) – how groups of people separated by distance were able to collaborate over the Internet to create new knowledge, specifically producing an accurate digital map of a part of the country, such as a village. Yet another example of science-in-the-making and which reinforces the point that the scientific enterprise is a complex social activity in which multiple institutions as well as the public participate is the article on 'War on cancer: US institutions race to map patients' genes', which was published on 23 April.

While judiciously infusing newspaper science into classroom teaching can spice up lessons, not all science teachers would have the necessary skill sets to do this effectively. Appropriate professional development programmes would be necessary to upskill them for the task. Given the demands of the school science curriculum and the administrative responsibilities of teachers as well as the perceived benefits of the nexus between school science and newspaper science, its use in lessons must not be construed as an add-on but as one which is integrated into classroom teaching. Otherwise, there is a risk that its introduction would be seen more in the context of an appendage that can be conveniently off-loaded for the resumption of normal teaching. A recent study conducted by Poon, Toh and Tan (2010) in a primary science classroom in Singapore used the format of a debate to extend what students learn about the interdependence of species in an ecosystem to an issue of contemporary significance reported in an international news media – the declining bee population; the students were able to obtain useful insights on the issue as a result of linking classroom content with real-world contexts.

It would appear that the science education curriculum may not yet be ready to prepare students formally for critical engagement of the science that appears in newspapers. However, there is a case for doing so, especially in this age of globalisation and new developments in science and technology. Indeed, the seminal report on 'Beyond 2000: Science education for the future' by Millar and Osborne (1998) has called for the science curriculum to be reformed so as to address the challenge of helping students to 'be able to understand and respond critically to media reports of issues with a science component' (p. 12). These ideas also resonate in the literature on 'science for citizenship' (Osborne, 2000) and lifelong learning (e.g., Solomon & Thomas, 1999). Jarman and McClune (2002) also opine that since articles in newspapers are generally pitched at the level of the 'nonscience audience', these would be especially useful for classroom teaching. It would be pertinent to conclude this section by quoting Elliott (2006) in his study of trainee teachers in a university in the UK:

> If even people such as these do not regularly follow scientific developments and debate in newspapers, what hope is there that less scientifically minded, scientifically educated, or academic peers do? This makes the case for introducing newspaper science in school classes even stronger. If school students can be given opportunities to read media reports of science and helped to understand and critically evaluate them, there is a chance that more will take up this habit into their adult lives. (p. 1260)

Limitations

The choice of just one 'continuous week' to explore the science content in the newspapers is a limitation of this study. Experts have recommended that there should be at least two 'constructed weeks' (Hansen et al., 1998; Riffe, et al., 2005) in order to study a year's coverage of articles in the press. As not all the articles published in the newspaper selected for this study are abstracted in online databases and as the university library does not store hard copies of newspapers beyond about 2 months, it was not possible to come up with two constructed weeks of newspapers on a random basis. As the present study was of an exploratory/pilot nature, it was felt that the choice of one continuous week was reasonable, given the constraints. It is also pertinent to add that the use of one continuous week of newspapers as the temporal horizon for analysis has also appeared in the literature (e.g., Wellington, 1991).

The choice of categories used in the classification of the articles is not mutually exclusive, and it is possible that a few articles could be placed in more than one category or that alternative categories beyond those described could emerge. This is a consequence of the interdisciplinary nature of science, which means that more than one category can often be apparent in the articles. The literature suggests that categories emerging from content analysis should have clearly demarcated domains into which content can be placed unequivocally (Chadwick, Bahar, & Albrecht, 1984). It has to be noted, however, that the recommendations by the authors were made in the context of social science research, from where the term 'content analysis' was derived, and its applicability to studies on science coverage in the press may have to be treated as a special case where exceptions would have to be made.

Column space (inclusive of photographs, where present), in terms of length and area, was measured with a simple ruler, and thus the accuracy of the values obtained is limited by the precision of the ruler used. However, they are not expected to influence the interpretations offered in this chapter.

It has to be stressed again that this study is of an exploratory nature and that the intent is to get some insights about the science portrayed in newspapers in Singapore as this is lacking in the literature. A more comprehensive research on the role of newspapers in contributing to public education about science in Singapore would be worthy of a funded study. Future work could build on the present study by expanding the time horizon over which science-based articles are surveyed and exploring more nuances in the framing of the articles as well as the salience of these frames. Other topics which could be explored are how controversial issues in science are presented, what the tone of coverage in the articles is and how the tensions between multiple perspectives in a topic are managed in the various articles.

Conclusion

In Singapore, public education about science comes under the mandate of a few agencies dedicated to informing the public on such matters. Contributing to and complementing these efforts are the efforts of other platforms. The role of science journalism via newspapers is such a platform and it has been explored in this study. It is shown that there is modest coverage of science in one major local newspaper, with the coverage being more pronounced for health/medicine and earth sciences for the period explored. As the newspaper medium has not been explored for science coverage in the Singapore context in the literature, this chapter is a small contribution in this respect.

References

Barrow, L. H. (2006). A brief history of inquiry: From Dewey to standards. *Journal of Science Teacher Education, 17*, 265–278.

Bucchi, M., & Mazzolini, R. G. (2003). Big science, little news: Science coverage in the Italian daily press, 1946–1997. *Public Understanding of Science, 12*, 7–14.

Caburnay, C. A., Kreuter, M. W., & Luke, D. A. (2003). The news on health behavior: Coverage of diet, activity and tobacco in local newspapers. *Health Education Behavior, 30*, 709–722.

Caleon, I. S., & Subramaniam, R. (2005). The impact of a cryogenics-based enrichment programme on attitude towards science and the learning of science concepts. *International Journal of Science Education, 27*, 679–704.

Caleon, I. S., & Subramaniam, R. (2007). Augmenting learning in an out-of-school context: The cognitive and affective impact of two cryogenics-based enrichment programmes on upper primary students. *Research in Science Education, 37*, 333–351.

Chadwick, B. A., Bahar, H. M., & Albrecht, S. L. (1984). Content analysis. In B. A. Chadwick, H. M. Bahar, & S. L. Albrecht (Eds.), *Social science research methods* (pp. 239–257). Englewood Cliffs, NJ: Prentice Hall.

Chiappetta, E. L. (1997). Inquiry-based science. *The Science Teacher, 64*(10), 22–26.

Clark, F., & Illman, D. L. (2006). A longitudinal study of the New York Times Science Times section. *Science Communication, 27*, 496–513.

Cooper, K. J. (2008). Politics and priorities: Inside the Egyptian press. *Arab Media and Society*. Retrieved June 10, 2013, from http://www.arabmediasociety.com/?article=689

Dairianathan, A., & Subramaniam, R. (2011). Learning about inheritance in an out-of-school setting. *International Journal of Science Education, 33*, 1079–1108.

Davidson, A. E., & Wallack, L. (2004). A content analysis of sexually transmitted diseases in the print news media. *Journal of Health Communication, 9*, 111–117.

Dearing, I., & Everest, R. (1996). Agenda-setting. In *Communication concepts* (Vol. 6). Thousand Oaks, CA: Sage.

Deprez, A., & Raeymaeckers, K. (2010). Framing the first and second Intifada: A longitudinal quantitative research design applied to the Flemish press. *European Journal of Communication, 25*(1), 3–23.

Dimopoulos, K., & Koulaidis, V. (2003). Science and technology education for citizenship: The potential role of the press. *Science Education, 127*, 241–256.

Elliott, P. (2006). Reviewing newspaper articles as a technique for enhancing the scientific literacy of student-teachers. *International Journal of Science Education, 28*, 1245–1265.

Friedman, S. M., Gorney, C. M., & Egolf, B. P. (1992). Chernobyl coverage: How the US media treated the nuclear industry. *Public Understanding of Science, 1*, 305–323.
Fursich, E., & Lester, E. P. (1996). Science journalism under scrutiny: A textual analysis of Science Times. *Critical Studies in Mass Communication, 13*, 24–43.
Granner, M. L., Sharpe, P. A., Burroughs, E. L., Fields, R., & Hallenbeck, J. (2010). Newspaper content analysis in evaluation of a community-based participatory project to increase physical activity. *Health Education Research, 25*, 656–667.
Hansen, A., Cottle, S., Negrine, R., & Newbold, C. (1998). *Mass communication research methods*. London: Macmillan.
Hijmans, E., Pleijter, A., & Wester, F. (2003). Covering scientific research in Dutch newspapers. *Science Communication, 25*, 153–176.
Jarman, R., & McClune, B. (2002). A survey of the use of newspapers in science instruction by secondary teachers in Northern Ireland. *International Journal of Science Education, 24*, 997–1020.
Jarman, R., & McClune, B. (2007). *Developing scientific literacy: Using news media in the classroom*. London: Open University Press.
Jarman, R., & McClune, B. (2010). Developing students' ability to engage critically with science in the news: Identifying elements of the 'media awareness' dimension. *The Curriculum Journal, 21*(1), 47–64.
Krippendorff, K. (1980). *Content analysis: An introduction to its methodology*. London: Sage.
Lederman, G. (1999). Teachers' understanding of the nature of science and classroom practice. *Journal of Research in Science Teaching, 36*, 916–929.
Logan, R., Peng, Z. J., & Wilson, N. F. (2000). Prevailing impressions in science and medical news: A content analysis of the Los Angeles Times and the Washington Post. *Science Communication, 22*, 27–45.
Massey, B. L. (2000). How three Southeast-Asian newspapers framed the haze of 1997–1998. *Asian Journal of Communication, 10*(1), 72–94.
McClune, B., & Jarman, R. (2010). Critical reading of science-based news reports: Establishing a knowledge, skills and attitudes framework. *International Journal of Science Education, 32*, 727–752.
Metcalfe, J., & Gascoigne, T. (1995). Science journalism in Australia. *Public Understanding of Science, 4*, 411–428.
Millar, R., & Osborne, J. E. (1998). *Beyond 2000: Science education for the future*. London: King's College London.
Miller, D. (1999). Mediating science: Promotional strategies, media coverage, public belief, and decision making. In E. Scanlon, E. Whitelegg, & S. Yates (Eds.), *Communicating science: Contexts and channels* (pp. 206–26). London: Routledge.
Miller, J. D. (1983). Scientific literacy – A conceptual and empirical review. *Daedalus, 112*(2), 29–48.
Morvillo, N., & Brooks, L. M. (1995). Headline science: Popular news stories spark student interest in biology. *The Science Teacher, 62*, 20–23.
Nelkin, D. (1995). *Selling science: How the press covers science and technology*. New York: W. H. Freeman.
Oon, P. T., & Subramaniam, R. (2010). Views of physics teachers on how to address the declining enrolment in physics at the university level. *Research in Science & Technological Education, 28*, 277–289.
Oon, P. T., & Subramaniam, R. (2011). On the declining interest in physics among students – From the perspective of teachers. *International Journal of Science Education, 33*, 727–746.
Osborne, J. (2000). Science for citizenship. In M. Monk & J. Osborne (Eds.), *Good practice in science teaching: What research has to say* (pp. 225–240). Buckingham, UK: Open University.
Pellenchia, M. G. (1997). Trends in science coverage: A content analysis of three US newspapers. *Public Understanding of Science, 6*, 49–68.

Poon, C. L., Toh, F. M., & Tan, A. L. (2010). Bees in the news: Connecting classroom science to real-life issues. *Science Scope, 33*(9), 64–68.

Riffe, D., Lacy, S., & Fico, F. G. (2005). *Analyzing media messages: Using quantitative content analysis in research* (2nd ed.). Mahwah, NJ: Erlbaum.

Scheufele, D. A., & Tewksbury, D. (2007). Framing, agenda-setting, and priming: The evolution of three media effects models. *Journal of Communication, 57*, 9–20.

Schooler, C., Sundar, S. S., & Flora, J. (1996). Effects of a Stanford five city project: Media advocacy program. *Health Education Quarterly, 23*, 346–364.

Shapin, S. (1992). Why the public ought to understand science-in-the-making. *Public Understanding of Science, 1*, 27–30.

Skinner, C. H., Rhymer, K. N., & McDaniel, E. C. (2000). Naturalistic direct observation in educational settings. In E. S. Shapiro & T. R. Kratoch (Eds.), *Conducting school-based assessments of child and adolescent behavior* (pp. 21–54). New York: The Guilford Press.

Solomon, J., & Thomas, J. (1999). Science education for the public understanding of science. *Studies in Science Education, 33*, 61–90.

Stamm, K. R., Clark, F., & Eblacas, P. R. (2000). Mass communication and public understanding of environmental problems: The case of global warming. *Public Understanding of Science, 9*, 219–237.

Tan, W. H. L., & Subramaniam, R. (1998). Developing countries need to popularise science. *New Scientist, 2139*, 52.

Tan, W. H. L., & Subramaniam, R. (1999). Scientific societies build better nations. *Nature, 399*, 633.

Tan, W. H. L., & Subramaniam, R. (2003). Science and technology centres as agents for promoting science culture in developing nations. *International Journal of Technology Management, 25*, 413–426.

Tan, W. H. L., & Subramaniam, R. (2009). Role of scientific academies and scientific societies in promoting science and technology: Experiences from Singapore. *International Journal of Technology Management, 46*, 38–50.

Tan, W. H. L., & Subramaniam, R. (2012). Environmental leadership in waste management: Experiences from Singapore. In D. R. Gallagher (Ed.), *Environmental leadership: A reference handbook* (pp. 704–710). Los Angeles: Sage.

Thomas, G., & Durant, J. (1987). Why should we promote the public understanding of science. In M. Shortland (Ed.), *Scientific literacy papers* (pp. 1–44). Oxford: Department of External Studies, University of Oxford.

Wallack, L., Woodruff, K., Dorfman, L., & Diaz, I. (1999). *News for a change: An advocate's guide to working with the media.* Thousand Oaks, CA: Sage.

Wellington, J. (1991). News paper science, school science: Friends or foes? *International Journal of Science Education, 13*, 363–372.

Wilkins, L. (1993). Between facts and values: Print media coverage of the greenhouse effect, 1987–1990. *Public Understanding of Science, 2*, 71–84.

Zamith, R., Pinto, J., & Villar, M. E. (2013). Constructing climate change in the Americas: An analysis of the news coverage in U.S. and South American newspapers. *Science Communication, 35*, 334–357.

Chapter 16
Learning Science Through Inquiry in Informal Contexts

John K. Gilbert

Keywords Learning by inquiry • Informal resources • Capabilities of resources • Science Centre Singapore • Possible future development

Supporting the Evolution of Science Education in Singapore

The meaning of the phrase 'learning science by inquiry' is opaque, leading to the use of a wide variety of words to partially characterise it, for example, 'scientific processes, scientific method, experimental approach, formulating hypotheses', and many others (Abd-El-Khalick & Akerson, 2004). As the focus of this volume is the introduction of inquiry-based science education in Singapore, it is perhaps most useful to consider its overall meaning in terms of the approach that it is supplementing or replacing. There seems to be a desire that a shift takes place from:

- a goal of providing science education for scientists, to providing science education for all;
- an image of science education as what we know, to science education as teaching science as a way of knowing;
- an image of science education that emphasises content and process goals to science education that stresses goals examining the relation between evidence and explanations;
- an emphasis on individual science lessons that demonstrate concepts, to science lesson sequences that promote reasoning with and about concepts;
- the study of science topics that examine current scientific thinking without regard for social context, to the study of scientific topics in social contexts;
- a view of science that emphasizes observation and experimentation, to a view that stresses theory and model building and their revision; and

J.K. Gilbert (✉)
King's College London, The University of Reading, Berkshire, UK
e-mail: John.K.gilbert@btinternet.com

A.-L. Tan et al. (eds.), *Inquiry into the Singapore Science Classroom*, Education Innovation Series, DOI 10.1007/978-981-4585-78-1_16,

- a view of scientific evidence principally derived from sense-perception (either direct or augmented), to a view that evidence is obtained from theory-driven observations (Duschl & Grandy, 2008, pp. 6–7).

The traditional approach of science education in Singapore has been very successful indeed, to judge from the excellent and sustained performance in international comparative assessments, Trends in International Mathematics and Science Study (TIMSS) and Programme for International Student Assessment (PISA), in which all sections of the population take a justified pride. However, the changes in purpose, style, content, and teaching technique that are implied by a substantial shift to inquiry-based science education are drastic. There is ample evidence from other major changes in the science curriculum attempted elsewhere that innovatory elements tend to become diluted or interpreted in terms of existing practice. This occurs as the progression of the new ideas takes place from their conception in the ideal curriculum, to their inclusion in the formal stated curriculum, to the curriculum as perceived by teachers, to the curriculum as taught in the classroom, to the curriculum as experienced by the students, to the attainment of the curriculum as reflected in student assessment (Van den Akker, 1998). If the major changes as envisaged in Singapore are to be successful, then this must be done so that the international comparative status of Singaporean science education is not allowed to degrade: public and political opinion would not tolerate that. As has been shown in earlier sections of this book, the Ministry of Education is providing support for change within schools on a large scale. However, such change will require all the support that it can get. This is where resources for informal science education can contribute.

Informal Science Education

In its purist form, informal science education can be defined as that which takes place in out-of-school environments where (1) both attendance and involvement are voluntary or free choice, rather than compulsory or coercive; (2) the curriculum, if any, and whether intended or not, has an underlying structure which is open, offers choices to learners, and tends not to be transmissive; (3) the activities in which the learners can be involved are nonevaluative and non-competitive, rather than assessed and graded; and (4) the social interaction is amongst groups likely to be heterogeneous with regard to age, rather than constrained between same-age peers and formalised with the teacher as the main adult (Stocklmayer, Rennie, & Gilbert, 2010, p. 10).

This pure form of informal science education is only likely to be found in two situations. First, in the education of adults as part of their lifelong learning, it is often undertaken in family or friendship groups. This situation may achieve a high salience if the Singapore government's project to substantially increase the population is followed through, for most of the newcomers are likely to come from countries with weaker educational systems. Second, in the education of school-age students undertaking voluntary self-improvement activities, it is often with their families and at weekends or in school holidays.

The great majority of use for informal resources will, however, be as a complement or supplement to the science education provided by schools. As such, students will be guided, or even directed, to particular resources. They will be required to undertake specific activities in a specific time frame, these activities being coordinated with the progression of their school-based work. These activities will be too expensive to provide on a school-by-school basis or involve the use of specialist knowledge, equipment, or pedagogic skills. Informal science education can be facilitated in a number of ways. It is to these that I now briefly turn.

The Capabilities of Informal Resources

A wide range of types of informal resources are now available. They can be grouped according to the medium in which they are provided: situation-based, activity-based, paper-based, conventional media-based, and computer-based. Taking these in turn:

Situation-Based Resources

The provision of one-off or short-series lectures, usually outside working hours, does allow aspects of science and technology that are not mandated in statutory curricula to be explored. While those attending such lectures will have chosen to do so, their limitation is that those attending have very few opportunities to engage in question-and-answer sessions with the presenter.

The majority of the situations that present opportunities for informal learning are provided by museums and science centres. Museums consist of collections of static objects placed in a spatial or temporal order to generate a narrative for the visitor. In recent decades, the notion of the 'interactive exhibit' has emerged, in which the visitor is invited to perform some action on the exhibit, the consequences of that action then being made available to the visitor. Museums and science centres are heavily used and the most extensively researched of approaches to the provision of informal science education (Bell, Lewenstein, Shouse, & Falmer, 2009; Paris, 2002).

Activity-Based Resources

Although science centres provide constrained opportunities for activity-based learning, the broadest opportunities for such learning are provided by full-scale inquiry projects where what is to be done, why, and how are entirely under the control of the student. Such projects provide excellent opportunities for students to work collaboratively with each other and to engage in questioning and argumentation. The use of laboratories where there is equipment not generally available in schools also falls into this category.

Paper-Based Resources

The most immediately available forms of paper-based resources are newspapers and magazines. However, the former tend not to carry many such articles and to usually treat the topics sensationally. There are some specialist magazines, e.g., '*New Scientist*' and '*Scientific American*', but their appeal is to people who are familiar with scientific ideas.

A substantial proportion of paper-based resources are provided by what can be called 'popular science' books. Readily available in many types of commercial outlet, e.g., supermarkets, railway stations, and airports, they are not textbooks but rather designed to communicate with diverse audiences (MacPherson & Della Sala, 2008). In a recent study of such books when focused on chemistry, it was found that the purposes for which they may be read vary a great deal (Afonso & Gilbert, 2013).

The use of 'popular science books' requires the reader to be literate. However, for those who are less secure in their skills of reading, the potential – as yet unrealised in science education – of the 'graphic novel' and other forms has much to commend them. Broadly defined as a story presented in comic-strip format as a book (Gertler & Lieber, 2004), the graphic novel approach is allied to that of the Japanese '*manga*' comic. In the latter case, the genre has been extensively used in the field of science fiction and therefore has potential for bona fide science education (Bosker, 2007). Where the significance of the text recedes in the face of that of the illustrations, the result is the 'comic', consisting of a sequential series of cartoons. The potential of the comic in science education has been reviewed (Tatalovic, 2009) and seems considerable.

Conventional Media-Based Resources

Radio has great potential to provide information to large populations, e.g., China, or to low-density populations, e.g., Australia (Mazzonetto, Merzagora, & Tola, 2005). This potential has largely been overlooked by the advent of television and video, which, although widely used in a range of formats, have been little researched (Coll, Gilbert, Pilot, & Streller, 2013).

Computer-Based Resources

There is no doubt that the explosive growth in the availability of the Internet is having an immense impact on informal learning. For example, anybody told by their physician that they have a particular condition can find out all about it within seconds. The very recent advent of the so-called social media, e.g., Facebook, offers great potential for the construction of knowledge by worldwide networks of people. It is perhaps inevitable that, given the vast amounts of effort being made to develop these computer-based forms of informal learning, almost no systematic, insightful, and generalisable research has taken place so far (Coll et al., 2013, p. 250).

Informal Resources to Provide Support for Inquiry-Based Learning

Each of the types of informal resource, outlined earlier, can make a specific contribution to supporting the changes needed to implement inquiry-based learning. Indeed, some of them can make several types of contribution.

Situation-Based Resources

These can provide:

- Access to specialist knowledge, for example, up-to-date scientific research and recent technological developments
- Learning about the nature of science

Activity-Based Resources

These can provide opportunities to:

- Physically and mentally explore phenomena of scientific interest, whether directly or indirectly
- Engage in genuine, sustained, inquiries, in particular how to create and test models

- Work collaboratively with other students, in contrast to the intensively competitive atmosphere that prevail in schools
- Develop the skills of questioning and argumentation
- Use specialist scientific laboratory equipment, for example scanning tunnel microscopy

Paper-Based Resources

These can provide opportunities to:

- Learn about the social, political, and economic implications of science and technology
- Learn about the nature of science

Conventional Media-Based Resources

These can provide:

- Access to specialist knowledge, e.g., up-to-date scientific research and recent technological developments
- Opportunities to learn about the social, political, and economic implications of science and technology
- Opportunities to learn about the nature of science

Computer-Based Resources

This diverse group can provide:

- Opportunities for teachers to learn how to support all the resources mentioned, in addition to their use by students

Current Support for Inquiry-Based Learning Provided by Informal Resources in Singapore

Taking each of the three chapters in this section in turn:

Chapter 13 (*Knowledge Building as a Boundary Object in Formal/Informal Learning*) presents a most engaging example of activity-based informal learning. A case study is presented in which opportunities to engage in four of the cardinal aspects of inquiry-based learning (i.e., 'conduct of a sustained inquiry', 'opportunity to explore a phenomenon', 'engagement in collaborative work', 'opportunity to engage in questioning and argumentation') are present. Most importantly, the activity reported is highly realistic in terms of the evolution of inquiry-based work from the established norm of a knowledge transmission approach to learning.

The case study is contained within a worthwhile theoretical framework in which the customary distinction between the knowledge and learning taking place within the school and the home are blurred. The case starts from normal school practical work – the growing of plants from seeds – and expands the scope of that by redefining the notion of 'homework' to permit the students to continue their inquiry at home and over several months. As the activity develops, the home makes an increasing contribution to knowledge: the ideas of the mother and grandmother are incorporated and the unanticipated involvement of family pets. These insights are shared with fellow students during class time.

Almost no evidence is presented of the changing perceptions of the students, their teacher, and the parents, on the nature of learning. We can infer, however, from the summary results contained in the two tables that this was substantial. In terms of the approach through 'activity theory' taken in this chapter, there must have been changes for the participants in terms of the notion of personal identity as reflected in the notion of 'learning', the ways in which the coordination of home- and school-based learning was perceived, the ways in which these changes were evaluated, and on the rate of transformation of educational practice.

Given the realities of the curriculum and the availability of time in school, inquiry-based learning will have to spread out into 'nonschool' time. Ways will have to be found to sustain such activities with children who are other-than-high-achievers and with less-supportive parents. If such approaches are to be spread across Singapore schools, detailed reports, with the inclusion of evidence, will be needed, for teachers are most likely to change their practice when they can see that their peers have successfully made desired changes. In short, more heavily funded and sustained projects, coupled to systematic dissemination, will be needed. However, this dissemination will have to be shorn of the abstract language of 'activity theory' and 'boundary objects' if it is to be acceptable to students, parents, and teachers.

Chapter 14 (*Science Centre Singapore as an Alternate Classroom*) is an account of the activities of Science Centre Singapore, the major systematic provider of informal and non-formal science and technology education in Singapore. Long established, much of its work has been focused on the augmentation of the traditional transmission teaching and factual learning approaches to science education prevalent in schools. Although all based to some extent on experiential approaches to learning, many of the major galleries fall into this tradition. Examples are the Exhibition Galleries (presenting demonstrations of high-technology scientific equipment), the Thematic Galleries (offering opportunity to use more commonplace scientific equipment), and the Problem-based Learning Gallery Pathways (worksheets closely linked to the school science curriculum). The emphasis of the work of the centre placed on these galleries will no doubt be augmented, if not replaced, by the progressive adoption of inquiry-based approaches to educational provision.

Some of the existing activities, for example, the Young Scientist Badge Scheme, the Student Attachment Scheme, and the National Science Challenge can readily be changed to allow for more student-initiated project work, albeit with very different degrees of adaptation. However, a number of distinctively inquiry-based, student-initiated project schemes have been introduced. Prominent amongst these are the Centre for Research and Applied Learning in Science (CRADLE), the Singapore Science and Engineering Fair (SSEF), and the Singapore Academy of Young Engineers and Scientists (SAYES) schemes.

If government educational policy remains fairly consistent, then we will see the Science Centre Singapore continuing to make a major contribution to informal and non-formal science education in Singapore. The significance of this contribution will be underpinned by a continuing, even greater, emphasis being placed on the provision of teacher attachment and resource development workshops. This is vital, for changes in teacher beliefs, attitudes, and skills are difficult to bring about (Van den Akker, 1998). This endeavour will be given a great impulse by the availability to teachers of the Masters of Science (Science Communication) jointly with the Australian National University.

All those associated with the Science Centre Singapore are to be congratulated on what they have achieved so far and on the rapidity of the evolution of their activities to meet the demands of inquiry-based science education. Two challenges remain: first, to take a clear focus on the science educational needs of the general public (these will be driven by personal, social, and economic needs, and not in any way constrained by formal science curricula) and, second, to move beyond an approach to evaluation which measure vague 'satisfaction' and 'interest'. Such measures provide no guidance for the future. Probing into what has been learnt and why things were interesting can only yield positive pointers to future activities.

Chapter 15 (*Public Education About Science in Singapore: Science Journalism via Newspapers*) is based on a well-researched inquiry into the treatment of science and technology in *The Straits Times* over 1 week. In addition to identifying when science-oriented articles appear (more at the weekend than in midweek) and their focus (more of health-related issues), the analysis produced some interesting quantitative measures: The number of column inches showed the relative extent of issues (but not their percentage to the whole newspaper), while the calculation of the Prominence Index suggested the status attributed by the editor of the newspaper to individual articles.

The chapter also points to some of the challenges that the presentation of science in newspapers face. The fact that only 22 of the 51 articles identified were generated locally may imply a relatively low-status reportage attributed to science by the editor of the newspaper: Only the analysis of other sections of the paper (e.g., sport, local politics) would enable this suggestion to be evaluated. It would have been interesting to know something of the professional backgrounds and experience of the authors of the 22 articles. These may or may not have included a necessary element of science. While wire-service articles may be relevant to local concerns, the lack of a particular context may make it difficult for readers to see the relevance to their lives and concerns of the other 29 articles.

The chapter draws attention to two issues: first, that only the meaning of some specialist words are spelt out in articles which meant that relying on memories of school science is likely to prove an inadequate resource for many and, second, the lack of attention given to the physical sciences. While the focus of much cutting-edge science is now on what can broadly be termed the biological sciences, it does seem very realistic, as Subramanian points out, for necessary physical science ideas to be inserted into articles.

This chapter does spell out what scientific or technological topics were dealt with, how, and to what extent, over the sample week chosen. While we can infer the nature and extent of the contribution of The Straits Times to informal science education in Singapore, the identification of contribution of the newspaper specifically to inquiry-based school science calls for additional research.

Filling the Gaps: The Possible Future Use/And or Development of Informal Resources in Singapore

Although prosperous, Singapore has a small population and therefore will find it difficult to justify the considerable outlay needed to develop new informal resources focused on science. So 'filling the gaps' will have to take place through adopting one or both of two options: collaboration with neighbouring countries and importing pre-existing materials.

In terms of collaborative developments, the most likely options involve nearby countries: those that share a similar climate and therefore can readily provide similar social and economic contexts around which to shape some of the many aspects of inquiry learning. Two that come to mind are the provision of freshwater and the improvement of agricultural production. Although a diversity of languages is spoken in those countries, English is both taught in all of them and could therefore be seen as a lingua franca.

Several of the media in which informal materials can be provided were not mentioned in any of the three chapters. The first of these are paper-based materials in the form of popular books about science. In this respect, Singapore is like most countries: The potential of this form has not been explored. This could readily be done, preferably in conjunction with the many retail outlets that sell newspapers, magazines, and the various genres of 'self-improvement' books. The second of these is the conventional media form, that is, the use of radio, television, and video. Again, I suspect that these are widely used but the extent and purpose for doing so has not been recognised and systematised. Third, and by far the most important, is the use of computer-based forms.

There is a great deal of evidence that computer-based informal learning is widespread across both subject areas and age groups (Hsi, 2007). The genre of 'video games' is long established and has been shown as possibly capable of supporting a wide range of learning behaviours (Gee, 2007). The recent and widespread use of portable wireless-connected mobile devices is increasingly being exploited for educational and entertainment purposes (Wong, 2012). This use is already well established in Singapore and, most helpfully for this chapter, is focused on inquiry-based learning (Song, Wong, & Looi, 2012). There are certainly educational implications for the exponential evolution of interactive websites and social networking sites (Schuck & Aubusson, 2010). The educational importance of this medium of provision seems bound to grow rapidly in importance over the next few decades.

While I would advocate increased attention to these forms of provision that have so far been relatively overlooked in Singapore, we should not lose sight of the fundamentals of informal learning about science: the need to ensure that students have opportunities to design, conduct, and report their own inquiries. This will entail them learning to collaborate with others and to engage in argumentation. This will only take place if not only science teachers but also the parents of students acquire the necessary knowledge and skills for this to be possible.

Ensuring That the Best Use Is Made of Informal Resources

I have written extensively about the shape of teacher education activities that actually lead to sustained professional development (Bell & Gilbert, 1996). They must (a) be sustained over a substantial period with incremental advances in the depth of consideration given to issues, (b) involve significant collaboration between the participants, (c) involve the trialling of ideas in everyday classrooms, and (d) be supported by the mentoring of experts in the field in question. The half- or one-day 'course' without preparation and follow-up, provided in lecture format and with only limited opportunities for interaction, is, quite frankly, a waste of public funds and teachers' time.

Providing systematic support for parents who wish to encourage the education of their offspring does not seem to be widely practised. 'Good practice' in so doing would perhaps emerge through an inquiry into the attitudes and actions of those parents who are also professional educators, e.g., teachers and university lecturers.

References

Abd-El-Khalick, F., & Akerson, V. L. (2004). Learning as conceptual change: Factors mediating the development of preservice elementary teachers' views of nature of science. *Science Education, 88*, 785–810.

Afonso, A., & Gilbert, J. K. (2013). The role of 'popular' books in informal chemical education. *International Journal of Science Education (B), 3*, 77–99.

Bell, B., & Gilbert, J. (1996). *Teacher development: A model from science education*. London: Falmer.

Bell, B., Lewenstein, B., Shouse, A. W., & Falmer, M. A. (2009). *Learning science in informal environments: People, places, and pursuits*. Washington, DC: The National Academies Press.

Bosker, B. (2007, August 31). Manga mania. *The Wall Street Journal*. Retrieved September 10, 2013, from http://online.wsj.com/article/SB118851157811713921.html?mod=googlenews_wsj

Coll, R., Gilbert, J., Pilot, A., & Streller, S. (2013). How to benefit from the informal and interdisciplinary dimensions of chemistry in teaching. In I. Eilks & A. Hofstein (Eds.), *Teaching chemistry: A study book* (pp. 241–268). Rotterdam, The Netherlands: Sense Publishers.

Duschl, R., & Grandy, R. (2008). Reconsidering the character and role of inquiry in school science: Framing the debates. In R. Duschl & R. Grandy (Eds.), *Teaching scientific inquiry* (pp. 1–37). Rotterdam, The Netherlands: Sense Publishers.

Gee, J. P. (2007). *What video games have to teach us about learning and literacy*. New York: Palgrave Macmillan.

Gertler, N., & Lieber, S. (2004). *The complete idiot's guide to creating a graphic novel*. New York: Alpha Books.

Hsi, S. (2007). Conceptual learning from the everyday activities of digital kids. *International Journal of Science Education, 29*, 1509–1529.

MacPherson, S., & Della Sala, S. (2008). Reviews of popular science books. *Cortex, 44*, 763.

Mazzonetto, M., Merzagora, M., & Tola, E. (2005). *Science in radio broadcasting*. Milan, Italy: Polimetrica.

Paris, S. (Ed.). (2002). *Perspectives on object-centred learning in museums*. Mahwah, NJ: Erlbaum.

Schuck, S., & Aubusson, P. (2010). Educational scenarios for digital futures. *Learning Media and Technology, 35*, 293–305.

Song, Y., Wong, L.-H., & Looi, C.-K. (2012). Fostering personalised learning in science inquiry supported by mobile technologies. *Educational Technology Research and Development, 60*, 679–701.

Stocklmayer, S., Rennie, J. L., & Gilbert, J. K. (2010). The roles of the formal and informal sectors in the provision of effective science education. *Studies in Science Education, 46*, 1–44.

Tatalovic, M. (2009). Science comics as tools for science education and communication: A brief exploratory study. *Journal of Science Communication, 8*(4), 1–16.

Van den Akker, J. (1998). The science curriculum: Between ideals and outcomes. In B. F. Fraser & K. G. Tobin (Eds.), *International handbook of science education* (pp. 421–448). Dordrecht, The Netherlands: Kluwer.

Wong, L.-H. (2012). A learner-centric view of mobile seamless learning. *British Journal of Educational Technology, 43*(1), 19–23.

Erratum to: Acknowledgements

Inquiry into the Singapore Science Classroom

A.-L. Tan et al. (eds.), *Inquiry into the Singapore Science Classroom*,
Education Innovation Series, DOI 10.1007/978-981-4585-78-1,

DOI 10.1007/978-981-4585-78-1

Production team had inadvertently misplaced the correction and change was made to Dr. Imelda Santos CALEON affiliation. The correct affiliation of Dr. Imelda Santos CALEON is given below:
"(National Institute of Education, Office of Education Research)"

The online version of the original book can be found at
http://dx.doi.org/10.1007/978-981-4585-78-1

A.-L. Tan et al. (eds.), *Inquiry into the Singapore Science Classroom*,
Education Innovation Series, DOI 10.1007/978-981-4585-78-1_17,

About the Authors

Tan-Ying Chin is a senior curriculum specialist at the Singapore Ministry of Education (MOE) Curriculum Planning and Development Division. She obtained her master's degree in education from NIE/NTU and her doctorate degree in science education from the Institute of Education, University of London. As a science specialist, she leads and works with schools and various partners to support science teaching and learning. Her research interest includes the design and implementation of curriculum, pedagogy, and assessment that support practice of science as inquiry in the primary classrooms. She believes in using research and practice to inform the design of curriculum, pedagogy, and assessment so that students enjoy, understand, and appreciate how science is interesting, relevant, and applicable in their everyday lives.

Barbara A. Crawford holds the position of full professor of science education at the University of Georgia in the USA, prior to her work at Cornell University. She was recently elected as fellow of the American Association for the Advancement of Science (AAAS). She has 16 years of experience teaching science in grades 7–12 (ages 12–18 years old). Her research agenda focuses on understanding how to teach children scientific inquiry and nature of science and critical thinking and viable ways to support teachers in teaching science as inquiry. How to teach children about carrying out and learning about scientific practices is a critical issue raised by many countries worldwide. Her research intersects with her teaching; in that understanding, how to create classrooms that engage children in carrying out science practices connects with how to support prospective and practising teachers in inquiry pedagogy. Connected with inquiry is an understanding of scientific models and modelling, how to support children in defending arguments, and appreciating how scientists work. The context for her research and teaching is authentic science environments. Her current funded project, Fossil Finders, aims to develop innovative resources and a model of teacher professional development centred in authentic science.

Anne Dairianathan joined the Science Centre Singapore in 1981 and has held various positions within the Education Division, in particular Life Science. She is currently the senior director for education programmes. She has a B.Sc. (Hons) in

A.-L. Tan et al. (eds.), *Inquiry into the Singapore Science Classroom*,
Education Innovation Series, DOI 10.1007/978-981-4585-78-1,

Microbiology from; the National University of Singapore and an M.A. from the Nanyang Technological University, Singapore. She has conducted enrichment programmes for students from preschool to junior college level, as well as for teachers, and been involved in exhibit development at SCS. She contributes as a member of the Genetic Modification Advisory Committee Subcommittee on Public Awareness and the National Book Development Council, Singapore. She co-authored *A Guide to Fruits and Seeds* published by SCS. She received an Education Award by the Japanese Chamber of Commerce and Industry (JCCI) Singapore Foundation in 2002 for her contribution to popularisation and promotion of science to the public. Her interests are in out-of-school learning and communication of science to various audiences.

John K. Gilbert took a B.Sc. in Chemistry at the University of Leicester (1962), a D.Phil. in Chemistry at the University of Sussex (1965), and a PGCE (Secondary Science) at the University of London (1968). He taught at King's School, Rochester, and at Banbury School until 1971 when he took up a Lectureship in Science Education at University of Keele, moving to the University of Surrey as a senior lecturer in 1974, subsequently becoming a reader in 1984. In 1988, he became professor of Science Education at the University of Reading, becoming professor emeritus in 2005. He is currently a visiting professor at King's College London and has been editor in chief of the International Journal of Science Education (A) since 1991 and co-editor of IJSE(B) since 2011. His initial research interests in children's 'alternative conceptions' evolved into a focus of 'models and modelling' and most recently to both 'visualisation and explanation' and to 'science communication'. In 2001 he received the NARST award for 'Distinguished Contributions to Science Education through Research'.

Mijung Kim is an assistant professor at the Department of Curriculum and Instruction, University of Victoria, Canada. Her research area includes teacher perception, dilemmas, and reflection on inquiry-based teaching, children's reasoning, and collaborative problem solving on socioscientific and environmental issues. Her current research focuses on understanding the relationship of students' scientific knowledge and decision-making in dialogical argumentation process.

Chor Yam Lau is a lecturer at the National Institute of Education, Nanyang Technological University. His current research interests are in the area of science inquiry, physics education, and lesson study. Prior to joining NIE, Chor Yam was a senior curriculum specialist in the Singapore Ministry of Education and has vast experience in curriculum review and in supporting schools with the implementation of new physics curriculum initiatives. He was a member of the Physics Syllabus Review Committee for the 2007 and 2013 syllabi. To support the implementation of the syllabi, he was involved in reviewing the physics textbooks and a workbook approved for use in schools, and he was also instrumental in facilitating the Physics by Inquiry workshops for teachers conducted by the University of Washington from 2005 to 2010.

Yew-Jin Lee was trained as a secondary school biology teacher in Singapore. Currently, he is a science teacher educator at the NIE with research interests in curriculum studies, scientific ways of knowing, and informal science learning environments. He also brings to his research cultural-historical concepts of learning as well as theoretical insights from the social sciences and humanities. Apart from being the co-editor of *Pedagogies: An International Journal* (Taylor & Francis), Yew Jin serves on the editorial boards of *Studies in Science Education* and *Research in Science Education*. He has a strong passion for development work and has travelled around the region for this purpose.

Shirley S. L. Lim is concurrently an associate dean at the Office of Graduate Studies and Professional Learning and associate professor in Natural Sciences and Science Education. A marine ecologist by training, she has interests in science inquiry. She was the president of the Singapore Institute of Biology (helming the professional society for biologists and biology educators for 5 years) and a member of the Singapore Science Centre Board and a council member of the Singapore National Academy of Science for 6 and 5 years, respectively. Shirley organised the 1st to 10th Singapore Biology Olympiads, trained, and led the national teams for competition at the International Biology Olympiads (IBO) for 11 years. She is one of the four steering committee members of the IBO fraternity and was re-elected for a second term in 2012. She organised the 23rd IBO hosted by Singapore in July 2012. Shirley is the Asian governor of the Crustacean Society (USA) and the vice president (Executive Board) of the International Union of Biological Science's Commission for Biological Education (USA).

Tit Meng Lim has been the chief executive at the Science Centre Singapore (SCS) since January 2010. He concurrently holds the position of associate professor at NUS Department of Biological Sciences. He has held many leadership positions at NUS, including that of vice dean from 2003 to 2008. He is the current president of the Singapore Association for the Advancement of Science (SAAS) and the first vice president of the Singapore National Academy. He is a board member in the Singapore Technologies Endowment Programme and the Singapore Millennium Foundation. He is also a director in the Board of the Association of Science-Technology Centers (ASTC) based in the USA. In 2008, he received the Outstanding Contributorship Award for the Ministry of Education (MOE) Science Research Programme for his services spanning 1988–2007. His other awards include the NUS Quality Service Award (2007), the MOE Service to Education Award (2009), and the Outstanding Science Entrepreneur Award (2010). Under his leadership, SCS has won many accolades locally and internationally, including the coveted Singapore Tourism Board Award for Best Education Experience in 2010, 2011, and 2012. He obtained his B.Sc. (Hons) in Zoology from the National University of Singapore and his PhD from University of Cambridge.

Chew-Leng Poon is the deputy director for Research and Evaluation at the Singapore Ministry of Education (MOE) Planning Division. She is concurrently appointed as MOE's principal specialist in Research and Curriculum. Dr Poon earned her master's degree in Curricular Studies at the Ohio State University and her Ph.D. from NIE/NTU, specialising in science inquiry pedagogy. She has taught Year 12 Chemistry, been in various school leadership roles, and spent 6 years in curriculum policy development before taking on her current roles. Currently, she is involved in several international studies—including the Programme for International Student Assessment (PISA), Trends in International Mathematics and Science Study (TIMSS), and Assessment and Teaching of 21st Century Skills (ATC21S). She also coleads a Network Learning Community of senior specialists, master teachers, and heads of Science Department in exploring inquiry science in action.

R. Subramaniam is an associate professor at the National Institute of Education in Nanyang Technological University. His principal research interests are in the areas of chemistry education, physics education, primary science education, and science communication. He has a deep interest in the use of creative and innovative approaches in the teaching of the physical sciences, the use of conceptual change strategies to remediate alternative conceptions in the sciences among students, and how students learn in informal science settings. He has completed a number of funded research projects, of which he was the principal investigator. More recently, he has become interested in the area of assessment.

Frederick Talaue is a doctoral candidate at the National Institute of Education, Nanyang Technological University. His research delves into teachers' positionality within discourses on science teaching and learning at the primary level in Singapore. His current research interests include inquiry-based teaching, language in the science classroom, and identity-based studies. Having been a science high school teacher, he also takes special interest in learning how one can teach various biology concepts more effectively.

Aik-Ling Tan is an associate professor at the National Institute of Education, Nanyang Technological University. Her research interests include the practice of science as inquiry in primary science classrooms as well as in biology laboratories. Of particular interest to her are the intricate classroom interactions between students and teachers that foster meaning learning in science. She is the principal investigator of a research group that aims to connect teachers' knowledge to students' learning. In this group, they examine teachers' and students' knowledge about science and science learning and create learning spaces that will positively enhance learning of science through inquiry.

Kim Chwee Daniel Tan started his career as a chemistry teacher in 1990. He has been a faculty member of the National Institute of Education, Nanyang Technological University, Singapore, since 1998. He held a concurrent appointment in 2006–2007 as an assistant director in the Educational Technology Division, Ministry of Education, as he wanted to experience and understand the processes involved in policymaking. He is currently the associate professor and deputy head (Teaching and Curriculum Matters) in the Natural Sciences and Science Education Academic Group. He teaches higher degree courses as well as chemistry pedagogy courses in the pre-service teacher education programmes. His research interests are chemistry curriculum, translational research, ICT in science education, students' understanding and alternative conceptions of science, practical work, and the use of representations in science.

Poh-Hiang Tan is a doctoral candidate at the National Institute of Education, Nanyang Technological University. Her research delves into primary science teachers' ideas about formative assessment, their eventual practices of formative assessment in their classroom, and their reflection on the tensions and dilemmas about formative assessment in the teaching of science as inquiry. She believes that this is an important aspect of research since our current knowledge of formative assessment is generally discipline independent and fuzzy. Examining formative assessment in the domain of science will allow the disciplinary characteristics of science to be infused. Poh Hiang was an assessment officer with the Singapore Examinations and Assessment Board for a decade, and she is currently a teaching fellow at the National Institute of Education.

Seng-Chee Tan is the head of the Learning Sciences and Technologies Academic Group of the National Institute of Education (Singapore). His research interests include computer-supported collaborative learning and knowledge creation in education. As a former high school science teacher, he has conducted many research studies in science classrooms, particularly those that involve the use of knowledge building pedagogy. He is also exploring conceptual change through social cultural perspectives.

Kenneth Tobin came to the Urban Education doctoral programme at the Graduate Center of CUNY in the fall semester of 2003. Prior to that, he had positions as tenured full professor at Florida State University (1987 to 1997) and the University of Pennsylvania (1997–2003). In Australia, he held university appointments at the Western Australian Institute of Technology (now Curtin University), Mount Lawley College, and Graylands College (now Edith Cowan University). Before becoming a university science educator in Australia in 1974, Tobin taught high school physics, chemistry, biology, general science, and mathematics for 10 years. He began a programme of research in 1973 that continues to the present day—teaching and learning of science and learning to teach science.

Darren Wong is a senior curriculum specialist at the Singapore Ministry of Education (MOE) Curriculum Planning and Development Division. He began his career as a physics teacher at a junior college in 1997 and served in MOE from 2005, where he was involved in a number of curriculum initiatives such as the design and development of innovative science curricula and the implementation and evaluation of school-based science practical assessment. He was later seconded to National Institute of Education, Nanyang Technological University, in 2009 as a science teacher educator, where he was involved in a number of funded research projects. His current research interests are in the areas of physics education research, inquiry-based learning in science, and teacher professional development. As a science curriculum specialist, he leads and works with teachers and various partners to support physics and science learning and teaching in schools.

Jennifer Yeo teaches physics education in Natural Sciences and Science Education Academic Group of National Institute of Education (Singapore). Her research interests include knowledge creation in science education, and students' construction of scientific explanation and the role of representations and argumentation in science inquiry learning at the primary science levels as well as secondary and junior college physics.

Benny H. W. Yung is an associate professor at the University of Hong Kong (HKU). He is also head of the Division of Science, Mathematics and Computing in the Faculty of Education. He was a founding member and the vice president of East-Asian Association for Science. Before joining HKU, he was a secondary school biology teacher, Biology Subject Officer at the Hong Kong Examinations Authority. Apart from scholarly books like *Assessment Reform in Science: Fairness and Fear*, he is also the author of a very popular biology textbook for the senior secondary curriculum in Hong Kong.

Acknowledgements

Please note the Erratum to this Acknowledgements at the last chapter of this book

We thank our reviewers who have volunteered their time and expertise to help read and comment on the chapters. (Names are listed in alphabetical order.)

Dr. Imelda Santos CALEON (*College of Education, Zhejiang University*)

Professor (Adjunct) S. GOPINATHAN (*National University of Singapore, Lee Kuan Yew School of Public Policy*)

Ms. Jennifer JOCZ (*National Institute of Education, Learning Sciences and Technologies*)

Dr. Michelle TAN (*National Institute of Education, Office of Education Research*)

Dr. Michael TAN (*National Institute of Education, Office of Education Research*)

Ast/P TANG Kok Sing (*National Institute of Education, Natural Sciences and Science Education*)

Dr. TEO Chew Lee (*Ministry of Education, Educational Technology Division*)

Ast/P TEO Tang Wee (*National Institute of Education, Natural Sciences and Science Education*)

A/P Benny H W YUNG (*University of Hong Kong, Faculty of Education*)

Dr. ZHAI Junqing (*College of Education, Zhejiang University*)

A.-L. Tan et al. (eds.), *Inquiry into the Singapore Science Classroom*,
Education Innovation Series, DOI 10.1007/978-981-4585-78-1,

Acknowledgements

Index

A.-L. Tan et al. (eds.), *Inquiry into the Singapore Science Classroom*,
Education Innovation Series, DOI 10.1007/978-981-4585-78-1,

MIX
Papier aus verantwortungsvollen Quellen
Paper from responsible sources
FSC® C105338

If you have any concerns about our products,
you can contact us on
ProductSafety@springernature.com

In case Publisher is established outside the EU,
the EU authorized representative is:
Springer Nature Customer Service Center GmbH
Europaplatz 3, 69115 Heidelberg, Germany

Printed by Libri Plureos GmbH
in Hamburg, Germany